Omics Approaches for Crop Improvement

Omics Approaches for Crop Improvement

Editors

Roxana Yockteng

Andrés J. Cortés

María Ángeles Castillejo

Basel • Beijing • Wuhan • Barcelona • Belgrade • Novi Sad • Cluj • Manchester

Editors

Roxana Yockteng
Corporación Colombiana de
Investigación Agropecuaria
(AGROSAVIA)
Tibaitatá
Colombia

Andrés J. Cortés
Corporación Colombiana de
Investigación Agropecuaria
(AGROSAVIA)
Universidad Nacional de
Colombia - Sede Medellín
Antioquia
Colombia

María Ángeles Castillejo
Department of Biochemistry
and Molecular Biology
University of Cordoba
Córdoba
Spain

Editorial Office
MDPI AG
Grosspeteranlage 5
4052 Basel, Switzerland

This is a reprint of articles from the Special Issue published online in the open access journal *Agronomy* (ISSN 2073-4395) (available at: www.mdpi.com/journal/agronomy/special_issues/crop_omics_genetic).

For citation purposes, cite each article independently as indicated on the article page online and as indicated below:

Lastname, A.A.; Lastname, B.B. Article Title. *Journal Name* **Year**, *Volume Number*, Page Range.

ISBN 978-3-7258-1494-7 (Hbk)
ISBN 978-3-7258-1493-0 (PDF)
doi.org/10.3390/books978-3-7258-1493-0

Contents

About the Editors

Roxana Yockteng

Dr. Yockteng is a biologist with a broad range of knowledge in plant genetics and evolution. She obtained her BSc and MSc diplomas from the Universidad de los Andes in Colombia, and her PhD from the Paris XI-Sud University in Orsay, France. She is an Associate Professor at the National Museum of Natural History in Paris, where she worked on the evolution of interaction between plants and organisms. She also worked as a researcher at the University of California at Berkeley, where she developed evolution and evo-devo studies in plants. Since October 2014, she has been a PhD researcher at AGROSAVIA, where she studies cacao genetic diversity and the genetic mechanisms involved in critical agronomical traits in this crop using omics data. She leads research projects on the characterization of the national plant germplasm bank to understand the conserved diversity and accelerate the selection of materials of crop species with economic importance in Colombia to improve resilience and productivity.

Andrés J. Cortés

Dr. Cortés holds an Associate Research position as a Geneticist at the Colombian Agricultural Research Corporation (AGROSAVIA), CI La Selva. Dr. Cortés graduated as a Plant Geneticist (PhD) from Uppsala University (Sweden), and as a Biologist (BSc Hons, MSc) from Universidad de los Andes (Colombia). His research experience dates back to the International Center for Tropical Agriculture (CGIAR – CIAT), the University of Fribourg (Switzerland), the Swiss Federal Institute for Forest, Snow and Landscape Research (WSL – SLF), and the Swedish University of Agricultural Sciences (SLU). Dr. Cortés is interested in investigating the genetic adaptive potential in plants and trees of agro-ecological interest using genomic, evolutionary, and ecological tools. In particular, he has explored abiotic stress tolerance and genetic diversity in beans, rootstock-mediated inheritance in avocado trees, genetic adaptation to climate change in willow species, and adaptive diversification across various plant lineages at the Páramo, a rapidly evolving ecosystem that provides water resources for the Neotropics.

María Ángeles Castillejo

Dr. Castillejo is a Professor at the University of Cordoba (Spain). She investigates the response of plants to biotic and abiotic stresses, including model systems, crop species, and forest species. Dr. Castillejo is focused on the study of the most representative species of the Mediterranean forest and the agrosilvopastoral system, the holm oak, from a multidisciplinary point of view. Through physiological and molecular studies (using biochemical and omic tools), her research aims to identify molecular markers against the main stresses that threaten the survival of this species: drought and decline syndrome. Dr. Castillejo has also initiated a research line that aims to activate defense responses in forest species through the use of biostimulants.

Preface

Ensuring food security and satisfying the nutritional needs of a constantly expanding global population is at the core of modern crop science. Traditional breeding methods have significantly contributed to agricultural productivity; nevertheless, climate change, evolving pests and diseases, and sustainable practices require novel approaches to be leveraged. Throughout this reprint, readers will grip the realm of 'omics', a repertoire of revolutionizing techniques transforming our understanding of plant biology and providing unprecedented potential for crop enhancement.

Omics technologies encompass genomics, transcriptomics, proteomics, metabolomics, and phenomics. Each field provides a unique lens through which the intricate biological processes underlying plant growth, development, and stress responses can be understood. By integrating data from these diverse sources, researchers can achieve a holistic view of the molecular networks and trade-offs underlying crop traits and pave the way for more precise and efficient breeding strategies.

The potential for revolutionizing varietal development through the integration of omics technology lies in its capacity to provide a comprehensive and dynamic understanding of the very same foundations of plant biology. It now becomes possible to customize crops for high yields resilient to environmental challenges and rich in critical nutrients by breaking down trade-offs and manipulating complex traits at the omics level. Moreover, the application of omics approaches supports sustainable agricultural practices by enabling the development of crops that require fewer inputs, such as water, fertilizers, and pesticides.

This reprint offers an exquisite overview of various omics technologies and their applications in crop improvement. It brings together contributions from leading experts in the field, offering insights into the latest advancements, methodologies, and case studies. By bridging the gap between fundamental research and practical applications, this fascinating compilation is a valuable resource for researchers, breeders, and students.

In the chapters that follow, readers will embark on a fascinating venture towards advances in plant omics, unveiling the transgressive potential these technologies hold for the future of crop improvement. As we embrace the era of precision agriculture, integrating omics approaches will undoubtedly play a pivotal role in meeting the global challenges of food security and sustainability.

Roxana Yockteng, Andrés J. Cortés, and María Ángeles Castillejo

Editors

Editorial

'Omics' Approaches for Crop Improvement

Andrés J. Cortés [1,*,†,‡], María Ángeles Castillejo [2,*,†] and Roxana Yockteng [3,4,*,†]

[1] Corporación Colombiana de Investigación Agropecuaria—AGROSAVIA, C.I. La Selva,
    Km 7 vía Rionegro—Las Palmas, Rionegro 054048, Colombia
[2] Agroforestry and Plant Biochemistry, Proteomics and Systems Biology—Department of Biochemistry
    and Molecular Biology, University of Cordoba, CO-CeiA3, 14014 Cordoba, Spain
[3] Corporación Colombiana de Investigación Agropecuaria—AGROSAVIA, C.I. Tibaitatá, Km 14 vía Mosquera,
    Mosquera 250047, Colombia
[4] Institut de Systématique, Evolution, Biodiversité-UMR-CNRS 7205, National Museum of Natural History,
    75005 Paris, France
* Correspondence: acortes@agrosavia.co (A.J.C.); bb2casam@uco.es (M.Á.C.); ryockteng@agrosavia.co (R.Y.)
† All authors contributed equally and share first authorship; their names are sorted alphabetically.
‡ Secondary address: Facultad de Ciencias Agrarias—Departamento de Ciencias Forestales,
    Universidad Nacional de Colombia—Sede Medellín, Medellín 050034, Colombia.

Citation: Cortés, A.J.; Castillejo, M.Á.; Yockteng, R. 'Omics' Approaches for Crop Improvement. *Agronomy* **2023**, *13*, 1401. https://doi.org/10.3390/agronomy13051401

Received: 12 May 2023
Revised: 17 May 2023
Accepted: 18 May 2023
Published: 19 May 2023

1. Introduction

The growing human population and climate change are imposing unprecedented challenges on the global food supply [1]. To cope with these pressures, crop improvement demands enhancing important agronomical traits beyond yield, such as adaptation, resistance, and nutritional value, by pivoting direct and indirect selection approaches [2]. The development of next-generation high-throughput screening technologies, referred to as 'omics', promises to speed up plant trait improvement [3] while producing more sustainable crops.

Large-scale techniques, such as genomics, transcriptomics, proteomics, metabolomics, and phenomics, have already provided large datasets for that purpose. Meanwhile, modern bioinformatic and machine-learning approaches are helping us to process this heterogeneous hyper-dimensional data [4] while ultimately understanding the mechanisms behind agronomic features within the contemporary plant breeding triangle (i.e., genomics vs. phenomics vs. enviromics) [5]. 'Omics' datasets are also being generated to study macro-scale interactions and deepen our knowledge of crop behavior across the microbial [6] and environmental [7,8] continua. However, despite these massive technological and computational developments [4], systemic efforts to integrate 'omics' studies to understand biochemical pathways and cellular networks of crop systems are in their infancy [9], especially in orphan species [10].

Therefore, this Special Issue envisions offering updated emergent views on large-scale 'omics'-based approaches. Specifically, the compilation explores the conceptual framework of the 'omics' paradigm [11], the practical uses of multiple 'omics' technologies, and their integration through trans-disciplinary bioinformatics as tools to improve qualitative and quantitative traits in a diverse panel of crop species.

2. Genomic-Enabled Crop Traceability and Improvement

Genomics is speeding up multiple steps in the breeding scheme (Table 1). For instance, in the downstream extreme of the breeding pipeline, Campuzano-Duque et al. [12] demonstrated the utility of high-throughput single nucleotide polymorphism (SNP) genotyping using SNP arrays to trace varietal purity of single plant selections (SPS). The authors successfully assessed the relationships and ancestry of plant selections from three inbreed origins (one original variety and two additional multi-lines) of forage oat (*Avena sativa*), and prioritized SNP candidates to ensure the genetic purity of these varieties. Meanwhile, in a more upstream introgression-breeding step, Pandit et al. [13] exemplified genomic-assisted selection within backcrossing schemes. The team pyramided three quantitative trait loci (QTLs) for submergence tolerance and grain yield in the rice (*Oryza sativa*) 'Maudamani' variety background, sourcing pyramided lines as novel cultivars or potential 'bridge' donors for further backcrossed generations.

Table 1. Collection of 10 studies in the Special Issue 'Omics Approaches for Crop Improvement'.

Plant Species	'Omics'	Research Goal	Sampling	Key Finding	Reference
Review					
Papaya (*Carica papaya*)	Transversal to 'omics' and systems biology	Review omics and bioinformatics advances for Papaya	Diverse cultivars and germplasm	'Omics' improved ripening, tolerance, and fruit quality	Zainal-Abidin et al. [11]
Genomics					
Forage Oat (*Avena sativa*)	iSelect 6K Bead-Chip	Evaluate the purity and relationships of SPS	AV-25 original, and AV25-T and AV25-S multi-lines	SNPs are a suitable tool to ensure genetic purity of oats	Campuzano-Duque et al. [12]
Rice (*Oryza sativa*)	Sub1, OsSPL14, and GW5 QTLs for tolerance and yield	Pyramid QTLs for submergence tolerance and yield	'Maudamani' variety background	Pyramided lines are useful as cultivars and as donors	Pandit et al. [13]
Transcriptomics					
Cacao (*Theobroma cacao*)	Phylogenetic, gene structure, and in silico expression	Report and characterize *tcGASA* genes in cacao	Cacao reference genome	*tcGASA* genes are target for resistant cacao varieties	Abdullah et al. [14]
Malvaceae family: Cacao, cotton, and jute fiber	Phylogenetic, synteny, and in silico expression	Characterize *MGT* genes in the Malvaceae family	*T. cacao, Gossypium hirsutum,* and *Corchorus capsularis*	MGTs interact with lipid/cell wall and photo-protection	Heidari et al. [15]
Rice (*Oryza sativa*)	In-house micro-array and mGCN	Unveil the mechanism of drought tolerance in *ABP57*	Drought-tolerant transgenic *Abp57*-OE line	MAPK, IAA and SA co-determine tolerance response	Abdullah-Zawawi et al. [16]
Proteomics					
Faba bean (*Vicia faba*)	2DE, MAL-DI-TOF/TOF, and zymography	Test leaf proteome effects to *Botrytis fabae* fungus	'Baraca' susceptible genotype, and resistant BPL710	Chloroplast PSII protein repair cycle linked to resistance	Castillejo et al. [17]
Tomato (*Solanum lycopersicum*)	TMT, HPLC, MS	Identify the effects of EFI vs. TSI on roots' protein level	Seedlings from the pure tomato cultivar 'Ouxiu-201'	EFI induces 513 DAPs adapted responses in roots	Wang et al. [18]
Phenomics					
Bean (*Phaseolus vulgaris*) × Tepary (*P. acutifolius*)	Multi-locality trials	Assess abiotic tolerance in inter-specific crosses	Interspecific backcross (86) between beans and Tepary	Interspecific backcrosses pyramid polygenic tolerance	Burbano-Erazo et al. [19]
Peanut (*Arachis hypogaea*)	HTP	Assess morphological variation in a CSSL	A total of 26 lines from a CSSL population	Chromosome segment from CWR sources variation	Gimode et al. [20]

Table is sorted bottom-up by 'omics' and species. ABP: auxin-binding protein, CSSL: chromosome segment substitution line, CWR: crop wild relatives, DAPs: differentially accumulated proteins, 2DE: two-dimensional gel electrophoresis, EFI: Ebb-and-flow sub-irrigation, GASA: gibberellic acid-stimulated *Arabidopsis*, HPLC: high-performance liquid chromatography, HTTP: high-throughput phenotyping, IAA: indole-3-acetic acid, MALDI-TOF/TOF: matrix-assisted laser desorption/ionization-time of flight, MAPK: mitogen-activated protein kinase, mGCN: modular gene co-expression network, MGT: magnesium (Mg) transporter, MS: mass-spectrometry, OE: overexpressed, SA: salicylic acid, SPS: single plant selections, TMT: tandem mass tag, TSI: top sprinkle irrigation.

3. Transcriptomic-Based Characterization and Validation

Studies at the genomic-transcriptomic interface prove insightful for gene-oriented phylogenetics [21], structural mapping [22], and functional characterizations [23]. In this regard, two studies from this Special Issue merged genome-based gene family screening with in silico expression analyses. First, Abdullah et al. [14] characterized *tcGASA* paralogous in cacao (*Theobroma cacao*) and identified targets to customize fungi-resistant varieties. Similarly, Heidari et al. [15] described *MGT* (O-6-Methylguanine-DNA Methyltransferase) genes also in cacao, and they expanded the search to orthologous in two other species of the Malvaceae family with economic importance, cotton (*Gossypium hirsutum*), and jute fiber (*Corchorus capsularis*). In silico expression analysis enabled the authors to pinpoint that MGT targets the network as part of the lipid/cell wall metabolism and photoprotection pathways. Both studies demonstrate how well consolidated genomic and transcriptomic resources can be combined in silico to source *ad hoc* gene and allelic mining.

Transcriptomic screenings are equally informative of functional gene validation and tissue/environment-conditioned expression profiling. For instance, in rice, Abdullah-Zawawi et al. [16] used in-house microarray technology to determine the subjacent regulatory machinery of drought tolerance in an *Abp57*-overexpressing transgenic line. The team recovered the MAPK, IAA, and SA pathways as co-determinants of the stress response. This way, transcriptomic resources offer detailed mechanistic understating underlying agronomical relevant phenotypes.

4. Proteomics Meets Orphan Species

Proteomics is a powerful tool that allows the identification of proteins that can be used as markers in breeding programs. The development of new methodologies, and genomics and transcriptomics databases has provided a rapid advance in plant proteomics in recent decades, including orphan species. The gel-free based techniques (shotgun or LC-MSMS), DDA (data dependent acquisition), and DIA (data-independent acquisition), and targeted strategies are the most frequently chosen methods.

Identifying proteins and derived prototypic peptides throughout shotgun proteomics has already lit up feasible paths to improve or select tolerant individuals to abiotic stresses, such as drought [24] and heavy metal toxicity [25]. By using a DIA strategy, a panel of peptides and proteins has been proposed as putative markers of resistance to *Peyronellaea pinodes* in peas [26]. However, classical gel-based proteomics techniques, such as 2DE, remain the method of choice in many experiments. Castillejo et al. [17] explored the proteomic consequences of biotic stresses using 2DE-MALDI/TOF MSMS analysis combined with protease activity assays. The authors evaluated leaf proteome responses to *Botrytis fabae* necrotrophic fungus in susceptible and resistant Faba bean (*Vicia faba*) genotypes, finding a predominant role in the chloroplast PSII protein repair cycle. More interestingly, these studies reinforce that proteomics advancements are already permeating orphan species [10], even in the forestry sector [3].

Meanwhile, in a slightly more studied crop system (*Solanum lycopersicum*), Wang et al. [18] traced the effects of ebb-and-flow sub-irrigation (EFI) at roots' protein level as compared to top sprinkle irrigation (TSI). The team identified 513 differentially accumulated proteins between treatments. Overall, these two studies are promising for un-leashing plant improvements via proteomics. We are looking forward to similar developments in proteomics and also in metabolomics on non-model crop species [27].

Meanwhile, a more challenging research gap remains open in the long term, regardless the crop species. An unanswered question until now is how upstream genomic, transcriptomic, proteomic, and metabolomic layers collide and jointly interact across climates and through time to finally shape downstream multi-dimensional phenomic expression in the field.

5. Phenomics Leverage Crop Wild Gene pools

A long-standing research gap in phenomic screening is its factual implementation in crop wild gene pools [28], which typically exhibit more environmentally dependent trait segregation [29,30]. Fortunately, in this collection, two studies have harnessed crop-wild diversity through phenomics. First, Gimode et al. [20] studied the morphological consequences of a chromosome segment in peanuts (*Arachis hypogaea*) inherited from a wild relative. The authors demonstrated that this chromosome segment sources valuable trait variation from the exotic gene pool into the cultivated background. This work is also outstanding because it sidesteps the main bottlenecks of interspecific crossing, which are recurrent species incompatibility [31] and polygenic variation [32]. To do so, the study relied on a panel of chromosome segment substitution lines (CSSL) that narrowed the introgression to discrete chromosome segments, which are easier to be retained within the recurrent parental species while conferring the desired phenotypic novelty (by definition, it would be highly desirable that the substituted chromosome segment matched a single haplotype block with strong internal linkage disequilibrIum (LD) to avoid spurious recombination events that may jeopardize its integrity and phenotypic determinism).

Another crop-wild innovation was the one by Burbano-Erazo et al. [19], who utilized multi-locality phenomics trials to characterize heat and drought tolerance in 86 interspecific backcross lines between the common bean (*Phaseolus vulgaris*) × the Tepary bean (*P. acutifolius*). The team managed to unlock interspecific adaptive variation, despite the natural incompatibility, throughout 'bridge' genotypes (those with comparatively lower incompatibility), and eventually delivered candidate introgressed lines capable of pyramiding polygenic abiotic tolerance. Other studies have also aimed to break interspecific barriers between the two species [33], genomically characterize the hybridization, and re-discover naturally occurring ecological adaptive variation for drought and heat stresses [34].

Overall, throughout this Special Issue, the works by Pandit et al. [13], Gimode et al. [20], and Burbano-Erazo et al. [19], respectively illustrated in rice, peanuts, and beans, the feasibility and power to update classical introgression breeding [35] with modern 'omics' approaches, such as genomics and phenomics. This integration enables guiding more rapidly and with better precision the pyramiding of exotic variation into elite commercial backgrounds. Alternatives to utilize interspecific genomic variability include grafting [36]. This ancient horticultural technique capable of physically merging two distinct species can be optimized for desirable trait variation using genomics [37], transcriptomics [38,39], phenomics, epigenomics [40,41], and beyond [42–44]. These promising examples amalgamate species diversity via introgression breeding and grafting, and update it to last-generation high-throughput standards. They corroborate the utility of 'omics' technologies for crop improvement without denying more classical, yet still very timing, schemes.

So far, the ultimate consequence of the bottom-up genomic, transcriptomic, proteomic, and metabolomic continuum is the phenotype. However, additional phenotypic modulation may be conferred by the emerging layers of the enviromics [7,8], epigenomics [45], and soil metagenomics [6] fields, as we envision in the next section.

6. Perspectives

Despite the effort of this Special Issue in compiling a diverse array of 'omics' sub-disciplines for crop improvement (Figure 1), high-throughput screening technologies have also permeated other promising fields that do not necessarily exhibit prominence in the present collection. For instance, as introduced in the previous section, scaling phenomics across the environmental continuum gradient would offer a more accurate prediction of the G × E interaction as part of the nascent enviromics framework [7,8], which ultimately merges [46] multi-environment phenomics screening [47] with genomic-based prediction [5,48,49]. Similarly, embracing an epigenomic footprint profiling [45] could also mechanistically disentangle a great proportion of crop phenotypic variance in reaction norms and plasticity gradients [50] naturally seen across climates [51,52].

SYNTHESIZED 'OMICS' APPROACHES FOR CROP IMPROVEMENT

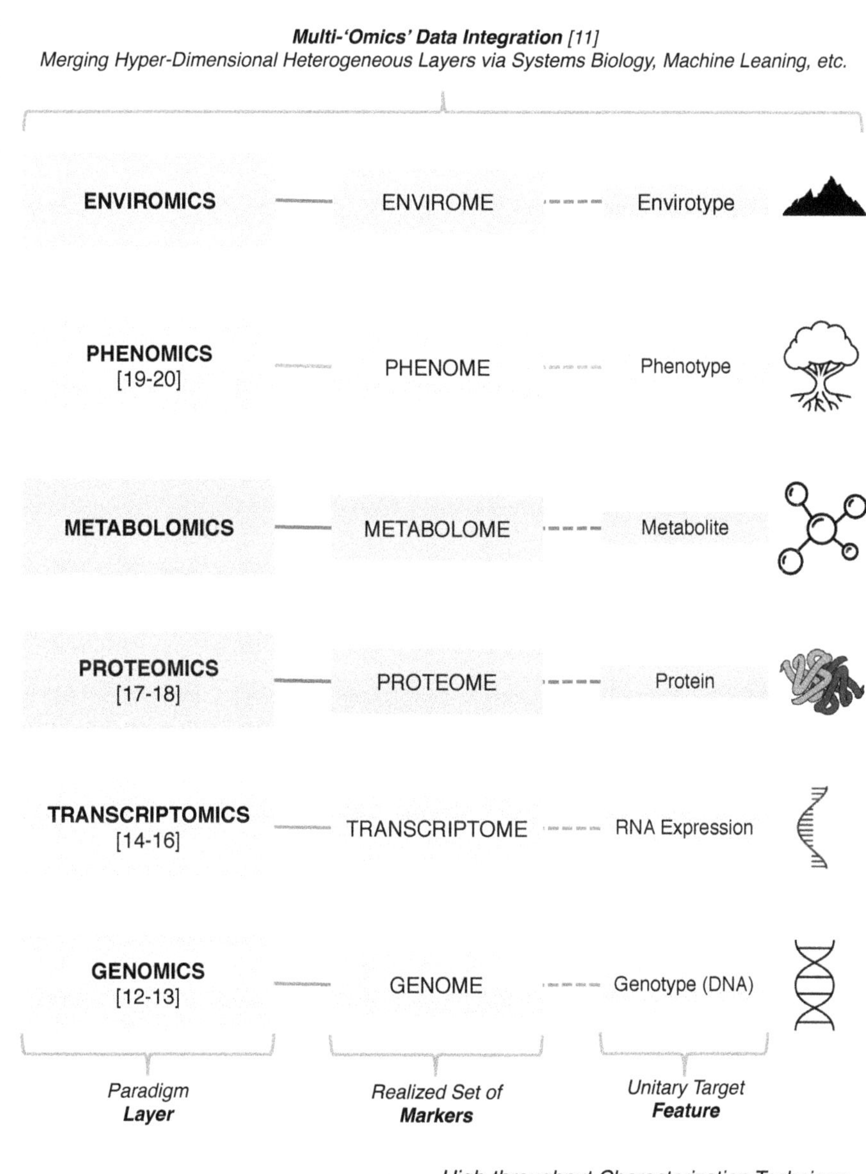

Figure 1. Synthesized 'Omics' Approaches for Crop Improvement. References from this Special Issue are indicated under Paradigm Layer within square brackets.

At another level, soil metagenomics on 'environmental' DNA [6] is boosting the retrieval of synergistic microorganisms for agriculture, given a growing sustainability requirement [53]. Backward 'omics' tools are also improving our understanding of crops' evolution [54,55] and their cultural heritage [56] by updating archeological records [57] with genomics [58] and phylogenomics [59,60] into modern paleogenomics [61,62].

While heterogeneous 'omics' data piles up across sub-disciplines, effective trans-disciplinary data merging and bioinformatics processing demands revolutionizing open-source record access [63,64], novel statistical algorithms [65], and unprecedented computational resources [66–68]. Speed breeding [69,70] through 'omics'-enabled [48,49,71], systems biology [9,72], and machine learning [4,73] predictions exemplify the promises of fast-forward customized crop breeding [2,74] by bridging the curse of dimensionality inherent from multi-'omics' data [10,75,76], while matching the modern seed delivery requirements [77]. This trend of analytical innovation may proceed further into the inflection point of the artificial general intelligence (AGI) hypothesis [78], eventually enabling human-unguided 'omics'-based plant improvement at an unforeseen pace.

Author Contributions: A.J.C., M.Á.C. and R.Y. conceived this Special Issue and together processed submissions. All authors have read and agreed to the published version of the manuscript.

Funding: A.J.C. received support from the British Council's Newton Fund during the execution of this Special Issue through grant 527023146. He was thanks Vetenskapsrådet (VR, 2016-00418/2022-04411) and Kungliga Vetenskapsakademien (KVA, BS2017-0036).

Data Availability Statement: Raw datasets were made available by each of the ten published articles [11–20] from the Special Issue on "Omics Approaches for Crop Improvement" (https://www.mdpi.com/journal/agronomy/special_issues/crop_omics_genetic (revised 18 May 2023).

Acknowledgments: A.J.C., M.Á.C. and R.Y., in their role as guest editors, thank the interest and determination of all authors, reviewers, and topic editors. Otherwise, this compilation on "Omics Approaches for Crop Improvement" would not have been possible. Agronomy's editorial team, editor-in-chief, and section assistant managing editor are as well thanked for encouraging and enabling A.J.C., M.Á.C. and R.Y. to host this Special Issue.

Conflicts of Interest: The authors declare no conflict of interest.

References

1. McCouch, S. Feeding the future. *Nature* **2013**, *499*, 23–24. [CrossRef] [PubMed]
2. Varshney, R.K.; Bohra, A.; Roorkiwal, M.; Barmukh, R.; Cowling, W.A.; Chitikineni, A.; Lam, H.-M.; Hickey, L.T.; Croser, J.S.; Bayer, P.E.; et al. Fast-forward breeding for a food-secure world. *Trends Genet.* **2021**, *37*, 1124–1136. [CrossRef] [PubMed]
3. Castillejo, M.A.; Pascual, J.; Jorrín-Novo, J.V.; Balbuena, T.S. Proteomics research in forest trees: A 2012–2022 update. *Front. Plant Sci.* **2023**, *14*, 1130665. [CrossRef] [PubMed]
4. Libbrecht, M.W.; Noble, W.S. Machine learning applications in genetics and genomics. *Nat. Rev. Genet.* **2015**, *16*, 321–332. [CrossRef]
5. Crossa, J.; Fritsche-Neto, R.; Montesinos-Lopez, O.A.; Costa-Neto, G.; Dreisigacker, S.; Montesinos-Lopez, A.; Bentley, A.R. The Modern Plant Breeding Triangle: Optimizing the Use of Genomics, Phenomics, and Enviromics Data. *Front. Plant Sci.* **2021**, *12*, 651480. [CrossRef]
6. Zhang, L.; Chen, F.; Zeng, Z.; Xu, M.; Sun, F.; Yang, L.; Bi, X.; Lin, Y.; Gao, Y.; Hao, H.; et al. Advances in Metagenomics and Its Application in Environmental Microorganisms. *Front. Microbiol.* **2021**, *12*, 3847. [CrossRef]
7. Resende, R.T.; Piepho, H.-P.; Rosa, G.J.M.; Silva-Junior, O.B.; e Silva, F.F.; de Resende, M.D.V.; Grattapaglia, D. Enviromics in breeding: Applications and perspectives on envirotypic-assisted selection. *Theor. Appl. Genet.* **2021**, *134*, 95–112. [CrossRef]
8. Costa-Neto, G.; Fritsche-Neto, R. Enviromics: Bridging different sources of data, building one framework. *Crop Breed. Appl. Biotechnol.* **2021**, *21*, e393521S12. [CrossRef]
9. Escandón, M.; Castillejo, M.; Jorrín-Novo, J.V.; Rey, M.-D. Molecular Research on Stress Responses in *Quercus* spp.: From Classical Biochemistry to Systems Biology through Omics Analysis. *Forests* **2021**, *12*, 364. [CrossRef]
10. Maldonado-Alconada, A.M.; Castillejo, M.; Rey, M.-D.; Labella-Ortega, M.; Tienda-Parrilla, M.; Hernández-Lao, T.; Honrubia-Gómez, I.; Ramírez-García, J.; Guerrero-Sanchez, V.M.; López-Hidalgo, C.; et al. Multiomics Molecular Research into the Recalcitrant and Orphan *Quercus ilex* Tree Species: Why, What for, and How. *Int. J. Mol. Sci.* **2022**, *23*, 9980. [CrossRef]
11. Zainal-Abidin, R.-A.; Ruhaizat-Ooi, I.-H.; Harun, S. A Review of Omics Technologies and Bioinformatics to Accelerate Improvement of Papaya Traits. *Agronomy* **2021**, *11*, 1356. [CrossRef]

12. Campuzano-Duque, L.F.; Bejarano-Garavito, D.; Castillo-Sierra, J.; Torres-Cuesta, D.R.; Cortés, A.J.; Blair, M.W. SNP Genotyping for Purity Assessment of a Forage Oat (*Avena sativa* L.) Variety from Colombia. *Agronomy* **2022**, *12*, 1710. [CrossRef]

13. Pandit, E.; Pawar, S.; Barik, S.R.; Mohanty, S.P.; Meher, J.; Pradhan, S.K. Marker-Assisted Backcross Breeding for Improvement of Submergence Tolerance and Grain Yield in the Popular Rice Variety 'Maudamani'. *Agronomy* **2021**, *11*, 1263. [CrossRef]

14. Abdullah; Faraji, S.; Mehmood, F.; Malik, H.M.T.; Ahmed, I.; Heidari, P.; Poczai, P. The GASA Gene Family in Cacao (*Theobroma cacao*, Malvaceae): Genome Wide Identification and Expression Analysis. *Agronomy* **2021**, *11*, 1425. [CrossRef]

15. Heidari, P.; Abdullah; Faraji, S.; Poczai, P. Magnesium transporter Gene Family: Genome-Wide Identification and Characterization in Theobroma cacao, Corchorus capsularis, and Gossypium hirsutum of Family Malvaceae. *Agronomy* **2021**, *11*, 1651. [CrossRef]

16. Abdullah-Zawawi, M.-R.; Tan, L.-W.; Ab Rahman, Z.; Ismail, I.; Zainal, Z. An Integration of Transcriptomic Data and Modular Gene Co-Expression Network Analysis Uncovers Drought Stress-Related Hub Genes in Transgenic Rice Overexpressing *OsAbp57*. *Agronomy* **2022**, *12*, 1959. [CrossRef]

17. Castillejo, M.Á.; Villegas-Fernández, Á.M.; Hernández-Lao, T.; Rubiales, D. Photosystem II Repair Cycle in Faba Bean May Play a Role in Its Resistance to *Botrytis fabae* Infection. *Agronomy* **2021**, *11*, 2247. [CrossRef]

18. Wang, K.; Ali, M.M.; Guo, T.; Su, S.; Chen, X.; Xu, J.; Chen, F. TMT-Based Quantitative Proteomic Analysis Reveals the Response of Tomato (*Solanum lycopersicum* L.) Seedlings to Ebb-and-Flow Subirrigation. *Agronomy* **2022**, *12*, 1880. [CrossRef]

19. Burbano-Erazo, E.; León-Pacheco, R.I.; Cordero-Cordero, C.C.; López-Hernández, F.; Cortés, A.J.; Tofiño-Rivera, A.P. Multi-Environment Yield Components in Advanced Common Bean (*Phaseolus vulgaris* L.) × Tepary Bean (*P. acutifolius* A. Gray) Interspecific Lines for Heat and Drought Tolerance. *Agronomy* **2021**, *11*, 1978. [CrossRef]

20. Gimode, D.; Chu, Y.; Holbrook, C.C.; Fonceka, D.; Porter, W.; Dobreva, I.; Teare, B.; Ruiz-Guzman, H.; Hays, D.; Ozias-Akins, P. High-Throughput Canopy and Belowground Phenotyping of a Set of Peanut CSSLs Detects Lines with Increased Pod Weight and Foliar Disease Tolerance. *Agronomy* **2023**, *13*, 1223. [CrossRef]

21. André, T.; Sass, C.; Yockteng, R.; Wendt, T.; Palma-Silva, C.; Specht, C.D. Deep genetic structure of a ground-herb along contrasting environments of seasonally dry understories in Amazonia and Cerrado as revealed from targeted genomic sequencing. *Bot. J. Linn. Soc.* **2021**, *199*, 196–209. [CrossRef]

22. Osorio-Guarín, J.A.; Berdugo-Cely, J.A.; Coronado-Silva, R.A.; Baez, E.; Jaimes, Y.; Yockteng, R. Genome-Wide Association Study Reveals Novel Candidate Genes Associated with Productivity and Disease Resistance to *Moniliophthora* spp. in Cacao (*Theobroma cacao* L.). *G3 Genes Genomes Genet.* **2020**, *10*, 1713–1725. [CrossRef] [PubMed]

23. López Hernández, F.; Cortés, A.J. Whole Transcriptome Sequencing Unveils the Genomic Determinants of Putative Somaclonal Variation in Mint (*Mentha* L.). *Int. J. Mol. Sci.* **2022**, *23*, 5291. [CrossRef] [PubMed]

24. San-Eufrasio, B.; Bigatton, E.D.; Guerrero-Sánchez, V.M.; Chaturvedi, P.; Jorrín-Novo, J.V.; Rey, M.-D.; Castillejo, M. Proteomics Data Analysis for the Identification of Proteins and Derived Proteotypic Peptides of Potential Use as Putative Drought Tolerance Markers for *Quercus ilex*. *Int. J. Mol. Sci.* **2021**, *22*, 3191. [CrossRef] [PubMed]

25. Salas-Moreno, M.; Castillejo, M.; Rodríguez-Cavallo, E.; Marrugo-Negrete, J.; Méndez-Cuadro, D.; Jorrín-Novo, J. Proteomic Changes in *Paspalum fasciculatum* Leaves Exposed to Cd Stress. *Plants* **2022**, *11*, 2455. [CrossRef]

26. Castillejo, M.Á.; Fondevilla-Aparicio, S.; Fuentes-Almagro, C.; Rubiales, D. Quantitative Analysis of Target Peptides Related to Resistance Against Ascochyta Blight (Peyronellaea pinodes) in Pea. *J. Proteome Res.* **2020**, *19*, 1000–1012. [CrossRef]

27. Tienda-Parrilla, M.; López-Hidalgo, C.; Guerrero-Sanchez, V.M.; Infantes-González, Á.; Valderrama-Fernández, R.; Castillejo, M.; Jorrín-Novo, J.V.; Rey, M.-D. Untargeted MS-Based Metabolomics Analysis of the Responses to Drought Stress in *Quercus ilex* L. Leaf Seedlings and the Identification of Putative Compounds Related to Tolerance. *Forests* **2022**, *13*, 551. [CrossRef]

28. Cortés, A.J.; Barnaby, J.Y. Editorial: Harnessing genebanks: High-throughput phenotyping and genotyping of crop wild relatives and landraces. *Front. Plant Sci.* **2023**, *14*, 1149469. [CrossRef]

29. Cortés, A.J.; Cornille, A.; Yockteng, R. Evolutionary Genetics of Crop-Wild Complexes. *Genes* **2022**, *13*, 1. [CrossRef]

30. Beebe, S.E.; Rao, I.M.; Cajiao, C.; Grajales, M. Selection for Drought Resistance in Common Bean Also Improves Yield in Phosphorus Limited and Favorable Environments. *Crop Sci.* **2008**, *48*, 582–592. [CrossRef]

31. Coyne, C.J.; Kumar, S.; von Wettberg, E.J.; Marques, E.; Berger, J.D.; Redden, R.J.; Ellis, T.N.; Brus, J.; Zablatzká, L.; Smýkal, P. Potential and limits of exploitation of crop wild relatives for pea, lentil, and chickpea improvement. *Legum. Sci.* **2020**, *2*, e36. [CrossRef]

32. Barghi, N.; Hermisson, J.; Schlötterer, C. Polygenic adaptation: A unifying framework to understand positive selection. *Nat. Rev. Genet.* **2020**, *21*, 769–781. [CrossRef]

33. Munoz, L.C.; Blair, M.W.; Duque, M.C.; Tohme, J.; Roca, W. Introgression in Common Bean × Tepary Bean Interspecific Congruity-Backcross Lines as Measured by AFLP Markers. *Crop Sci.* **2004**, *44*, 637–645. [CrossRef]

34. Buitrago-Bitar, M.A.; Cortés, A.J.; López-Hernández, F.; Londoño-Caicedo, J.M.; Muñoz-Florez, J.E.; Muñoz, L.C.; Blair, M.W. Allelic Diversity at Abiotic Stress Responsive Genes in Relationship to Ecological Drought Indices for Cultivated Tepary Bean, *Phaseolus acutifolius* A. Gray, and Its Wild Relatives. *Genes* **2021**, *12*, 556. [CrossRef]

35. Migicovsky, Z.; Myles, S. Exploiting Wild Relatives for Genomics-assisted Breeding of Perennial Crops. *Front. Plant Sci.* **2017**, *8*, 460. [CrossRef]

36. Warschefsky, E.J.; Klein, L.L.; Frank, M.H.; Chitwood, D.H.; Londo, J.P.; von Wettberg, E.J.B.; Miller, A.J. Rootstocks: Diversity, Domestication, and Impacts on Shoot Phenotypes. *Trends Plant Sci.* **2016**, *21*, 418–437. [CrossRef]

37. Loupit, G.; Cookson, S.J. Identifying Molecular Markers of Successful Graft Union Formation and Compatibility. *Front. Plant Sci.* **2020**, *11*, 610352. [CrossRef]

38. Gautier, A.T.; Chambaud, C.; Brocard, L.; Ollat, N.; A Gambetta, G.; Delrot, S.; Cookson, S.J. Merging genotypes: Graft union formation and scion–rootstock interactions. *J. Exp. Bot.* **2019**, *70*, 747–755. [CrossRef]

39. Guillaumie, S.; Decroocq, S.; Ollat, N.; Delrot, S.; Gomès, E.; Cookson, S.J. Dissecting the control of shoot development in grapevine: Genetics and genomics identify potential regulators. *BMC Plant Biol.* **2020**, *20*, 70. [CrossRef]

40. Kapazoglou, A.; Tani, E.; Avramidou, E.V.; Abraham, E.M.; Gerakari, M.; Megariti, S.; Doupis, G.; Doulis, A.G. Epigenetic Changes and Transcriptional Reprogramming Upon Woody Plant Grafting for Crop Sustainability in a Changing Environment. *Front. Plant Sci.* **2021**, *11*, 613004. [CrossRef]

41. Tsaballa, A.; Xanthopoulou, A.; Madesis, P.; Tsaftaris, A.; Nianiou-Obeidat, I. Vegetable Grafting From a Molecular Point of View: The Involvement of Epigenetics in Rootstock-Scion Interactions. *Front. Plant Sci.* **2021**, *11*, 621999. [CrossRef] [PubMed]

42. Albacete, A.; Martínez-Andújar, C.; Martínez-Pérez, A.; Thompson, A.; Dodd, I.C.; Pérez-Alfocea, F. Unravelling rootstockxscion interactions to improve food security. *J. Exp. Bot.* **2015**, *66*, 2211–2226. [CrossRef] [PubMed]

43. Goldschmidt, E.E. Plant grafting: New mechanisms, evolutionary implications. *Front. Plant Sci.* **2014**, *5*, 727. [CrossRef] [PubMed]

44. Wang, J.; Jiang, L.; Wu, R. Plant grafting: How genetic exchange promotes vascular reconnection. *New Phytol.* **2017**, *214*, 56–65. [CrossRef] [PubMed]

45. Mehrmohamadi, M.; Sepehri, M.H.; Nazer, N.; Norouzi, M.R. A Comparative Overview of Epigenomic Profiling Methods. *Front. Cell Dev. Biol.* **2021**, *9*, 714687. [CrossRef]

46. Cooper, M.; Messina, C.D. Can We Harness "Enviromics" to Accelerate Crop Improvement by Integrating Breeding and Agronomy? *Front. Plant Sci.* **2021**, *12*, 735143. [CrossRef]

47. Costa-Neto, G.; Fritsche-Neto, R.; Crossa, J. Nonlinear kernels, dominance, and envirotyping data increase the accuracy of genome-based prediction in multi-environment trials. *Heredity* **2020**, *126*, 92–106. [CrossRef]

48. Desta, Z.A.; Ortiz, R. Genomic selection: Genome-wide prediction in plant improvement. *Trends Plant Sci.* **2014**, *19*, 592–601. [CrossRef]

49. Crossa, J.; Pérez-Rodríguez, P.; Cuevas, J.; Montesinos-López, O.; Jarquín, D.; de los Campos, G.; Burgueño, J.; González-Camacho, J.M.; Pérez-Elizalde, S.; Beyene, Y.; et al. Genomic Selection in Plant Breeding: Methods, Models, and Perspectives. *Trends Plant Sci.* **2017**, *22*, 961–975. [CrossRef]

50. Costa-Neto, G.; Crossa, J.; Fritsche-Neto, R. Enviromic Assembly Increases Accuracy and Reduces Costs of the Genomic Prediction for Yield Plasticity in Maize. *Front. Plant Sci.* **2021**, *12*, 717552. [CrossRef]

51. Cortés, A.J.; López-Hernández, F.; Blair, M.W. Genome–Environment Associations, an Innovative Tool for Studying Heritable Evolutionary Adaptation in Orphan Crops and Wild Relatives. *Front. Genet.* **2022**, *13*, 1562. [CrossRef]

52. Hancock, A.M.; Brachi, B.; Faure, N.; Horton, M.W.; Jarymowycz, L.B.; Sperone, F.G.; Toomajian, C.; Roux, F.; Bergelson, J. Adaptation to Climate Across the *Arabidopsis thaliana* Genome. *Science* **2011**, *334*, 83–86. [CrossRef]

53. Nwachukwu, B.C.; Babalola, O.O. Metagenomics: A Tool for Exploring Key Microbiome With the Potentials for Improving Sustainable Agriculture. *Front. Sustain. Food Syst.* **2022**, *6*, 886987. [CrossRef]

54. Meyer, R.S.; Purugganan, M.D. Evolution of crop species: Genetics of domestication and diversification. *Nat. Rev. Genet.* **2013**, *14*, 840–852. [CrossRef]

55. Cortés, A.J.; Skeen, P.; Blair, M.W.; Chacón-Sánchez, M.I. Does the Genomic Landscape of Species Divergence in Phaseolus Beans Coerce Parallel Signatures of Adaptation and Domestication? *Front. Plant Sci.* **2018**, *9*, 1816. [CrossRef]

56. Przelomska, N.A.S.; Armstrong, C.G.; Kistler, L. Ancient Plant DNA as a Window Into the Cultural Heritage and Biodiversity of Our Food System. *Front. Ecol. Evol.* **2020**, *8*, 74. [CrossRef]

57. Purugganan, M.D.; Fuller, D.Q. Archaeological data reveal slow rates of evolution during plant domestication. *Evolution* **2011**, *65*, 171–183. [CrossRef]

58. Purugganan, M.D.; Jackson, S.A. Advancing crop genomics from lab to field. *Nat. Genet.* **2021**, *53*, 595–601. [CrossRef]

59. Wang, M.; Zhang, L.; Zhang, Z.; Li, M.; Wang, D.; Zhang, X.; Xi, Z.; Keefover-Ring, K.; Smart, L.B.; DiFazio, S.P.; et al. Phylogenomics of the genus *Populus* reveals extensive interspecific gene flow and balancing selection. *New Phytol.* **2020**, *225*, 1370–1382. [CrossRef]

60. Cheng, L.; Li, M.; Han, Q.; Qiao, Z.; Hao, Y.; Balbuena, T.S.; Zhao, Y. Phylogenomics Resolves the Phylogeny of Theaceae by Using Low-Copy and Multi-Copy Nuclear Gene Makers and Uncovers a Fast Radiation Event Contributing to Tea Plants Diversity. *Biology* **2022**, *11*, 1007. [CrossRef]

61. Barrera-Redondo, J.; Piñero, D.; Eguiarte, L.E. Genomic, Transcriptomic and Epigenomic Tools to Study the Domestication of Plants and Animals: A Field Guide for Beginners. *Front. Genet.* **2020**, *11*, 742. [CrossRef] [PubMed]

62. Swarts, K.; Gutaker, R.M.; Benz, B.; Blake, M.; Bukowski, R.; Holland, J.; Kruse-Peeples, M.; Lepak, N.; Prim, L.; Romay, M.C.; et al. Genomic estimation of complex traits reveals ancient maize adaptation to temperate North America. *Science* **2017**, *357*, 512–515. [CrossRef]

63. McCouch, S.R.; Wright, M.H.; Tung, C.-W.; Maron, L.G.; McNally, K.L.; Fitzgerald, M.; Singh, N.; DeClerck, G.; Agosto-Perez, F.; Korniliev, P.; et al. Open access resources for genome-wide association mapping in rice. *Nat. Commun.* **2016**, *7*, 10532. [CrossRef] [PubMed]

64. Spindel, J.E.; McCouch, S.R. When more is better: How data sharing would accelerate genomic selection of crop plants. *New Phytol.* **2016**, *212*, 814–826. [CrossRef] [PubMed]

65. Cortés, A.J.; López-Hernández, F. Harnessing Crop Wild Diversity for Climate Change Adaptation. *Genes* **2021**, *12*, 783. [CrossRef]

66. Wang, P.; Schumacher, A.M.; Shiu, S.-H. Computational prediction of plant metabolic pathways. *Curr. Opin. Plant Biol.* **2022**, *66*, 102171. [CrossRef]

67. Tirnaz, S.; Zandberg, J.; Thomas, W.J.W.; Marsh, J.; Edwards, D.; Batley, J. Application of crop wild relatives in modern breeding: An overview of resources, experimental and computational methodologies. *Front. Plant Sci.* **2022**, *13*, 1008904. [CrossRef]
68. Montesinos-López, O.A.; Vallejo, F.J.M.; Crossa, J.; Gianola, D.; Hernández-Suárez, C.M.; Montesinos-López, A.; Juliana, P.; Singh, R. A Benchmarking Between Deep Learning, Support Vector Machine and Bayesian Threshold Best Linear Unbiased Prediction for Predicting Ordinal Traits in Plant Breeding. *G3 Genes Genomes Genet.* **2019**, *9*, 601–618. [CrossRef]
69. Varshney, R.K.; Barmukh, R.; Roorkiwal, M.; Qi, Y.; Kholova, J.; Tuberosa, R.; Reynolds, M.P.; Tardieu, F.; Siddique, K.H.M. Breeding custom-designed crops for improved drought adaptation. *Adv. Genet.* **2021**, *2*, e202100017. [CrossRef]
70. Bohra, A.; Kilian, B.; Sivasankar, S.; Caccamo, M.; Mba, C.; McCouch, S.R.; Varshney, R.K. Reap the crop wild relatives for breeding future crops. *Trends Biotechnol.* **2022**, *40*, 412–431. [CrossRef]
71. Arenas, S.; Cortés, A.J.; Mastretta-Yanes, A.; Jaramillo-Correa, J.P. Evaluating the accuracy of genomic prediction for the management and conservation of relictual natural tree populations. *Tree Genet. Genomes* **2021**, *17*, 12. [CrossRef]
72. Myburg, A.A.; Hussey, S.; Wang, J.P.; Street, N.R.; Mizrachi, E. Systems and Synthetic Biology of Forest Trees: A Bioengineering Paradigm for Woody Biomass Feedstocks. *Front. Plant Sci.* **2019**, *10*, 775. [CrossRef]
73. Schrider, D.R.; Kern, A.D. Supervised Machine Learning for Population Genetics: A New Paradigm. *Trends Genet.* **2018**, *34*, 301–312. [CrossRef]
74. Grattapaglia, D.; Silva-Junior, O.B.; Resende, R.T.; Cappa, E.P.; Müller, B.S.F.; Tan, B.; Isik, F.; Ratcliffe, B.; El-Kassaby, Y.A. Quantitative Genetics and Genomics Converge to Accelerate Forest Tree Breeding. *Front. Plant Sci.* **2018**, *9*, 1693. [CrossRef]
75. Guerrero-Sánchez, V.M.; López-Hidalgo, C.; Rey, M.-D.; Castillejo, M.Á.; Jorrín-Novo, J.V.; Escandón, M. Multiomic Data Integration in the Analysis of Drought-Responsive Mechanisms in *Quercus ilex* Seedlings. *Plants* **2022**, *11*, 3067. [CrossRef]
76. López-Hidalgo, C.; Guerrero-Sánchez, V.M.; Gómez-Gálvez, I.; Sánchez-Lucas, R.; Castillejo-Sánchez, M.A.; Alconada, A.M.M.; Valledor, L.; Jorrín-Novo, J.V. A Multi-Omics Analysis Pipeline for the Metabolic Pathway Reconstruction in the Orphan Species Quercus ilex. *Front. Plant Sci.* **2018**, *9*, 935. [CrossRef]
77. Peláez, D.; Aguilar, P.A.; Mercado, M.; López-Hernández, F.; Guzmán, M.; Burbano-Erazo, E.; Denning-James, K.; Medina, C.I.; Blair, M.W.; De Vega, J.J.; et al. Genotype Selection, and Seed Uniformity and Multiplication to Ensure Common Bean (*Phaseolus vulgaris* L.) var. Liborino. *Agronomy* **2022**, *12*, 2285. [CrossRef]
78. Khan, M.H.U.; Wang, S.; Wang, J.; Ahmar, S.; Saeed, S.; Khan, S.U.; Xu, X.; Chen, H.; Bhat, J.A.; Feng, X. Applications of Artificial Intelligence in Climate-Resilient Smart-Crop Breeding. *Int. J. Mol. Sci.* **2022**, *23*, 11156. [CrossRef]

Review

A Review of Omics Technologies and Bioinformatics to Accelerate Improvement of Papaya Traits

Rabiatul-Adawiah Zainal-Abidin [1], Insyirah-Hannah Ruhaizat-Ooi [2] and Sarahani Harun [2,*

[1]   Biotechnology and Nanotechnology Research Centre, Malaysian Agricultural Research and Development Institute (MARDI), Serdang 43400, Malaysia; rabiatul@mardi.gov.my
[2]   Centre for Bioinformatics Research, Institute of Systems Biology (INBIOSIS), Universiti Kebangsaan Malaysia, Bangi 43600, Malaysia; P109077@siswa.ukm.edu.my
*   Correspondence: sarahani@ukm.edu.my

Abstract: Papaya (*Carica papaya*) is an economically important fruit crop that is mostly planted in tropical and subtropical regions. Major diseases of papaya, such as the papaya dieback disease (PDD), papaya ringspot virus (PRSV) disease, and papaya sticky disease (PSD), have caused large yield and economic losses in papaya-producing countries worldwide. Postharvest losses have also contributed to the decline in papaya production. Hence, there is an urgent need to secure the production of papaya for a growing world population. Integration of omics resources in crop breeding is anticipated in order to facilitate better-designed crops in the breeding programme. In papaya research, the application of omics and bioinformatics approaches are gradually increased and are underway. Hence, this review focuses on addressing omics technologies and bioinformatics that are used in papaya research. To date, four traits of the papaya have been studied using omics and bioinformatics approaches, which include its ripening process, abiotic stress, disease resistance, and fruit quality (i.e., sweetness, fruit shape, and fruit size). This review also highlights the potential of genetics and genomics data, as well as the systems biology approach that can be applied in a papaya-breeding programme in the near future.

Keywords: bioinformatics; comparative genomics; molecular markers; next-generation sequencing; omics; papaya

Citation: Zainal-Abidin, R.-A.; Ruhaizat-Ooi, I.-H.; Harun, S. A Review of Omics Technologies and Bioinformatics to Accelerate Improvement of Papaya Traits. *Agronomy* **2021**, *11*, 1356. https://doi.org/10.3390/agronomy11071356

Academic Editors: Roxana Yockteng, Andrés J. Cortés and María Ángeles Castillejo

Received: 30 May 2021
Accepted: 17 June 2021
Published: 1 July 2021

Publisher's Note: MDPI stays neutral with regard to jurisdictional claims in published maps and institutional affiliations.

1. Introduction

Papaya (*Carica papaya*) is one of the tropical fruits that is widely grown in tropical and subtropical countries such as Australia, Brazil, Malaysia, Thailand, and South America. In worldwide fruit production, the papaya is ranked in third place after mango and pineapple fruit production [1]. The papaya has a high nutritional content and medicinal value, which make it widely planted and globally popular. For instance, the fruit is rich in vitamin A, vitamin C, folate, potassium, and niacin, and its leaves, stems, and roots are suitable for alternative medicine [2].

Papaya has nine pairs of chromosomes and is a diploid. The papaya, which has a genome size of 372 Mb, belongs to the Brassicales order and Caricaceae family. Among the well-known papaya cultivars in Hawaii are Solo, Sunrise, SunUp, and Rainbow [3], while Eksotika and Sekaki are well-known cultivars in Malaysia [3]. To date, six papaya cultivars, namely Eksotika, Eksotika II, Sekaki, Three Pillars, Frangi, and Viorica have been registered with the Department of Agriculture, Malaysia [4]. Correspondingly, Japan also produces papaya cultivars, namely Ishigaki Sango and Ishigaki Wondrous [5]; in India, the well-known papaya cultivars are Co1 and Co2 [3].

As a perennial herb, the breeding cycle of the papaya takes about seven to nine months until a ripe fruit is produced. The breeding cycle of the papaya is divided into three stages that are represented by seed germination, flowering, and fruit setting to fruit harvesting (Figure 1) [6,7]. The papaya seed germinates within two to four weeks after sowing [8].

Then, after being transplanted, its vegetative state starts to grow into single-stemmed trees, each bearing a rosette of large, deeply lobed leaves at the apex about two to three months after transplantation [6–8]. The papaya plant produces its flowers at the age of three to six months, and produces ripe fruits at seven to nine months [6,7].

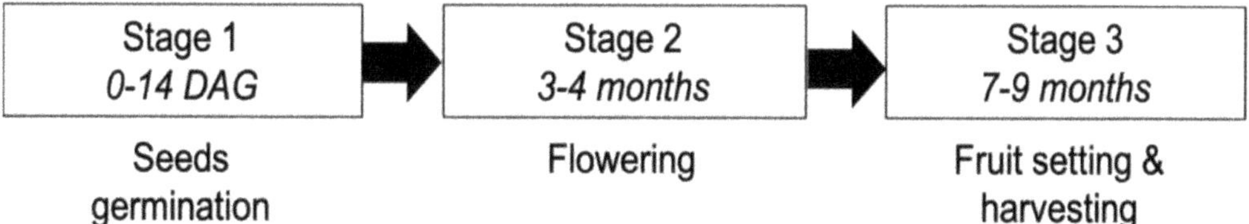

Figure 1. The breeding cycle of papaya. DAG denotes day after germination.

Papaya is cultivated in fertile and well-drained soils with a pH of 5.5 to 6.5 [8,9]. Nutrients for the papaya are essential in order to increase the productivity and quality of papaya production. Nitrogen (N) and phosphorus (P) are essential nutrients during the early growth of the papaya in order to ensure an optimum growth of foliage, trunks, and roots, as well as induce higher productivity [10]. Potassium (K) is needed at the papaya fruiting stage because it is essential for improving the fruit quality (i.e., sweetness and pulp thickness) [11,12]. Hence, these agronomic aspects are contributed to ensure that breeding of papaya is successful. The breeding of papaya has produced more papaya cultivars with desired agronomic traits such as high yield, fruit shape, fruit size, and sweetness. However, the papaya is easily exposed to pathogens and being infected by diseases, (i.e., the papaya dieback disease (PDD), papaya ringspot virus (PRSV) disease, and papaya sticky disease (PSD)), which have resulted in a decline in papaya production. In addition, high temperature and water stress in papaya plantation are being affected by climate change, which is a challenge to the papaya industry, especially for its production yield [13–15]. The high temperature (i.e., above 35 °C) causes flowers to drop and sex changes in female and hermaphrodite flowers [13]. Although high temperature (i.e., 28 °C) promotes the fast growth of papaya, low pollen viability and early maturation result in imperfect quality fruits and a low yield [13]. Interestingly, high temperature coupled with a higher moisture content produces higher total soluble solids (TSS) in papaya [13]. A sufficient amount of water is essential to papaya because it determines the fruit size and fruit quality [15]. For instance, in dry conditions, the fruits are smaller with a hard texture when ripe.

One major way to overcome the constraints due to climate change and papaya diseases is by breeding for new and improved papaya that has been enhanced with desired traits, such as resistance to disease, resistance to abiotic stress, delayed ripening, and sweetness. Recent trends in crop improvements have shown the integration of omics approaches (i.e., genomics, transcriptomics, proteomics, and metabolomics) and bioinformatics in breeding programmes [16–19]. The use of omics and bioinformatic approaches in crop breeding helps to obtain a holistic understanding of the genetic and genomic bases of the crop, as well as to understand the molecular interaction among genes, proteins, and metabolites, especially regarding complex traits. Thus, the integration of omics and bioinformatic approaches in crop-breeding programmes is anticipated to facilitate the development of climate-resilient crop varieties and efficient germplasm screening, as well as to accelerate the rate of the genetic gain in a crop [18,19].

To date, papaya-breeding programmes have focused on improving yield and quality, resistance to abiotic stress condition, resistance to disease, as well as delayed ripening. Remarkable success has been achieved in the genetic engineering of papaya: the first transgenic papaya cultivar SunUp developed resistance to papaya ring spot virus (PRSV) disease [20]. In 2008, the first papaya genome, from the cultivar SunUp, was sequenced using whole-genome shotgun Sanger sequencing [21]. The papaya cultivar SunUp is a transgenic papaya and was the first papaya genome sequenced [21]. The SunUp genome sequence was annotated, and yielded 27,793 protein-coding transcripts. To date, SunUp has been used as a reference genome for various comparative genomics analyses of the papaya genome. Previous efforts have exploited the genome sequences, molecular markers, and physical and genetic maps for improvement of papaya traits [22–25]. The domestication and genetics of the papaya have also been discussed [26]; while Dhekney et al. [8] and Palei et al. [27] have summarised the use of biotechnology tools and the progress of research that has been conducted on papaya, they have not covered all the omics and bioinformatic approaches that have been used in the improvement of papaya traits.

Hence, this review paper aims to examine the recent progress of the improvement of papaya traits using the omics and bioinformatic approaches, and their application in breeding for new and improved papaya cultivars (Figure 2). Each omics approach is shown in Figure 2, outlining how most of the generated molecular or omics data are analysed and visualised using bioinformatics approaches. The integration of systems biology with the analysed omics data analysis will cater to the identification of molecular markers, as well as candidate genes, proteins, and metabolites, for application in papaya-breeding programmes. In this review paper, the future perspective of using omics approaches in improving the desired papaya traits are also discussed.

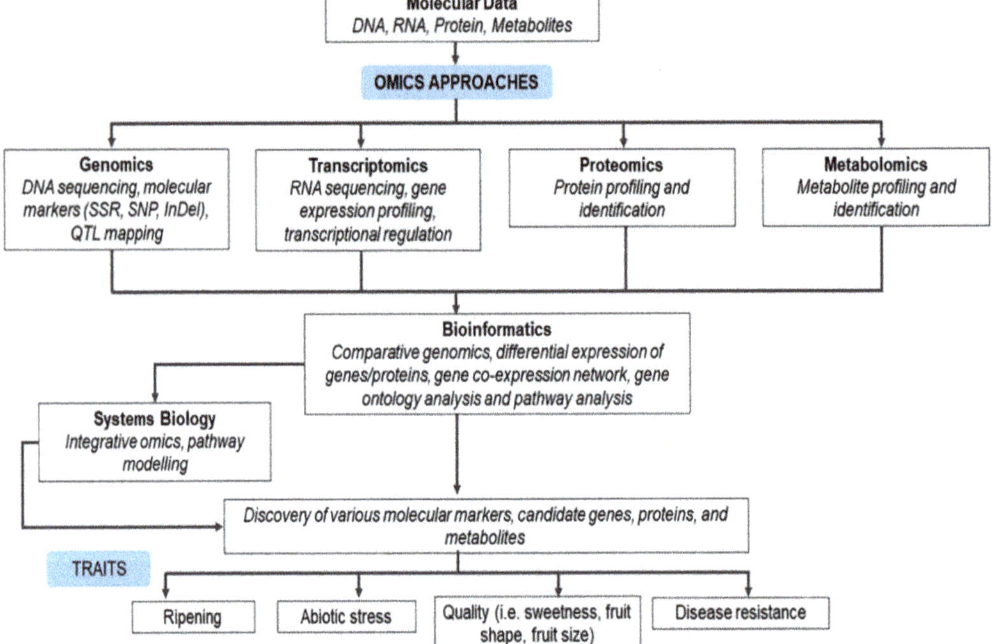

Figure 2. A summary of the omics approaches that are used in the improvement of papaya traits. Four traits have been studied using omics and bioinformatic approaches, including ripening, abiotic stress, disease resistance, and fruit quality.

2. The Role of Bioinformatics in Analysing Omics Data

Bioinformatics is an interdisciplinary field of various research backgrounds that comprises biology, computer science, mathematics, and statistics in order to extract, analyse, integrate, and visualise the biological data that are generated from omics platform technologies [28]. Several bioinformatics tools and biological databases have been developed and are accessible to other researchers in similar fields [19,28]. The development of useful bioinformatics tools is also important for crop breeders in order to inform them of the target selection of traits effortlessly [19,28]. The continuous effort in building and managing the integrated databases would contribute to the translational research in which multiomics data analysis has become an integral part of systems biology.

Comparative Genomics Analysis of Carica papaya

The completion of the SunUp papaya reference genome sequence provides the opportunity for researchers to perform a comparative genomics analysis of the papaya genome. Comparative genomics, which is also known as comparative genome-wide analysis, is an area of study in which the structure and function of genomes from different species or varieties are compared [29]. Using computational approaches, the comparison of genomes and their contents among the cultivars or species enable researchers to identify sequence features (i.e., genes and proteins) that are conserved among the species [30].

Such examples include the comparative genome-wide analysis of the papaya genome in order to identify, annotate, and classify the genes into several gene and transcription-factor families that are associated with abiotic stress [31], disease resistance [24,32], ripening [33,34], and flower development [35]. The identification and annotation of the selected gene families in the papaya genome have also provided information on their gene structure, and phylogenetic tree relationships in relation to orthologs or paralogs from other species. Consequently, the potential genes and transcription factors were validated to further investigate their expression patterns under different conditions [31–35].

The papaya is susceptible to multiple pathogens, such as *Erwinia mallotivora* [36], ringspot virus [37], and papaya meleira virus (PMeV) [38]. Hence, in order to understand the basis of papaya resistance and susceptibility, the method is to perform a comparative genome-wide analysis of the papaya genome. The genome-wide analysis of papaya sequences has revealed that the papaya has fewer (0.2%) disease-resistance genes (i.e., nucleotide binding site (NBS)-containing *R* genes) than *Arabidopsis thaliana* (0.68%) and *O. sativa* (1.38%), which makes conventional breeding for resistance difficult [24]. Similarly, Praza-Echeverria et al. [32] investigated the potential of *NPR1*, a pathogenesis-related gene, in papaya resistance against a pathogen by comparing the *NPR1* in *Arabidopsis* and tomato genome sequences. This could lead to the application of the identified gene in genetic engineering for crop improvement.

The understanding of the papaya ripening process is important in order to reduce the occurrence of postharvest losses in the papaya industry. To improve the ripening trait in papaya, it is crucial to identify the genes that are responsible for the ripening process. Coupled with bioinformatics analysis, Liu et al. [33] identified 14 potential SQUAMOSA protein-binding protein-like (SPL) genes, whereas Xu et al. [34] investigated 18 potential auxin/indole-3-acetic acid (Aux/IAA) genes in the papaya ripening process. The involvement of SPL and Aux/IAA gene families in the papaya ripening process is not well known. Hence, these efforts provide an opportunity to understand the roles of auxin-responsive genes and SPL in the papaya ripening process.

Using the genome sequence of the papaya, the basic helix–loop–helix (bHLH) transcription factors that were associated with abiotic stress were identified [31]. A total of 73 candidate genes from the bHLH family were detected using comparative genome-wide analysis. The quantitative real time PCT (qRT-PCR) experiment also revealed the role of candidate bHLHs that might be responsible for abiotic stress responses (i.e., salt, drought, and cold stresses) in papaya. Using a similar approach, Liu et al. [35] identified 11 potential genes in the auxin response factor (ARF) transcription factor family, and their role in papaya flower development.

In summary, a comparative genomics analysis of papaya can be initiated by retrieving the papaya genome sequence from public databases. To date, five databases can be used to retrieve the papaya genome sequence; namely, the NCBI (https://ftp.ncbi.nlm.nih.gov/genomes/all/annotation_releases/3649/100/GCF_000150535.2_Papaya1.0/ accessed on 22 June 2021) [39], PlantGDB [40], Phytozome [41], EnsemblPlants [42], and PLAZA [43]. HMMER [44] and BLAST [45] were developed to perform similarity and homolog sequence analyses. InterProScan [46], Pfam [47], and SMART [48] tools were designed to identify the gene family, motif sequence, and domain association, respectively. PLACE [49] and PlantCARE [50] are tools that are used for the analysis of cis-regulatory elements and promoter identification of the genes of interest. The Gene Structure Display Server (GSDS) [51] annotates gene structures; i.e., exon, intron, 3′UTR, and 5′UTR. To construct a phylogenetic tree, multiple sequence alignment must be performed using the gene sequences containing the domain or motif of interest as input data. ClustalX [52] and MAFFT [53] are multiple sequence alignment programmes that are used for aligning the gene sequences, and MEGA is a tool that is used for constructing a phylogenetic tree [54].

Figure 3 summarises the workflow of the comparative genomics analysis of the papaya gene families using bioinformatic approaches that are related to disease resistance, ripening, flower development, and abiotic stress.

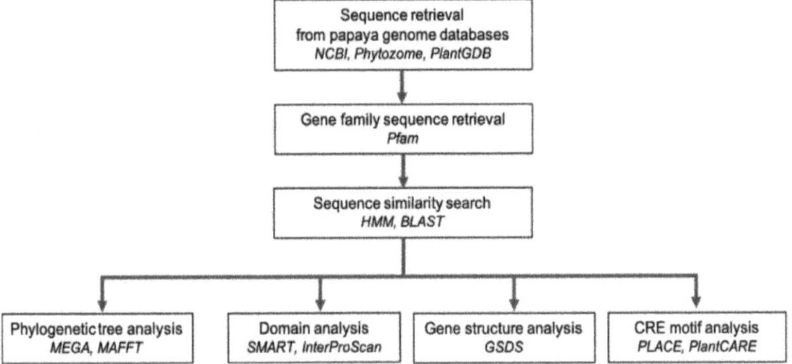

Figure 3. Workflow for comparative genomics analysis of the papaya gene families.

3. Application of Omics Technologies in *Carica papaya*

The suffix 'omics' can be described as the screening of the whole molecular data from a living organism for as genes (genomics), mRNA (transcriptomics), proteins (proteomics), and metabolites (metabolomics). As seen by the expanding number of publications over the years, these omics methods, as well as systems and synthetic biology, are becoming increasingly popular [55]. As compared to other model fruit crops such as the tomato and grape, the multiomics study of papaya is still in its infancy. However, various omics data of papaya have been generated from each omics platform, which can lead to the understanding of the complex phenotypic variations that can facilitate papaya breeding strategies in the improvement of papaya traits (Table 1) [56].

3.1. Genomics and Molecular Markers

Advancements in next-generation sequencing technology (NGS) have resulted in the different types of sequencing platforms (i.e., Illumina, PacBio, and Oxford Nanopore Technologies), which produce high-quality sequences, including longer sequence reads and fewer sequence error rates [56]. In addition, NGS has offered an affordable cost, facilitating the sequencing of the genomes of various plant species and cultivars [57]. For instance, the whole-genome resequencing of the papaya cultivars Eksotika and Sekaki from Malaysia [58], as well as the Sunset cultivar from Hawaii [59], have been carried out using the Illumina platform. The whole genome resequencing of papaya cultivars has led to the identification of a large number of variants that have been annotated into coding genes, where the identified SNPs could be developed as molecular markers in the application of a marker-assisted selection of papaya breeding.

The large volumes of data that have been generated using next-generation sequencing technology have increased the efficiency of the development of molecular markers, such as simple sequence repeat (SSR) and single nucleotide polymorphism (SNP). SSR and SNP are types of molecular markers that are widely used to improve the agronomic traits of fruit trees [5]. Molecular markers, such as restriction fragment length polymorphism (RFLP), amplified fragment length polymorphism (AFLP), and random amplified polymorphic DNA (RAPD), are often limited by low reproducibility, laborious techniques, and time-consuming processes [60]; using SSR and SNP will greatly expedite the detection by using a high-throughput genotyping platform.

Previous efforts have identified 116,43 SSRs in the SunUp genome sequence [61]. In addition, 73 SSRs were found in 25 genes that were related to fruit ripening. A similar study focused on the development of high polymorphic SSR from the whole-genome resequencing of Sunrise Solo (Hawaiian cultivar) and RB2 (Australian cultivar) [62]. The developed SSR markers can be used to differentiate the elite cultivars, and can be used in papaya breeding selection.

Genomics variants such as SNP and insertion–deletion (InDel) are highly abundant in plant genomes [63]. Previous efforts have discovered putative SNPs in the genome-wide analysis of papaya genome resequencing [20,21,48]. Zainal-Abidin et al. [58] sequenced the genome of the papaya cultivars Eksotika and Sekaki in order to identify and annotate SNPs. The identified SNPs will be useful for genotyping to develop a papaya SNP panel. Similarly, the whole-genome resequencing of papaya cultivars such as SunUp (transgenic) and SunSet (nontransgenic) has identified SNPs, InDels, and structural variations [59]. These variants' data have been used to perform a comparative genomics analysis between transgenic and nontransgenic papaya. Recently, Bohry et al. [64] carried out genome-wide identification of SNP and InDels between the two parental lines of UC10 hybrids, the Formosa elite lines Sekati and JS-12, using the Illumina MiSeq (Illumina Inc., San Diego, CA, USA) platform. Interestingly, the putative variants that were located in the ripening-related genes (RRGs) were suggested to be validated through functional analysis and the genotyping platform. Subsequently, these candidate SNPs that are associated with RRGs can be applied in diversity and genetic-mapping studies, as well as for application in marker-assisted selection (MAS) of papaya.

Recent efforts have been conducted to identify the structural variations (SVs) in the genomes of 25 wild and 42 cultivated papaya [65]. The SVs (i.e., insertions, deletions, inversions, transposable elements, and copy number variations) contain relatively long DNA changes as compared to SNPs and InDels [63,65]. These SV data have been used to unravel the effect of SVs on the papaya phenotype and its adaptation during the domestication process [65]. Detailed GO enrichment between the wild and cultivated papaya has identified genes that are artificially selected during papaya domestication. Environmental adaptability, sexual reproduction, and essential traits such as pistil development, embryonic development, flowering duration, crop yield, pedicel elongation, defence response, and disease response are all influenced by these genes. This study would facilitate the understanding of the genes that are involved during the process of papaya domestication, and provide potential SV data to develop molecular markers in papaya breeding programmes.

The establishment of a bioinformatics pipeline in discovering large genomic variants (i.e., SNP, InDel) has made it possible to unravel the genomic variants in the papaya genome from the various cultivars. The identified and annotated SNPs in disease-resistance and ripening-related genes could be applied in the MAS of the papaya as a tool to aid the selection process in papaya germplasm, as well as in the study of its genetic diversity.

The genetic map is useful for dissecting the genetic components of complex traits [66]. The first genetic map of papaya was developed between Sunrise Solo x Line UH536, which comprised 61 RAPD markers and was distributed in 11 linkage groups (LG) [22]. Then, Deputy et al. [67] developed the genetic map of Kapoho x SunUp that comprised 1498 AFLP in 12 LG; while Chen et al. [23] developed the genetic map of AU9 x SunUp, which comprised 706 SSR in 12 LGs. Blas et al. identified 14 QTL controlling fruit size and shape of papaya [68]. However, progress in the development of papaya QTL has been limited by a lack of genetic and genomic information. Genome-wide identification of SNPs in the papaya genome sequence has allowed the development of a high-density genetic map. This approach has been applied for unravelling the fruit quality of papaya, which has been developed in the F2 population from the RB2 x Sunrise Solo cross using the genotyping-by-sequencing approach (GBS) [69]. The highlights include the candidate genes that are related to the fruit quality traits, such as fruit size, fruit shape, sweetness, length, and firmness. These QTL data that are associated with the genes, and which are related to the fruit quality traits, can be used as candidates for gene exploration in the selection of SNPs and InDels using bioinformatics analysis [64]. Notably, candidate genes with associated SNP markers represent a valuable resource for the future of strategic selective breeding of elite papaya cultivars.

Although the genome sequences of papaya cultivars have been determined, no work has been carried out on the pangenome analysis of *C. papaya*. Pangenome analysis enables us to capture the entire set of genes from papaya cultivars, as well as to overcome the limitations of relying on a single reference genome [70,71]. The identification of SVs can represent pangenome analysis that would enable us to capture the entire set of genes from papaya cultivars, as well as to overcome the limitations of relying on a single reference genome [70,71]. In addition, more new candidate genes (i.e., genes for disease resistance and ripening) or gene pools can be identified from the wild germplasm and molecular markers, and these can be developed to screen for resistant varieties in the field [70,71].

3.2. Transcriptomics

The transcriptome is defined as the whole set of transcripts in a cell and the quantification of its specific developmental stage or physiological condition [72]. Hence, transcriptome plays a role in estimating the expression of genes, as well as in deciphering the regulation of genes in tissues and organs. Unravelling the transcriptome of tissues and organs of the species that are of interest can be carried out using RNA sequencing technology (RNA-seq), which has been shown to be highly reproducible and enables the simultaneous study of expressed gene samples [73–79]. These RNA-seq features have made RNA-seq experiments widely used in most transcriptome studies, including the transcriptome analysis of papaya plants.

Previous studies on papaya transcriptomes have focused on the analysis of several papaya traits, such as drought effects [73], fruit quality [74,75], sex determination [76,77], and disease mechanisms [78,79] (Table 1). The molecular response of papaya plants can be observed by analysing the tissue-specific differentially expressed genes (DEGs) using a gene-enrichment analysis. Gene-enrichment analysis using AgriGO [80] or ShinyGO [81] provides information on the biological processes that are regulated in response to the desired traits in papaya.

A transcriptome study of the delayed sticky disease symptoms in papaya has revealed the involvement of stress-responses genes in tolerance mechanisms at the pre-flowering stage [79]. In addition, the authors found that the salicylic-activated genes (i.e., *PR1, PR2, PR5*, WRKY) contributed to the delayed symptoms, while the activation of candidate genes such as *NPR1*, UDP-glucuronosyltransferase (UGT), and ethylene limit salicylic acid allowed the PSD symptoms to develop. The nutrient transporter gene family (i.e., nitrate, ammonium, potassium, sodium, phosphate, and sulfate) was the upregulated gene in the host during the infection, and had been shown to act as sensors for plant immunity [78,79].

Shen et al. [75] used RNA-seq technology to elucidate the fruit-colouration process in the ripening condition of papaya. A total of 13 candidate genes, including beta-carotene hydroxylase (CHYB), carotene ε-monooxygenase (LUT1), violaxanthin de-epoxidase (VDE), phytoene synthases (PSY1, PSY2), phytoene desaturases (PDS1, PDS2), zeta-carotene desaturae (ZDS), lycopene cyclases (CYCB, LCYB1, LCYB2, LCYE), and zeaxanthin epoxidase (ZEP), were detected in the papaya fruit transcriptome, which showed that these genes were involved in the carotenoid biosynthetic pathway.

A transcriptome study of the papaya male flower and male-to-hermaphrodite sex-reversal flower demonstrated the involvement of 1756 differentially expressed genes in sex determination [76]. Of these, four papaya homologous genes, including three PIN1 and one PIN3, were found to be upregulated in the male-to-hermaphrodite sexual-reversal flowers.

In addition, the authors found the phytohormone-biosynthesis and signal-transduction pathways in the male-to-hermaphrodite sex-reversal flowers. Similarly, Zhou et al. [77] used RNA-seq technology to investigate the underlying mechanism in DNA methylation contribution to the male and female flowers. Dissecting the molecular mechanism that regulates sex expression in the papaya provides valuable resources that facilitate an understanding of sexual differentiation in the papaya.

In the papaya disease-resistance trait, the transcriptomes of SunUp and Sunset leaves were sequenced to investigate the expression changes in both of the papaya cultivars, and were compared to the regulated genes in the resistant and susceptible varieties [78]. This comparative transcriptomic analysis found that there were few disease-resistance and hormone-related genes in SunUp, indicating the PRSV resistance in SunUp transgenic papaya. The finding of this study also provided evidence that genetically modified papaya is not harmful.

Using transcriptome data, gene clusters can be observed by performing a gene coexpression network analysis (Figure 4). This analysis identifies gene–gene interaction, regulation of biological processes, and molecular functions that occur within a cluster of genes. In a transcriptome study of papaya leaves, sap, and roots that were under mild and severe drought, several transcription factors (i.e., WRKY, MYB, bHLH) that were commonly linked to abiotic stress conditions were identified [73]. These transcription factors can be potential regulators in the leaves and roots of papaya plants under drought-stress conditions.

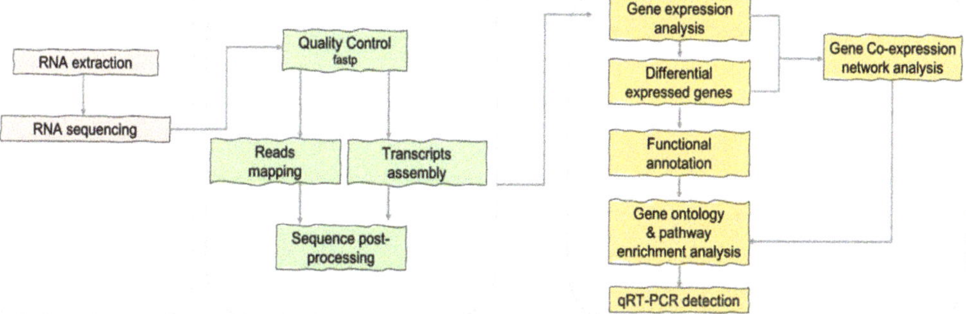

Figure 4. Schematic representation of the papaya RNA-seq analysis using a bioinformatics approach. The potential genes and transcription factor can be validated using qRT-PCR detection, which can be applied in precision breeding through gene editing or marker-assisted selection (MAS).

3.3. Proteomics

The two-dimensional differential gel electrophoresis (2D-DIGE), isobaric-tags for relative and absolute quantification (iTRAQ), and matrix-assisted laser desorption/ionisation time-of-flight mass spectrometry (MALDI-TOF/TOF MS) platforms have been widely used to estimate the expressed and abundance of proteins in plants, animals, and humans [82]. In papaya, the use of a proteomics platform has focused on the ripening process [83,84] and disease mechanism [85,86] (Table 1).

The use of a proteomics platform to study the underlying mechanism in papaya fruit ripening was first reported in 2011. Nogueira et al. [83] performed a comparative proteomic analysis of climacteric and preclimacteric papaya cultivars using the 2D-DIGE platform. Several proteins that were closely related to metabolic changes (i.e., cell wall, ethylene biosynthesis, climacteric respiratory burst, stress response, and chromoplast differentiation) in the ripening papaya were found, suggesting that these candidate proteins might be involved in the fruit-ripening process. The 2D-DIGE protein analysis, which was performed in order to identify the differentially expressed proteins during papaya ripening, also suggested the role of 1-methylcyclopropene (1-MCP) in affecting the fruit-ripening process [83]. In addition, several expressed proteins were associated with sugar and cell-wall metabolism, signalling, defence and stress responses, folding and protein stability, and ion channels. Proteins related to pectinesterases, SODs, and the dienelactone hydrolase family deserve further attention, as these proteins might be involved in the ripening processes [84]. In a proteomic study that was applied to different ripening stages in papaya, differential accumulated proteins (DAPs) were quantified using HPLC fractionation and LC-MS/MS [85]. Interestingly, unsaturated fatty acids (i.e., methyl palmitoleate and methyl alpha-octadecatrienoic) were increased during the ripening, indicating that they might be associated with volatile formation in papaya fruit.

Rodrigues et al. [86] used 2-DE and DIGE to identify and distinguish proteins that had accumulated during sticky disease infection, which was caused by the papaya meleira virus (PeMV). A proteomics study of the compatible reaction of *C. papaya* cv. Eksotika in response to *E. mallotivora* attack was carried out using an iTRAQ mass spectrometry analysis, and showed differentially expressed proteins that were related to metabolic processes, defence response, and response to stress [87]. Similarly, iTRAQ mass spectrometry was used to quantify the effector proteins in *E. mallotivora*, and suggested the type III secretion system (T3SS) as an important protein that contributed to the bacterial pathogenicity and virulance [88]. A high-throughput proteomic study using an LC-MS/MS-based label-free proteomics approach was used to assess the protein expression between PMeV-infected preflowering *C. papaya* and control plants [89]. It was suggested that the increased modulation of photosynthesis, the 26S proteasome, and cell-wall remodelling-associated proteins were involved in the initiation of papaya plant immunity.

In general, identifying differentially expressed proteins provides valuable resources for selecting essential proteins in improving *C. papaya* agronomic traits (i.e., disease resistance and ripening). To encourage reproducibility of proteomics data, the MS-based proteomics data can be uploaded to a public database such as the ProteomeXchange Consortium [90]. This effort adds additional value of the data in efforts towards the improvement of papaya traits.

3.4. Metabolomics

Metabolomics is the study of small molecules such as metabolites, substrates, and metabolic production in an organ, tissue, or cell [5]. Metabolomics has been used to identify and measure differentially expressed metabolites and to gain an insight into biochemical composition under different environmental conditions [91]. Papaya is rich in secondary metabolites that serve as a source of nutrients for human health. High-performance liquid chromatography or gas chromatography tandem mass spectrometry (LC/GC-MS) and liquid chromatography-quadrupole time-of-flight (LC-QTOF) have been widely used in studies of plant metabolomics [92]. Determining the standard chemical compounds of metabolites and their MS spectrum data is essential in a metabolomic-profiling analysis.

Previous studies performed a comparative analysis of metabolites among papaya cultivars [93,94] (Table 1). The identification of metabolites among these papaya cultivars is important in order to identify the cultivar with the essential metabolites that are associated with quality traits (i.e., sweetness and ripening) that subsequently are to be applied in papaya breeding programmes. For example, carotenoids, tocopherols, and glucosinolates are highly abundant in papaya, and all of them are sources of antioxidants [95,96]. The identification of unique metabolites in a papaya cultivar has highlighted the importance of the chemical marker in authenticating papaya-based food products [94].

Chilling injury is a part of the constraint in maintaining the freshness of papaya, especially for exported papaya. A study by Wu et al. [97] identified different metabolite profiles in the papaya peel at a temperature of 4 °C. The metabolites were associated with aroma traits, such as organic acid, amino acids, hexanal, carbonic acid, pentadecyl propyl ester, and methyl geranate, in papaya peel [97]. The elucidation of the metabolite profile that involves chilling stress at 4 °C can be applied to regulate the storage temperature of the papaya to prevent chilling injury, and to extend its storage period.

Apart from the fruit, studies of metabolite profiling in other parts of the papaya, including its leaves and seeds, have been conducted to detect the metabolites' phytomedicinal properties [98,99]. Papaya leaves are rich in metabolites (i.e., phenolics, flavonoid, saponins, and tocopherol) that have potential antimicrobial, anticancer, antioxidant, and pesticide properties [100]. Hence, papaya leaves have been widely used in the pharmacological industry for drug development.

Metabolomics that has been combined with other high-throughput omics technologies, such as transcriptomics and proteomics, is referred to as integrated metabolomics, which is sometimes used in studies that are aimed at understanding metabolism as the phenotype of genome function [92]. However, little is known on the integration between metabolomics and transcriptomics to understand the mechanism during the fruit-ripening process. It has been reported that metabolites (i.e., flavonoids, terpenoids, organic acids, phenolic acids, and alkaloids) are closely related to the ripening disorder of fruits [101–103]. The elucidation of these potential metabolites during papaya-ripening processes will provide valuable information for developing a strategy for postharvest storage and improving the fruit quality.

Table 1. A summary of the recent omics and bioinformatics approaches that are used in the improvement of papaya traits.

Type of Omics Platform	Traits/Conditions	Descriptions	Approach	Reference
Genomics	-	Whole-genome sequences of papaya cultivar SunUp. Development of first papaya reference genome sequences.	Whole-genome shotgun Sanger sequencing	[2]
	-	Whole-genome resequencing of papaya cultivars Eksotika and Sekaki to identify putative SNPs. The identified SNPs between Eksotika and Sekaki located in genes of interest could be suggested for validation using a genotyping platform.	Whole-genome resequencing using Illumina HiSeq2000 (Illumina Inc., CA, USA) and bioinformatic analysis	[58]
	-	Whole-genome resequencing of papaya cultivar SunUp (transgenic) and Sunset (nontransgenic) to identify SNPs and InDels, and used in comparing transgenic and nontransgenic papaya. The identified SNPs and InDels that were located in high-impact genes could be applied in marker-assisted PRSV disease-resistance breeding in papaya.	Whole-genome resequencing using Illumina HiSeq2000 (Illumina Inc., CA, USA) and bioinformatic analysis	[59]
	-	Whole-genome resequencing of wild-type and cultivated papaya to detect structural variations in papaya, and used in understanding the process of papaya domestication.	Whole-genome resequencing using Illumina HiSeq2500 (Illumina Inc., CA, USA) and bioinformatic analysis	[65]
	Ripening	Gene-based SSR marker development focusing on genes related to fruit ripening.	Bioinformatics and genotyping	[61]
		Polymorphic SSR marker development for marker-assisted breeding in papaya.	Whole-genome resequencing using Illumina HiSeq4000 (Illumina Inc., Foster City, CA, USA), bioinformatics, and genotyping	[62]
		Genome-wide identification of SNPs and InDels using whole-genome resequencing of two papaya cultivars, namely Sekaki and JS-12. The SNPs that were located in RRGs are potential SNPs to be converted in PCR markers, and could be applied in papaya genetic mapping and diversity studies, as well as marker-assisted selection.	Whole-genome resequencing using Illumina Miseq (Illumina Inc., Foster City, CA, USA)	[64]

Table 1. Cont.

Type of Omics Platform	Traits/Conditions	Descriptions	Approach	Reference
	Abiotic stress	Genome-wide analysis of basic helix–loop–helix (bHLH) transcription factors. Candidate bHLH genes that might be responsible for abiotic stress.	Comparative genomics and quantitative real-time PCR (qRT-PCR)	[31]
	Disease resistance	Genome-wide analysis of NBS resistance gene family. Candidate resistance (R) genes potentially responsible for disease-resistance mechanism.	Comparative genomics and quantitative real-time PCR (qRT-PCR)	[24]
	Disease resistance	Genome-wide analysis of NPR1 family. Candidate pathogenesis-related genes that might be responsible for a disease-resistance mechanism.	Comparative genomics and quantitative real-time PCR (qRT-PCR)	[32]
	Ripening	Genome-wide analysis of SQUAMOSA promoter binding protein-like gene family in papaya. Candidate ripening- and development-related genes.	Comparative genomics and quantitative real-time PCR (qRT-PCR)	[33]
	Ripening	Genome-wide analysis of Aux/IAA gene family. Candidate ripening-related genes in papaya.	Comparative genomics and quantitative real-time PCR (qRT-PCR)	[34]
	Flower development	Genome-wide analysis of auxin response factor (ARF) family genes related to flower and fruit development in papaya. Candidate genes related to flower and fruit development.	Comparative genomics and Quantitative real-time PCR (qRT-PCR)	[35]
Transcriptomics	Drought tolerance	Coexpression network analysis to identify genes and transcription factors related to abiotic stress.	Transcriptome sequencing using Illumina NextSeq500 (Illumina Inc., Foster City, CA, USA) and coexpression network analysis	[73]
	Ripening mechanism	Identification of potential regulatory genes during papaya ripening underlying 1-MCP treatment.	Transcriptome sequencing using Hiseq Xten (Illumina Inc., Foster City, CA, USA)	[74]
	Fruit colouration	Identification of potential TF regulating the carotenoid biosynthetic pathway.	Transcriptome sequencing using Illumina HiSeq2500 (Illumina Inc., Foster City, CA, USA)	[75]
	Sex determination	Differential expressed genes in sex determination of papaya, in male-to-hermaphrodite and male flowers.	Transcriptome sequencing using Illumina HiSeq2500 (Illumina Inc., Foster City, CA, USA)	[76]
	Disease resistance	Identification of disease-resistance genes in PRSV-resistant and susceptible cultivars.	Transcriptome sequencing using Illumina HiSeq2500 (Illumina Inc., Foster City, CA, USA)	[78]
	Disease resistance	Identification of stress-response genes and nutrient upregulated genes in tolerance mechanism of papaya sticky disease.	Transcriptome sequencing using Illumina HiSeq2000 (Illumina Inc., Foster City, CA, USA)	[79]
Proteomics	Ripening mechanism	Comparative proteomic analysis of climacteric and preclimacteric papaya cultivars.	2-DGE and LC-MS/MS	[83]

Table 1. *Cont.*

Type of Omics Platform	Traits/Conditions	Descriptions	Approach	Reference
	Ripening mechanism	Differentially expressed proteins during papaya ripening.	2-DGE and QTRAP hybrid tandem mass spectrometer	[84]
	Ripening mechanism	Differentially accumulated proteins (DAPs) during papaya ripening.	HPLC and LC-MS/MS	[85]
	Disease mechanism	Identification of differentially expressed proteins in healthy and PMev disease leaf samples in the Golden cultivar. Metabolism-related proteins were downregulated, and stress-responsive proteins were upregulated.	MALDI-TOF-MS/MS and DIGE/LC-IonTrap-MS/MS	[86]
	Disease mechanism	Differentially expressed proteins of compatible reaction between Eksotika papaya and *E. mallotivora*	iTRAQ mass spectrometry	[87]
	Disease mechanism	Protein expression between PMeV-infected preflowering *C. papaya* and control plants	LC-MS/MS-based label-free proteomics	[88]
Metabolomics	Fruit ripening	Comparative analysis of metabolite profiling between Eksotika and Sekaki cultivars.	GC-MS	[93]
	Fruit ripening	Profiling analysis of bioactive and volatile compounds in two papaya cultivars, namely Sel-42 and Tainung.	HPLC-ESI-MS/MS	[94]
	Fruit ripening	Comparative profiling of carotenoids and volatile in yellow and red flashed between Sui huang and Sui hong cultivars.	HPLC-ApCI-MS	[95]
	Fruit ripening	Identification of genes and metabolites regulating fruit ripening and softening in papaya cultivar Suiyou-2.	Transcriptome sequencing using Illumina Hiseq Xten (Illumina Inc., Foster City, CA, USA) and metabolomics profiling using HPLC-ESI-MS/MS	[103]
	Chilling injury	Elucidating of primary metabolites and volatile changes in papaya peel in response to chilling stress.	GC-MS /MS	[97]
	Bioactive properties	Metabolite profiling in papaya leaves.	UPLC-ESI-MS and GC-MS/MS	[98–100]

4. Future Perspective

The progress of the improvement of papaya traits has been limited due to the lack of genetic and genomic information on papaya. The outcome of recent omics studies of papaya plants suggests that there is potential for using these valuable genetic and genomic resources as a breeding tool to improve the desired traits in papaya. Incorporating omics data in papaya breeding programmes with a focus on abiotic stress, disease resistance, delayed ripening, and sweetness offers a promising strategy for developing high-quality traits in papaya cultivars without compromising their yields or agronomic traits. A previous study in Mexico conducted a network analysis of the interaction between viruses in papaya orchards. This viral metagenomics study, which was coupled with a network analysis, could contribute to the understanding of the host–pathogen interactions, which would cater to the management strategies against PRSV and non PRSV symptoms in papaya [104]. The potential of computational approaches in understanding these biological systems has been employed in crop improvement. The computational models that were constructed integrate genome and phenome information, which led to new experimental strategies in improving crop production [105].

In papaya, pangenome analysis has not received much attention. Using whole-genome resequencing of papaya cultivars from diverse germplasms enables researchers to perform pangenome analyses, which would facilitate the identification of core and variable genes in various papaya genomes. This effort has been performed in soy bean [106] and *B. rapa* [107]. A further area of interest is to screen for favourable alleles of diverse resistance genes sourced from the wild germplasm of papaya.

Another direction in papaya-trait improvement is employing the genome-editing approach. Functional analysis of candidate genes in the papaya–pathogen system can be performed using the clustered regularly interspaced short palindromic repeats (CRISPR)/CRISPR-associated protein 9(Cas9) system. The CRISPR-Cas9 system was successfully applied in banana [108], apple [109], and kiwifruit [110].

5. Conclusions

The use of omics data plays a role as an advanced breeding tool that will enable the faster and more accurate selection of key consumer-driven traits. Integration of various high-throughput omics platforms may accelerate the research on papaya crop improvement. In addition, the application of computational approaches is key in revealing and filling data gaps, which will be valuable in the designing of new experimentation and measurement strategies that would result in enhanced papaya quality, as well as the ability of it to be sustained under various environmental conditions.

Author Contributions: R.-A.Z.-A. and S.H. conceived the idea; R.-A.Z.-A. wrote the original draft; S.H. and I.-H.R.-O. reviewed and edited the manuscript. All authors have read and agreed to the published version of the manuscript.

Funding: This work was funded by the Government of Malaysia under the 12th Malaysia Plan Development through Malaysian Agricultural Research and Development Institute (MARDI), grant number (PRB502).

Conflicts of Interest: The authors declare no conflict of interest.

References

1. Food and Agriculture Organization. *Major Tropical Fruits, Market Review*; Food and Agriculture Organization: Rome, Italy, 2020; pp. 1–5.
2. Ming, R.; Yu, Q.; Moore, P.H.; Paull, R.E.; Chen, N.J.; Wang, M.L.; Zhu, Y.J.; Schuler, M.A.; Jiang, J.; Paterson, A.H. Genome of papaya, a fast growing tropical fruit tree. *Tree Genet. Genomes* **2012**, *8*, 445–462. [CrossRef]
3. Chan, Y. Breeding papaya (*Carica papaya* L.). In *Breeding Plantation Tree Crops. Tropical Species*, 1st ed.; Jain, S.M., Priyadarshan, P.M., Eds.; Springer: New York, NY, USA, 2009; Volume 1, pp. 121–159.
4. Plant Variety Protection Malaysia. Available online: http://pvpbkkt.doa.gov.my/ (accessed on 10 June 2021).
5. Ogata, T.; Yamanaka, S.; Shoda, M.; Urasaki, N.; Yamamoto, T. Current status of tropical fruit breeding and genetics for three tropical fruit species cultivated in Japan: Pineapple, mango, and papaya. *Breed. Sci.* **2016**, *66*, 69–81. [CrossRef] [PubMed]
6. Storey, W.B. Genetics of papaya. *J. Hered.* **1953**, *44*, 70–78. [CrossRef]
7. Tamaki, M.; Urasaki, N.; Motomura, K.; Nakamura, I.; Adaniya, S. Shortening the breeding cycle of papaya (*Carica papaya* L.) by culturing embryos treated with ethrel. *Plant Cell Tissues Organ Cult.* **2011**, *106*, 225–233. [CrossRef]
8. Dhekney, S.A.; Kandel, R.; Bergey, D.R.; Sitther, V.; Soorianathasundaram, K.; Litz, R.E. Advances in papaya biotechnology. *Biocatal. Agric. Biotechnol.* **2016**, *5*, 133–142. [CrossRef]
9. Vos, C.; Arancon, N. Soil and plant nutrient management and fruit production of papaya (*Carica papaya*) in Keaau, Hawaii. *J. Plant Nutr.* **2019**, *43*, 384–395. [CrossRef]
10. Cruz, A.F.; de Oliveira, B.F.; de Carvalho Pires, M. Optimum Level of Nitrogen and Phosphorus to Achieve Better Papaya (*Carica papaya* var. Solo) Seedlings Growth and Mycorrhizal Colonization. *Int. J. Fruit Sci.* **2019**, *17*, 1–11. [CrossRef]
11. Fallas-Corralesa, R.G.; van der Zee, S.E.A.T.M. Diagnosis and management of nutrient constraints in papaya. In *Fruit Crops: Diagnosis and Management of Nutrient Constraints*, 1st ed.; Srivastava, A.K., Hu, C., Eds.; Elsevier: Amsterdam, The Netherlands, 2020; Volume 1, pp. 607–628.
12. Santos, E.M.C.; Cavalcante, Í.H.L.; da Silva Junior, G.B.; Albano, F.G. Impact of nitrogen and potassium nutrition on papaya (pawpaw) fruit quality. *Biosci. J. Uberlândia* **2015**, *31*, 1341–1348. [CrossRef]
13. Dinesh, M.R.; Reddy, B.M.C. Physiological Basis of Growth and Fruit Yield Characteristics of Tropical and Sub-tropical Fruits to Temperature. In *Tropical Fruit Tree Species and Climate Change*; Sthapit, S.R., Scherr, S.J., Eds.; Bioversity International: Rome, Italy, 2012; Volume 1, pp. 45–70.
14. Salinas, I.; Hueso, J.J.; Cuevas, J. Active Control of Greenhouse Climate Enhances Papaya Growth and Yield at an Affordable Cost. *Agronomy* **2021**, *11*, 378. [CrossRef]

15. Ramírez, A.G.; Rodríguez, L.M.P.; Erosa, F.E.; Fuentes, G.; Santamaría, J.M. Identification of the SHINE clade of AP2/ERF domain transcription factors genes in Carica papaya; Their gene expression and their possible role in wax accumulation and water deficit stress tolerance in a wild and a commercial papaya genotypes. *Environ. Exp. Bot.* **2021**, *183*, 104341. [CrossRef]
16. Shiratake, K.; Suzuki, M. Omics studies of citrus, grape and rosaceae fruit trees. *Breed. Sci.* **2016**, *66*, 122–138. [CrossRef]
17. Ali, M.W.; Borrill, P. Applying genomic resources to accelerate wheat biofortification. *Heredity* **2020**, *125*, 386–395. [CrossRef]
18. Roorkiwal, M.; Jain, A.; Kale, S.M.; Doddamani, D.; Chitikineni, A.; Thudi, M.; Varshney, R.K. Development and evaluation of high-density Axiom®CicerSNP Array for high-resolution genetic mapping and breeding applications in chickpea. *Plant Biotechnol. J.* **2018**, *16*, 890–901. [CrossRef]
19. Raza, A.; Tabassum, J.; Kudapa, H.; Varshney, R.K. Can omics deliver temperature resilient ready-to- grow crops? *Crit. Rev. Biotechnol.* **2021**, 1–24. [CrossRef]
20. Fitch, M.M.M.; Manshardt, R.M.; Gonsalves, D.; Slightom, J.L.; Sanford, J.C. Virus Resistant Papaya Plants Derived from Tissues Bombarded with the Coat Protein Gene of Papaya Ringspot Virus. *Bio Technol.* **1992**, *10*, 1466–1472. [CrossRef]
21. Ming, R.; Hou, S.; Feng, Y.; Yu, Q.; Dionne-Laporte, A.; Saw, J.H.; Senin, P.; Wang, W.; Ly, B.V.; Lewis, K.L.T.; et al. The draft genome of the transgenic tropical fruit tree papaya (*Carica papaya* Linnaeus). *Nature* **2008**, *452*, 991–996. [CrossRef]
22. Sondur, S.N.; Manshardt, R.M.; Stiles, J.I. A genetic map of papaya based on random amplified polymorphic DNA markers. *Theor. Appl. Genet.* **1996**, *93*, 547–553. [CrossRef]
23. Chen, C.; Yu, Q.; Hou, S.; Li, Y.; Eustice, M.; Skelton, R.L.; Veatch, O.; Herdes, R.E.; Diebold, L.; Saw, J.; et al. Construction of a sequence-tagged high-density genetic map of papaya for comparative structural and evolutionary genomics in Brassicales. *Genetics* **2007**, *177*, 2481–2491. [CrossRef]
24. Porter, B.W.; Paidi, M.; Ming, R.; Alam, M.; Nishijima, W.T.; Zhu, Y.J. Genome-wide analysis of *Carica papaya* reveals a small NBS resistance gene family. *Mol. Genet. Genom.* **2009**, *281*, 609–626. [CrossRef]
25. Yu, Q.; Tong, E.; Skelton, R.L.; Bowers, J.E.; Jones, M.R.; Murray, J.E.; Hou, S.; Guan, P.; Acob, R.A.; Luo, M.C.; et al. A physical map of the papaya genome with integrated genetic map and genome sequence. *BMC Genom.* **2009**, *10*, 371. [CrossRef]
26. Chávez-Pesqueira, M.; Núñez-Farfán, J. Domestication and genetics of papaya: A review. *Front. Ecol. Evol.* **2017**, *5*. [CrossRef]
27. Palei, S.; Dash, D.K.; Rout, G.R. Biology and biotechnology of papaya, an important fruit crop of tropics: A review. *Vegetos* **2018**, *31*, 1–15. [CrossRef]
28. Marsh, J.I.; Hu, H.; Gill, M.; Batley, J.; Edwards, D. Crop breeding for a changing climate: Integrating phenomics and genomics with bioinformatics. *Theor. Appl. Genet.* **2021**, *2021*, 1–14. [CrossRef]
29. Ellegren, H. Comparative genomics and the study of evolution by natural selection. *Mol. Ecol.* **2008**, *17*, 4586–4596. [CrossRef]
30. Morrell, P.L.; Buckler, E.S.; Ross-Ibarra, J. Crop genomics: Advances and applications. *Nat. Rev. Genet.* **2012**, *13*, 85–96. [CrossRef]
31. Yang, M.; Zhou, C.; Yang, H.; Kuang, R.; Huang, B.; Wei, Y. Genome-wide analysis of basic helix-loop-helix transcription factors in papaya (*Carica papaya* L.). *PeerJ* **2020**, *8*, e9319. [CrossRef]
32. Praza-Echeverria, S.; Santamaría, J.M.; Fuentes, G.; De los Angeles Menéndez-Cerón, M.; Vallejo-Reyna, M.A.; Herrera-Valencia, V.A. The NPR1 family of transcription cofactors in papaya: Insights into its structure, phylogeny and expression. *Genes Genom.* **2012**, *34*, 379–390. [CrossRef]
33. Liu, K.; Yuan, C.; Feng, S.; Zhong, S.; Li, H.; Zhong, J.; Shen, C.; Liu, J. Genome-wide analysis and characterization of Aux/IAA family genes related to fruit ripening in papaya (*Carica papaya* L.). *BMC Genom.* **2017**, *18*, 1–12. [CrossRef]
34. Xu, Y.; Xu, H.; Wall, M.M.; Yang, J. Roles of transcription factor SQUAMOSA promoter binding protein-like gene family in papaya (*Carica papaya*) development and ripening. *Genomics* **2020**, *112*, 2734–2747. [CrossRef]
35. Liu, K.; Yuan, C.; Li, H.; Lin, W.; Yang, Y.; Shen, C.; Zheng, X. Genome-wide identification and characterization of auxin response factor (ARF) family genes related to flower and fruit development in papaya (*Carica papaya* L.). *BMC Genom.* **2015**, *16*, 901. [CrossRef]
36. Amin, M.N.; Bunawan, H.; Redzuan, R.A.; Jaganath, I.B.S. *Erwinia mallotivora* sp., a new pathogen of papaya (*Carica papaya*) in peninsular Malaysia. *Int. J. Mol. Sci.* **2011**, *12*, 39–45. [CrossRef] [PubMed]
37. Purcifull, D.E.; Edwardson, J.R.; Hiebert, E.; Gonsalves, D. Papaya Ringspot Virus. In *CMI/AAB Description of Plant Viruses*; Coronel, R.E., Ed.; Wageningen University: Wageningen, The Netherlands, 1984; Volume 2.
38. Maciel-Zambolim, E.; Kunieda-Alonso, S.; Matsuoka, K.; De Carvalho, M.G.; Zerbini, F.M. Purification and some properties of Papaya meleira virus, a novel virus infecting papayas in Brazil. *Plant Pathol.* **2003**, *52*, 389–394. [CrossRef]
39. National Center for Botechnology Information (NCBI). Available online: https://ftp.ncbi.nlm.nih.gov/genomes/all/annotation_releases/3649/100/GCF_000150535.2_Papaya1.0/ (accessed on 29 May 2021).
40. Duvick, J.; Fu, A.; Muppirala, U.; Sabharwal, M.; Wilkerson, M.D.; Lawrence, C.J.; Lushbough, C.; Brendel, V. PlantGDB: A resource for comparative plant genomics. *Nucleic Acids Res.* **2008**, *36*, 959–965. [CrossRef] [PubMed]
41. Goodstein, D.M.; Shu, S.; Howson, R.; Neupane, R.; Hayes, R.D.; Fazo, J.; Mitros, T.; Dirks, W.; Hellsten, U.; Putnam, N.; et al. Phytozome: A comparative platform for green plant genomics. *Nucleic Acids Res.* **2012**, *40*, D1178–D1186. [CrossRef]
42. Yates, A.; Akanni, W.; Amode, M.R.; Barrell, D.; Billis, K.; Carvalho-Silva, D.; Cummins, C.; Clapham, P.; Fitzgerald, S.; Gil, L.; et al. Ensembl 2016. *Nucleic Acids Res.* **2016**, *44*, D710–D716. [CrossRef]
43. Van Bel, M.; Diels, T.; Vancaester, E.; Kreft, L.; Botzki, A.; Van De Peer, Y.; Coppens, F.; Vandepoele, K. PLAZA 4.0: An integrative resource for functional, evolutionary and comparative plant genomics. *Nucleic Acids Res.* **2018**, *46*, 1190–1196. [CrossRef]
44. Prakash, A.; Jeffryes, M.; Bateman, A.; Finn, R.D. The HMMER Web Server for Protein Sequence Similarity Search. *Curr. Protoc. Bioinform.* **2017**, *60*, 3.15.1–3.15.23. [CrossRef]

45. Altschul, S.F.; Gertz, E.M.; Agarwala, R.; Schäffer, A.A.; Yu, Y.K. PSI-BLAST pseudocounts and the minimum description length principle. *Nucleic Acids Res.* **2009**, *37*, 815–824. [CrossRef]
46. Jones, P.; Binns, D.; Chang, H.Y.; Fraser, M.; Li, W.; McAnulla, C.; McWilliam, H.; Maslen, J.; Mitchell, A.; Nuka, G.; et al. InterProScan 5: Genome-scale protein function classification. *Bioinformatics* **2014**, *30*, 1236–1240. [CrossRef]
47. El-Gebali, S.; Mistry, J.; Bateman, A.; Eddy, S.R.; Luciani, A.; Potter, S.C.; Qureshi, M.; Richardson, L.J.; Salazar, G.A.; Smart, A.; et al. The Pfam protein families database in 2019. *Nucleic Acids Res.* **2019**, *47*, D427–D432. [CrossRef]
48. Letunic, I.; Bork, P. 20 years of the SMART protein domain annotation resource. *Nucleic Acids Res.* **2018**, *46*, D493–D496. [CrossRef]
49. Higo, K.; Ugawa, Y.; Iwamoto, M.; Korenaga, T. Plant cis-acting regulatory DNA elements (PLACE) database: 1999. *Nucleic Acids Res.* **1999**, *27*, 297–300. [CrossRef]
50. Lescot, M.; Déhais, P.; Thijs, G.; Marchal, K.; Moreau, Y.; Van De Peer, Y.; Rouzé, P.; Rombauts, S. PlantCARE, a database of plant cis-acting regulatory elements and a portal to tools for in silico analysis of promoter sequences. *Nucleic Acids Res.* **2002**, *30*, 325–327. [CrossRef]
51. Hu, B.; Jin, J.; Guo, A.Y.; Zhang, H.; Luo, J.; Gao, G. GSDS 2.0: An upgraded gene feature visualization server. *Bioinformatics* **2015**, *31*, 1296–1297. [CrossRef]
52. Thompson, J.D.; Gibson, T.J.; Higgins, D.G. Multiple Sequence Alignment Using ClustalW and ClustalX. *Curr. Protoc. Bioinform.* **2003**, 2.3.1–2.3.22. [CrossRef]
53. Katoh, K.; Misawa, K.; Kuma, K.I.; Miyata, T. MAFFT: A novel method for rapid multiple sequence alignment based on fast Fourier transform. *Nucleic Acids Res.* **2002**, *30*, 3059–3066. [CrossRef]
54. Kumar, S.; Stecher, G.; Li, M.; Knyaz, C.; Tamura, K. MEGA X: Molecular evolutionary genetics analysis across computing platforms. *Mol. Biol. Evol.* **2018**, *35*, 1547–1549. [CrossRef]
55. Aizat, W.M.; Ismail, I.; Noor, N.M. Recent Development in Omics Studies. In *Omics Applications for Systems Biology*, 1st ed.; Aizat, W., Goh, H.H., Baharum, S., Eds.; Springer International Publishing: Cham, Switzerland, 2018; Volume 1102, pp. 1–9. [CrossRef]
56. Scossa, F.; Alseekh, S.; Fernie, A.R. Integrating multi-omics data for crop improvement. *J. Plant Physiol.* **2021**, *257*, 1–16. [CrossRef]
57. Choi, H.K. Translational genomics and multi-omics integrated approaches as a useful strategy for crop breeding. *Genes Genom.* **2019**, *41*, 133–146. [CrossRef]
58. Zainal-Abidin, R.; Abu-Bakar, N.; Yusof, M.M.F.; Abdullah, N. Sequence information on single nucleotide polymorphism (SNP) through genome sequencing analysis of *Carica papaya* variety Eksotika and Sekaki. *J. Trop. Agric. Food Sci.* **2016**, *44*, 219–228.
59. Fang, J.; Fang, J.; Fang, J.; Fang, J.; Wood, A.M.; Chen, Y.; Chen, Y.; Yue, J.; Ming, R.; Ming, R. Genomic variation between PRSV resistant transgenic SunUp and its progenitor cultivar Sunset. *BMC Genom.* **2020**, *21*. [CrossRef]
60. Nadeem, M.A.; Nawaz, M.A.; Shahid, M.Q.; Doğan, Y.; Comertpay, G.; Yıldız, M.; Hatipoğlu, R.; Ahmad, F.; Alsaleh, A.; Labhane, N.; et al. DNA molecular markers in plant breeding: Current status and recent advancements in genomic selection and genome editing. *Biotechnol. Biotechnol. Equip.* **2018**, *32*, 261–285. [CrossRef]
61. Vidal, N.M.; Grazziotin, A.L.; Ramos, H.C.C.; Pereira, M.G.; Venancio, T.M. Development of a gene-centered SSR atlas as a resource for papaya (*Carica papaya*) marker-assisted selection and population genetic studies. *PLoS ONE* **2014**, *9*, e112654. [CrossRef]
62. Nantawan, U.; Kanchana-udomkan, C.; Drew, R.; Ford, R. Development of polymorphic simple sequence repeat (SSR) markers from genome re-sequencing of *Carica papaya* L. 'Sunrise Solo' and 'RB2' for marker-assisted breeding. *Acta Hortic.* **2018**, *1205*, 687–695. [CrossRef]
63. Saxena, R.K.; Edwards, D.; Varshney, R.K. Structural variations in plant genomes. *Brief. Funct. Genom. Proteom.* **2014**, *13*, 296–307. [CrossRef]
64. Bohry, D.; Ramos, H.C.C.; dos Santos, P.H.D.; Boechat, M.S.B.; Arêdes, F.A.S.; Pirovani, A.A.V.; Pereira, M.G. Discovery of SNPs and InDels in papaya genotypes and its potential for marker assisted selection of fruit quality traits. *Sci. Rep.* **2021**, *11*, 292. [CrossRef]
65. Liao, Z.; Zhang, X.; Zhang, S.; Lin, Z.; Zhang, X.; Ming, R. Structural variations in papaya genomes. *BMC Genom.* **2021**, *22*, 335. [CrossRef]
66. Collard, B.C.Y.; Jahufer, M.Z.Z.; Brouwer, J.B.; Pang, E.C.K. An introduction to markers, quantitative trait loci (QTL) mapping and marker-assisted selection for crop improvement: The basic concepts. *Euphytica* **2005**, *142*, 169–196. [CrossRef]
67. Deputy, J.C.; Ming, R.; Ma, H.; Liu, Z.; Fitch, M.M.M.; Wang, M.; Manshardt, R.; Stiles, J.I. Molecular markers for sex determination in papaya (*Carica papaya* L.). *Theor. Appl. Genet.* **2002**, *106*, 107–111. [CrossRef]
68. Blas, A.L.; Yu, Q.; Veatch, O.J.; Paull, R.E.; Moore, P.H.; Ming, R. Genetic mapping of quantitative trait loci controlling fruit size and shape in papaya. *Mol. Breed.* **2012**, *29*, 457–466. [CrossRef]
69. Nantawan, U.; Kanchana-Udomkan, C.; Bar, I.; Ford, R. Linkage mapping and quantitative trait loci analysis of sweetness and other fruit quality traits in papaya. *BMC Plant Biol.* **2019**, *19*, 449. [CrossRef] [PubMed]
70. Tao, Y.; Zhao, X.; Mace, E.; Henry, R.; Jordan, D. Exploring and Exploiting Pan-genomics for Crop Improvement. *Mol. Plant* **2019**, *12*, 156–169. [CrossRef] [PubMed]
71. Neik, T.X.; Amas, J.; Barbetti, M.; Edwards, D.; Batley, J. Understanding host–pathogen interactions in brassica napus in the omics era. *Plants* **2020**, *9*, 1336. [CrossRef] [PubMed]
72. Wang, Z.; Gerstein, M.; Snyder, M. RNA-Seq: A revolutionary tool for transcriptomics. *Nat. Rev. Genet.* **2009**, *10*, 57–63. [CrossRef]
73. Gamboa-Tuz, S.D.; Pereira-Santana, A.; Zamora-Briseño, J.A.; Castano, E.; Espadas-Gil, F.; Ayala-Sumuano, J.T.; Keb-Llanes, M.Á.; Sanchez-Teyer, F.; Rodríguez-Zapata, L.C. Transcriptomics and co-expression networks reveal tissue-specific responses and regulatory hubs under mild and severe drought in papaya (*Carica papaya* L.). *Sci. Rep.* **2018**, *8*, 14539. [CrossRef]

74. Zhu, X.; Ye, L.; Ding, X.; Gao, Q.; Xiao, S.; Tan, Q.; Huang, J.; Chen, W.; Li, X. Transcriptomic analysis reveals key factors in fruit ripening and rubbery texture caused by 1-MCP in papaya. *BMC Plant Biol.* **2019**, *19*, 1–23. [CrossRef]

75. Shen, Y.H.; Lu, B.G.; Feng, L.; Yang, F.Y.; Geng, J.J.; Ming, R.; Chen, X.J. Isolation of ripening-related genes from ethylene/1-MCP treated papaya through RNA-seq. *BMC Genom.* **2017**, *18*, 1–13. [CrossRef]

76. Lin, H.; Liao, Z.; Zhang, L.; Yu, Q. Transcriptome analysis of the male-to-hermaphrodite sex reversal induced by low temperature in papaya. *Tree Genet. Genomes* **2016**, *12*, 94. [CrossRef]

77. Zhou, P.; Zhang, X.; Fatima, M.; Ma, X.; Fang, H.; Yan, H.; Ming, R. DNA methylome and transcriptome landscapes revealed differential characteristics of dioecious flowers in papaya. *Hortic. Res.* **2020**, *7*, 81. [CrossRef]

78. Fang, J.; Lin, A.; Qiu, W.; Cai, H.; Umar, M.; Chen, R.; Ming, R. Transcriptome profiling revealed stress-induced and disease resistance genes up-regulated in PRSV resistant transgenic papaya. *Front. Plant Sci.* **2016**, *7*, 855. [CrossRef]

79. Madroñero, J.; Rodrigues, S.P.; Antunes, T.F.S.; Abreu, P.M.V.; Ventura, J.A.; Fernandes, A.A.R.; Fernandes, P.M.B. Transcriptome analysis provides insights into the delayed sticky disease symptoms in *Carica papaya*. *Plant Cell Rep.* **2018**, *37*, 967–980. [CrossRef]

80. Tian, T.; Liu, Y.; Yan, H.; You, Q.; Yi, X.; Du, Z.; Xu, W.; Su, Z. AgriGO v2.0: A GO analysis toolkit for the agricultural community, 2017 update. *Nucleic Acids Res.* **2017**, *45*, W122–W129. [CrossRef]

81. Ge, S.X.; Jung, D.; Yao, R. ShinyGO: A graphical gene-set enrichment tool for animals and plants. *Bioinformatics* **2020**, *36*, 2628–2629. [CrossRef]

82. Aslam, B.; Basit, M.; Nisar, M.A.; Khurshid, M.; Rasool, M.H. Proteomics: Technologies and their applications. *J. Chromatogr. Sci.* **2017**, *55*, 182–196. [CrossRef]

83. Nogueira, S.B.; Labate, C.A.; Gozzo, F.C.; Pilau, E.J.; Lajolo, F.M.; Oliveira do Nascimento, J.R. Proteomic analysis of papaya fruit ripening using 2DE-DIGE. *J. Proteom.* **2011**, *75*, 1428–1439. [CrossRef]

84. Huerta-Ocampo, J.Á.; Osuna-Castro, J.A.; Lino-López, G.J.; Barrera-Pacheco, A.; Mendoza-Hernández, G.; De León-Rodríguez, A.; Barba de la Rosa, A.P. Proteomic analysis of differentially accumulated proteins during ripening and in response to 1-MCP in papaya fruit. *J. Proteom.* **2012**, *75*, 2160–2169. [CrossRef]

85. Jiang, B.; Ou, S.; Xu, L.; Mai, W.; Ye, M.; Gu, H.; Zhang, T.; Yuan, C.; Shen, C.; Wang, J.; et al. Comparative proteomic analysis provides novel insights into the regulation mechanism underlying papaya (*Carica papaya* L.) exocarp during fruit ripening process. *BMC Plant Biol.* **2019**, *19*. [CrossRef]

86. Rodrigues, S.P.; Ventura, J.A.; Aguilar, C.; Nakayasu, E.S.; Almeida, I.C.; Fernandes, P.M.B.; Zingali, R.B. Proteomic analysis of papaya (*Carica papaya* L.) displaying typical sticky disease symptoms. *Proteomics* **2011**, *11*, 2592–2602. [CrossRef]

87. Abu Bakar, N.; Badrun, R.; Shaharuddin, N.A.; Labuh, R. iTRAQ Proteins Analysis of Early Infected Papaya Plants with Papaya Dieback Pathogen. *Asian J. Plant Biol.* **2015**, *3*, 1–7.

88. Abu Bakar, N.; Badrun, R.; Rozano, L.; Ahmad, L.; Raih, M.M.; Tarmizi, A.A. Identification and validation of putative *Erwinia mallotivora* effectors via quantitative proteomics and Real Time Analysis. *J. Agric. Food. Technol.* **2017**, *7*, 10–21.

89. Soares, E.A.; Werth, E.G.; Madroñero, L.J.; Ventura, J.A.; Rodrigues, S.P.; Hicks, L.M.; Fernandes, P.M.B. Label-free quantitative proteomic analysis of pre-flowering PMeV-infected *Carica papaya* L. *J. Proteom.* **2016**, *151*, 275–283. [CrossRef] [PubMed]

90. Vizcaíno, J.A.; Deutsch, E.W.; Wang, R.; Csordas, A.; Reisinger, F.; Ríos, D.; Dianes, J.A.; Sun, Z.; Farrah, T.; Bandeira, N.; et al. ProteomeXchange provides globally coordinated proteomics data submission and dissemination. *Nat. Biotechnol.* **2014**, *32*, 223–226. [CrossRef] [PubMed]

91. Kusano, M.; Yang, Z.; Okazaki, Y.; Nakabayashi, R.; Fukushima, A.; Saito, K. Using metabolomic approaches to explore chemical diversity in rice. *Mol. Plant* **2015**, *8*, 58–67. [CrossRef] [PubMed]

92. Okazaki, Y.; Saito, K. Integrated metabolomics and phytochemical genomics approaches for studies on rice. *Gigascience* **2016**, *5*, 1–8. [CrossRef]

93. Sanimah, S.; Sarip, J. Metabolomic analysis of *Carica papaya* variety Eksotika and Sekaki. *J. Trop. Agric. Food Sci.* **2015**, *43*, 103–117.

94. Kelebek, H.; Selli, S.; Gubbuk, H.; Gunes, E. Comparative evaluation of volatiles, phenolics, sugars, organic acids and antioxidant properties of Sel-42 and Tainung papaya varieties. *Food Chem.* **2015**, *173*, 912–919. [CrossRef]

95. Jing, G.; Li, T.; Qu, H.; Yun, Z.; Jia, Y.; Zheng, X.; Jiang, Y. Carotenoids and volatile profiles of yellow- and red-fleshed papaya fruit in relation to the expression of carotenoid cleavage dioxygenase genes. *Postharvest Biol. Technol.* **2015**, *109*, 114–119. [CrossRef]

96. Santana, L. Nutraceutical Potential of *Carica papaya* in Metabolic syndrome. *Nutrients* **2019**, *5*, 1608. [CrossRef]

97. Wu, Q.; Li, Z.; Chen, X.; Yun, Z.; Li, T.; Jiang, Y. Comparative metabolites profiling of harvested papaya (*Carica papaya* L.) peel in response to chilling stress. *J. Sci. Food Agric.* **2019**, *99*, 6868–6881. [CrossRef]

98. Gogna, N.; Hamid, N.; Dorai, K. Metabolomic profiling of the phytomedicinal constituents of *Carica papaya* L. leaves and seeds by 1H NMR spectroscopy and multivariate statistical analysis. *J. Pharm. Biomed. Anal.* **2015**, *115*, 74–85. [CrossRef]

99. Harini, R.S.; Saivikashini, A.; Keerthana, G. Profiling metabolites of Carica papaya Linn. variety CO7 through GC-MS analysis. *J. Pharmacogn. Phytochem.* **2016**, *5*, 200–203.

100. Vuong, Q.V.; Hirun, S.; Roach, P.D.; Bowyer, M.C.; Phillips, P.A.; Scarlett, C.J. Effect of extraction conditions on total phenolic compounds and antioxidant activities of Carica papaya leaf aqueous extracts. *J. Herb. Med.* **2013**, *3*, 104–111. [CrossRef]

101. Oikawa, A.; Otsuka, T.; Nakabayashi, R.; Jikumaru, Y.; Isuzugawa, K.; Murayama, H.; Saito, K.; Shiratake, K. Metabolic profiling of developing pear fruits reveals dynamic variation in primary and secondary metabolites, including plant hormones. *PLoS ONE* **2015**, *10*, 1–18. [CrossRef]

102. Yun, Z.; Li, T.; Gao, H.; Zhu, H.; Gupta, V.K.; Jiang, Y. Integrated transcriptomic, proteomic and metabolomics analysis reveals peel ripening of harvested banana under natural condition. *Biomolecules* **2019**, *9*, 167. [CrossRef]
103. Zheng, S.; Hao, Y.; Fan, S.; Cai, J.; Chen, W.; Li, X.; Zhu, X. Metabolomic and transcriptomic profiling provide novel insights into fruit ripening and ripening disorder caused by 1-MCP treatments in papaya. *Int. J. Mol. Sci.* **2021**, *22*, 916. [CrossRef]
104. Alcalá-Briseño, R.I.; Casarrubias-Castillo, K.; López-Ley, D.; Garrett, K.A.; Silva-Rosales, L. Network analysis of the papaya orchard virome from two agroecological regions of Chiapas, Mexico. *mSystems* **2020**, *5*, 1–17. [CrossRef]
105. Benes, B.; Guan, K.; Lang, M.; Long, S.P.; Lynch, J.P.; Marshall-Colón, A.; Peng, B.; Schnable, J.; Sweetlove, L.J.; Turk, M.J. Multiscale computational models can guide experimentation and targeted measurements for crop improvement. *Plant J.* **2020**, *103*, 21–31. [CrossRef]
106. Li, Y.H.; Zhou, G.; Ma, J.; Jiang, W.; Jin, L.G.; Zhang, Z.; Guo, Y.; Zhang, J.; Sui, Y.; Zheng, L.; et al. De novo assembly of soybean wild relatives for pan-genome analysis of diversity and agronomic traits. *Nat. Biotechnol.* **2014**, *32*, 1045–1052. [CrossRef]
107. Golicz, A.A.; Bayer, P.E.; Barker, G.C.; Edger, P.P.; Kim, H.R.; Martinez, P.A.; Chan, C.K.K.; Severn-Ellis, A.; McCombie, W.R.; Parkin, I.A.P.; et al. The pangenome of an agronomically important crop plant Brassica oleracea. *Nat. Commun.* **2016**, *7*, 1–8. [CrossRef] [PubMed]
108. Kaur, N.; Alok, A.; Shivani; Kaur, N.; Pandey, P.; Awasthi, P.; Tiwari, S. CRISPR/Cas9-mediated efficient editing in phytoene desaturase (PDS) demonstrates precise manipulation in banana cv. Rasthali genome. *Funct. Integr. Genom.* **2018**, *18*, 89–99. [CrossRef]
109. Charrier, A.; Vergne, E.; Dousset, N.; Richer, A.; Petiteau, A.; Chevreau, E. Efficient targeted mutagenesis in apple and first time edition of pear using the CRISPR-Cas9 system. *Front. Plant Sci.* **2019**, *10*, 40. [CrossRef] [PubMed]
110. Wang, Z.; Wang, S.; Li, D.; Zhang, Q.; Li, L.; Zhong, C.; Liu, Y.; Huang, H. Optimized paired-sgRNA/Cas9 cloning and expression cassette triggers high-efficiency multiplex genome editing in kiwifruit. *Plant Biotechnol. J.* **2018**, *16*, 1424–1433. [CrossRef] [PubMed]

Article

SNP Genotyping for Purity Assessment of a Forage Oat (*Avena sativa* L.) Variety from Colombia

Luis Fernando Campuzano-Duque [1], Diego Bejarano-Garavito [2], Javier Castillo-Sierra [2], Daniel Ricardo Torres-Cuesta [2], Andrés J. Cortés [3,†] and Matthew Wohlgemuth Blair [4,*]

[1] Corporación Colombiana de Investigación Agropecuaria (AGROSAVIA)-CI La Libertad, Villavicencio 230002, Colombia; lcampuzano@agrosavia.co
[2] Corporación Colombiana de Investigación Agropecuaria (AGROSAVIA)-CI Tibaitatá, Bogotá 111321, Colombia; dbejarano@agrosavia.co (D.B.-G.); jcastillos@agrosavia.co (J.C.-S.); dtorres@agrosavia.co (D.R.T.-C.)
[3] Corporación Colombiana de Investigación Agropecuaria (AGROSAVIA)-CI La Selva, Rionegro 054048, Colombia; acortes@agrosavia.co
[4] Department of Agricultural and Environmental Sciences, Tennessee State University, Nashville, TN 37209, USA
* Correspondence: mblair@tnstate.edu; Tel.: +1-615-963-7467
† Secondary Address: Departamento de Ciencias Forestales, Facultad de Ciencias Agrarias, Universidad Nacional de Colombia, Sede Medellín, Medellín 050034, Colombia.

Abstract: Single nucleotide polymorphism (SNP) markers have multiple applications in plant breeding of small grains. They are used for the selection of divergent parents, the identification of genetic variants and marker-assisted selection. However, the use of SNPs in varietal purity assessment is under-reported, especially for multi-line varieties from the public sector. In the case of variety evaluation, these genetic markers are tools for maintaining varietal distinctness, uniformity and stability needed for cultivar release of multi-line or pure-line varieties of inbred crops. The objective of this research was to evaluate the purity and relationships of one original (AV-25) and two multi-line sub-populations (AV25-T and AV25-S) of the inbreeding species, oats (*Avena sativa* L.). Both subpopulations could be useful as forages in the central highland region of Colombia (>2000 masl), such as in the departments of Boyacá and Cundinamarca, even though they were derived from an original composite mixture widely used in the mountainsides of the southern department of Nariño named Avena 25. Representative single plant selections (SPS) from the two sub-populations were grown together with SPS harvests from off-type plants (early and late) and plants from the original AV25 composite mixture, to determine their genetic similarity. Plants were genotyped by DNA extraction of a plateful of 96 individual plant samples and SNPs were detected for an Illumina Infinium 6K Chip assay. The data were used for the analysis of genetic structure and population relationships. The grouping observed based on the genetic data indicated that AV25-T and AV25-S were homogeneous populations and somewhat divergent in their genetic profile compared to the original AV25-C mix. In addition, to the two commercial, certified oat varieties (Cajicá and Cayuse) were different from these. The early and late selections were probable contaminants and could be discarded. We concluded that the use of SNP markers is an appropriate tool for ensuring genetic purity of oat varieties.

Keywords: composite mix; genetic structure; multi-line variety; single nucleotide polymorphism markers; varietal purity

Citation: Campuzano-Duque, L.F.; Bejarano-Garavito, D.; Castillo-Sierra, J.; Torres-Cuesta, D.R.; Cortés, A.J.; Blair, M.W. SNP Genotyping for Purity Assessment of a Forage Oat (*Avena sativa* L.) Variety from Colombia. *Agronomy* **2022**, *12*, 1710. https://doi.org/10.3390/agronomy12071710

Academic Editor: Andreas Katsiotis

Received: 20 May 2022
Accepted: 13 July 2022
Published: 20 July 2022

Publisher's Note: MDPI stays neutral with regard to jurisdictional claims in published maps and institutional affiliations.

1. Introduction

Cultivated oat (*Avena sativa* L.) is a hexaploid small grain species (2n = 6x = 42) with a large genome (11.3–14.0 Gb) that originated in the Fertile Crescent region known as the Near East in western Asia and from there spread around the world [1]. Oats arrived in the Americas with European conquests and are currently among the most important cereals in the world after maize (*Zea mays* L.), rice (*Oryza sativa*), wheat (*Triticum aestivum*), sorghum (*Sorghum bicolor*), millets (*Panucum, Pennisetum, Setaria* spp.) and barley (*Hordeum vulgare*) but more popular than rye (*Secale cereale*). Unlike some of these cereals, oats are confined to temperate areas in northern latitudes or cool wet climates, such as those in highlands of the subtropics.

Major producers of oats include Canada, Russia, Finland, Poland and the United Kingdom in the northern hemisphere, and Australia and Brazil in the southern hemisphere. In highland (>2000 m above sea level) areas of the Andes, from Bolivia to Colombia, Ecuador and Peru, oats are used as a fodder source, in pasture, hay or silage, alone or with forage legumes [2] and is considered a food alternative for the diet of dairy cows to improve milk production and quality, especially in times of forage shortages by drought or frost [3].

Molecular markers are well known tools for plant breeding [4] with many methods developed since the advent of the polymerase chain reaction (PCR), which allowed the development of easy-to-use genetic assays for varied purposes adaptable to many of the different stages of genetic improvement of plants and animals [5–7]. These include the pre-breeding steps of biodiversity analysis, conservation, utilization [6,8], breeding analysis of the reproductive system of a species, characterization and selection of parents in crossing plans and monitoring of segregating populations [9]. Markers have been used as genetic tools for analysis of qualitative and quantitative characters and introgression of genes from wild species (e.g., [10]). They have also been used successfully in other lines of plant research, such as phylogeny [8], and the diagnosis of pests or pathogens [11]. In post-breeding steps, molecular markers are useful for confirming the identity, purity and homogeneity of varieties and in the tracking of genetically modified organisms (GMOs) [6,9]

In the last decade, single nucleotide polymorphisms (SNPs) have become the most common type of biallelic sequence-based molecular marker used in plant and animal genomes [7,12,13]. In addition to having stable inheritance from generation to generation, SNP markers target loci where only two alleles are observed within a population that arise from mutations or mismatch repair [9]. The use of SNPs is widespread in genetic studies of plants, which includes genomic diversity [4,14], construction of linkage maps [15,16], genomic selection [17,18], population structure analysis [19–21], genetic mapping [5] and genome–environment associations [22].

The target number of SNPs for a crop depends on its genome size, level of ploidy, research investments, population stratification and overall linkage disequilibrium patterns [23]. As examples, the capacity of currently available SNP arrays from Illumina go up to 820,000 (820K) in the large-genome hexaploid wheats (*Triticum aestivum*), 700K in rice (*Oryza sativa*) [4], 487K in triploid apples (*Malus domestica*) [24], 345K in the dodecaploid genome of sugar cane (*Saccharum officinarum*) [25], 90K in octoploid strawberries (*Fragaria vesca*) [26] and 58K in tetraploid peanuts (*Arachis hypogaeae*) [27,28]. Diploid species require less SNPs for coverage.

In the case of common oats, the first SNP array developed with Illumina technology contained 3072 assays and was applied to build a consensus map with 985 mapped loci [29]. The SNP chip was later expanded to the 6K Illumina oat BeadChip containing nearly 6000 assays, with a success rate of 86.6% [6,7]. Recently, this SNP array has been transformed into a 6K BeadChip layout containing 257 Infinium I and 5486 Infinium II features corresponding to 5743 SNPs.

The goal of this study was to determine the variety purity, plant to plant diversity and early or late season off-types of a multi-line (composite mix) population of forage oats (AV25-C) commonly grown in the highlands (>2000 masl) of southern Colombia along with the diversity of two derived sub-populations (AV25-S and AV25-T) selected for similar altitudes in central Colombian departments of Boyacá and Cundinamarca by using SNP fingerprinting based on the Infinium assay and two control genotypes along with selection of phenological extremes and multi-plant sampling. This is a first use of SNP chips for evaluating varietal purity in a forage oat multi-line. We plan to utilize SNPs to determine the genetic underpinning of the phenotypic variants of cereal oats.

2. Materials and Methods

2.1. Plant Materials Used

The main forage oat (*Avena sativa* L.) variety used in this study was AV25, a multi-line population developed in southern Colombia early in the 21st century by the Colombian Agricultural Research Corporation (AGROSAVIA). As a candidate population for varietal release in 2014, the multi-line was grown in a plot for foundation seed at the Obonuco Research Station of AGROSAVIA in Pasto, Colombia ($1°11'56''$ N, $77°41'44''$ W). For the next generation, dormancy was broken over a six-month storage period. Then, 100 spike-to-row selections were established at two field sites in 2015: (1) in the Tibaitatá Experimental station (near the town of Mosquera in the province of Cundinamarca) and (2) on the Finca Sutacón near the town of Susacón in the municipality of Paipa (Province Boyacá). Mass selection across and between rows was applied at each location, resulting in two new composite sub-populations, called AV25-T and AV25-S from Tibaitatá and Susacón, respectively. The original mixture: AV25-C (the population from Obonuco) was compared to the two subpopulations: AV25-T (Tibaitatá selection) and AV25-S (Sutacón selection) as well as to two off-types selections, namely AV25-P (early selection from Obonuco) and AV25-Ta (late selection from Obonuco). Two commercial varieties (Cajicá and Cayuse) were grown as checks.

2.2. Agronomic Locations Used

The grow out was carried out in the second half of 2016 at the Tibaitatá experimental station where 400 rows each of single plant selections (SPS) from the populations AV25-C, AV25-T, AV25-S were grown. In addition, 25 SPS rows for each of the off-type oats, AV-25-P, AV25-Ta and each of the commercial varieties Cajica and Cayuse were established there. Each row was 4.0 m long, and rows were planted 0.3 m apart. A planting density of 60 kg seed ha^{-1} was used, and the fertilizer regime and weed control followed the recommendation of AGROSAVIA [3], with the goal of obtaining approximately 60 t ha^{-1} in forage yield. Harvesting was set at stage Z7.0 as in Zadoks et al. [30].

2.3. Leaf Tissue and DNA Extraction

A total of 96 single plant DNA samples were taken from the flag leaves of an equal number of plants that were selected based on representation of the full SPS plot when they had reached the heading stage (Z6.0) according to the maturity scale of Zadoks et al. [30]. The leaf samples included 22 plants from AV25-C, 30 each from AV25-T and AV25-S; 4 each from Cajicá and Cayuse, and 3 each from AV25-P and AV25-Ta. DNA was extracted from approximately 4 cm of leaf tissue placed in individual 1.5 mL Eppendorf tubes on ice. These tissue samples were then ground to a powder with liquid nitrogen and a plastic mortar. The Mo-Bio® kit for plant tissue (PowerPlant® ProDNA Isolation, San Francisco, CA, USA) was used to obtain DNA from ground leaves. The quantity and quality of the DNA was evaluated with a NanoDrop® 2000 spectrophotometer (Thermo Fisher Scientific Inc., Wilmington, DE, USA). Samples with a 260/280 UV wavelength ratio close to 1.8, and a 260/230 ratio close to 2.0 with a minimum DNA concentration of 50 ng µL^{-1} were obtained after dilution.

2.4. SNP Genotyping

The iSelect 6K BeadChip specific to oats [6] was used to evaluate the 96 DNA samples describe above. The process was carried out in the molecular genetics laboratory at the Tibaitatá research facility of AGROSAVIA, and the procedure for array hybridization was according to the Infinium-II assay instructions (Illumina, Inc., San Diego, CA, USA). SNP genotypes were analyzed with the software GenomeStudio v2011.1 with a GeneCaller setting of 0.15. Additionally, a quality control process was carried out on the genotyping data, according to the rules described by Wiggans et al. (2010), using PLINK v1.9 software [31].

2.5. Data Analysis

The 'genome function' of the PLINK was used to perform the grouping analysis and to evaluate diversity patterns. We used a grouping approach based on an identity matrix estimated from the allelic frequencies. The distance matrix was also calculated and used for (a) principal component analysis (PCA), and (b) neighbor joining (NJ) dendogram reconstruction also in PLINK. These analyses allowed the visualization of direct relationships and the grouping of samples from the different varieties, which were later quantified explicitly by means on an analysis of molecular variance (AMOVA) test.

In addition, the population structure was determined with the ADMIXTURE software version 1.3.0 [32]. Five independent simulations were run for values from $K = 2$ to $K = 4$, under the parameters established by default (100,000 burn-ins and 200,000 iterations in the MCMC analysis). The best K value was determined based on the PCA diagrams, AMOVA analysis, cross-run cluster stability, and with the cross-validation likelihood procedure. This methodology allowed us to determine the genetic profile of each sample and establish the degree of purity or admixture of the individual plants. To graph the results of these analyses, we used R software (R Core Team, 2017).

Based on the optimum level of clustering, pairwise relative divergence (F_{ST}) scores, according to Weir and Cockerham (1984), were computed per marker for each pair of clusters using customized R scripts. Bidirectional gene flow among pairs of clusters was also estimated as the number of migrants per generation (N_em) following Beerli and Felsenstein [33]. Networks depicting pairwise relative divergence and bidirectional migration rates were drawn using the R package *qgraph*. Finally, in order to describe genetic patterns of diversity and identify F_{ST}-outlier SNP markers, we further computed per-marker expected (H_E) and observed (H_O) heterozygosity, nucleotide diversity as measured by π (Nei, 1987), Watterson's theta (θ) estimator [34], and Tajima's D [35] using the software Tassel v.5 [36] and customized R scripts. We compared these statistics among them via Pearson's correlations (*cor.test* function), and F_{ST} against the π score, using customized R codes (R Core Team, 2017).

3. Results

3.1. Oat Selections Based on Harvest Date

During the grow-out for AV-25-C, phenotypic variation was observed especially for days to flowering. About 20% of plants were late flowering (110 days after planting, dap), 34% were early (55 dap) and 46% were intermediate (80 dap). Correspondingly, harvests were either late (starting at 150 dap), intermediate (135 dap) or early (100 dap). Selections (Allard, 1967) of off-type plants were harvested and threshed to produce the AV25-P (early) and AV25-Ta (late) mass selections from the original AV25-C population. Meanwhile, the AV25-S and AV25-T selections made in Sutacón and Tibaitatá locations also from the AV25-C composite were made over two years (2015–2017) to be more uniform. These were both intermediate in phenology, with flowering around 55 dap and harvest around 135 dap (Table S1).

3.2. Oat SNP Chip Success Rate

The iSelect 6K BeadChip array from Illumina® for oats allowed the genotyping of 4975 SNPs distributed throughout the A. sativa genome as described by Tinker et al. [6]. The initial screen was across the 96 DNA samples representing subsampling of each oat varietal selection. SNPs were excluded if the call rate was less than 90%, or if they presented an extreme deviation from a Hardy–Weinberg equilibrium (*p*-value < 0.01). In addition, any SNPs that were monomorphic or having a minimum allele frequency (MFA) below 0.05% were removed. After applying these criteria for quality control, 1672 SNP loci were retained that met the parameters for further analysis. In addition to SNP locus validation, there were eight DNA samples which had a call rate lower than 90%, perhaps indicating worse quality DNA, and were eliminated. These included three lines from AV25-C (13.6% of total), one from AV25-P (33.3%), one from AV25-S (3.3%) and one from AV25-Ta (33.3%). The study, therefore had 88 genotype samples for population analysis that were sufficient for evaluating structure with the main sub-populations (AV25-C, AV25-S and AV25-T), while noticing trends on the off-types (AV25-P and AV25-Ta). The latter were of less interest to us because they were not being considered for varietal release as the former were.

3.3. Clustering of Individual Plants

The analysis of the subpopulation structure made it possible to determine that there was genetic differentiation between the evaluated oat genotypes (Figure 1). With a value of $K = 2$, a subgroup with the largest number of genotypes (red bars) corresponded mostly to the purified selections for AV25-T and AV25-S and some genotypes from the mixed original population, AV25-C. The other subgroup observed (yellow bars) included the samples of the commercial varieties, the AV25-P and AV25-Ta plants and some of the AV25-C samples, mainly those that correspond to the early maturing phenotype. At $K = 3$ value, there was a subgroup (yellow color) with a similar genetic profile as at $K = 2$, including all the samples of the purified AV25-S and AV25-T genotypes and most but not all the AV25-C samples. The second group (green color) included commercial varieties and plants with early and late phenotypes, while the third genetic subgroup (red color) was present only in some of the AV25-C samples. It is evident that within AV25-C there is a significant degree of mixture and some late off-type genotypes that were eliminated by mass selection. These results coincide with field observations, since this cultivar presents plants of different phenologies (early, intermediate and late). With the $K = 4$ value, we still found a common genetic profile for Cayuse and Cajicá and AV25-P and AV25-Ta genotypes.

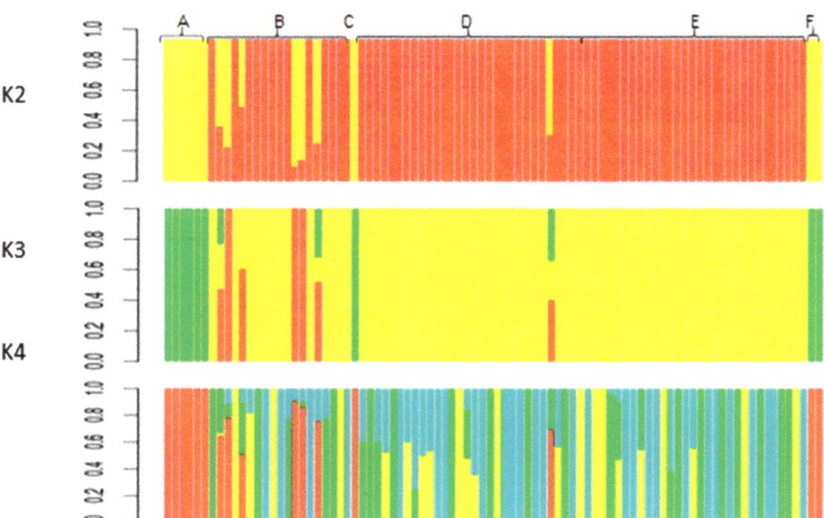

Figure 1. Genetic structure of 88 plants representing seven forage oat genotypes (A = Cajicá and Cayuse, commercial varieties), B = AV25 C, the mixed source population, C = AV25-P, the early selection D = AV25-S, mass selection samples from Sutacon, E = AV25-T mass selection samples from Tibaitatá and F = AV25-Ta, the late selection. Each individual panel is divided into subgroups coded in colors based on clustering K-value from 2 to 4 being the number of groups assumed. Length of the bar segment represents the estimated proportion of sample membership.

3.4. Principal Component Analysis

Figure 2 shows the principal component analysis for the 88 samples analyzed. The first two components of the PCA analysis showed three clearly defined groups, with a compact first group located to the left of the graph (Figure 2a) corresponding to the purified genotypes of AV25-S (29 genotypes with SNP profiles), all the AV25-T (30 genotypes) and some of the AV25-C samples (13 genotypes).

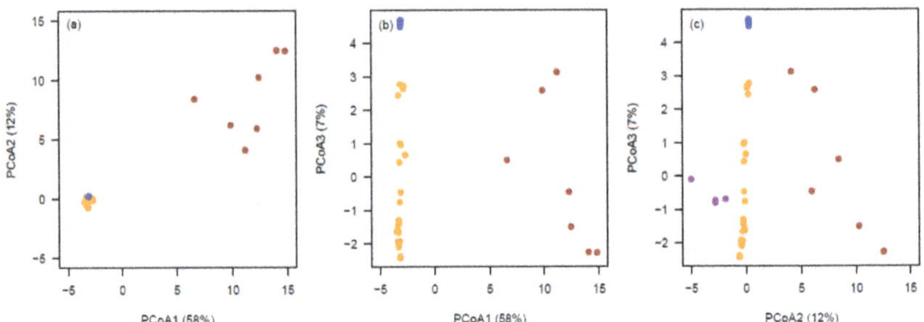

Figure 2. Analysis of main components (PCoA) for estimating genetic structure patterns of forage oat genotypes used in this study: with (**a**) first two components, (**b**) first and second components, and (**c**) second and third components. The percentage of variation explained by each component is shown within parenthesis in the label of the corresponding axis. Axes are drawn to the same scale to make comparisons. Different colors represent groups.

When expanding the area of the graph to the third component (Figure 2b,c), there was evidence of variation between samples for the AV-25 group: a second group located in the upper right part of the graph included the AV25-P samples and a single AV25-S sample. The variation that existed within this group was greater than that evidenced in the other groups, which suggests a possible mixture with the other genotypes. The third group identified in the PCA was in the lower right part of the graph with the first two components (Figure 2a) and corresponded to the samples of commercial varieties and AV25-P and AV25-Ta, which presented a well-defined, distinct genetic profile.

Different color dots were used to represent the groupings with the majority of the AV25-S lines in the orange group, the AV25-T lines in the blue group, the AV25-P lines in the brown group and the AV25-Ta lines in the purple group. The genotypes from the original mixed sample AV25-C were together with other groups.

The patterns were broadly concordant with the distance-based dendogram using the neighbor joining algorithm (Supplementary Figure S1) where dots on the tree and in the PCA were in same grouping colors. Nonetheless, this approach exhibited less resolution at distinguishing genetic differences among individual plants when compared to the PCA. An AMOVA test relying on the grouping from the NJ tree and from the PCA indicated that 63.6% of the genetic variation could be explained by genetic cluster, while 36.4% was found within genotype (p-value < 0.001). Finally, unsupervised Bayesian clustering allowed for a more detailed reconstruction of the admixture level within individual plants from the selections compared to the original population of AV25-C (Supplementary Figure S2). In $K = 2$, most AV25-T and AV25-S genotypes grouped together with some samples of AV25-C. The other group included the samples of the commercial varieties, the AV25-P and AV25-Ta plants and some of the AV25-C samples, mainly those that corresponded to the early phenotype.

Given that PCA diagrams, AMOVA analysis, and cross-run cluster stability had substructure beyond the hierarchical level of $K = 2$, we explored population stratification and admixture at two additional K strata. When using a $K = 3$ value there was a subgroup with a similar genetic profile to $K = 2$, which corresponded to all the samples of the purified AV25-T genotype and most of the AV25-S samples, except for one sample showing a mixed genetic profile. This group also included AV25-C samples. The second group was made up of commercial varieties and plants with extreme phenotypes (early and late), while the third genetic profile was present only in some of the AV25-C samples. It was evident that within AV25-C there was a significant degree of admixture, sharing genetic profiles with the other groups evaluated.

3.5. Relative Divergence and Genetic Patterns of Diversity

Based on the unsupervised genetic clustering described above, pairwise relative divergence (F_{ST}) scores were estimated between groups and ranged from 0.06 to 0.25 (Figure 3a). Pairwise F_{ST} scores were higher for all three comparisons of blue and red clusters against the brown cluster for AV25-P samples and a single AV25-S versus purple cluster with commercial varieties.

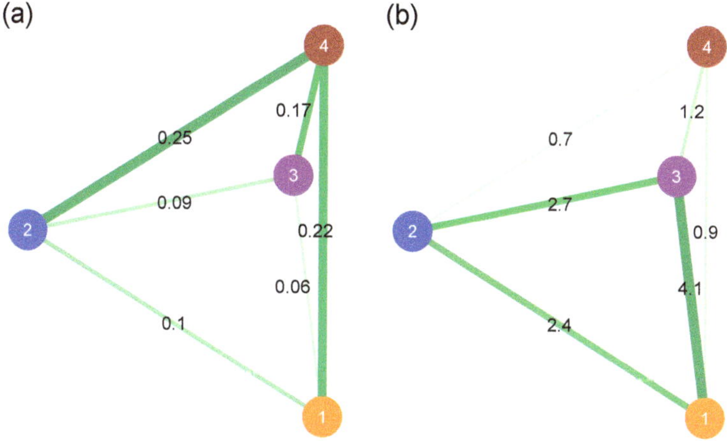

Figure 3. Networks depicting (**a**) relative divergence scores (F_{ST}), and (**b**) bidirectional gene flow (computed by migrants per generation, $N_e m$) among four forage oat genetic clusters (circles). Clusters were determined following the unsupervised clustering approaches of the previous section: where the orange (1) and blue (2) clusters included samples of the purified genotypes AV25-S (29 lines), AV25-T (30) and some of the AV25-C samples (13); the purple cluster (3) included samples of commercial varieties, and AV25-P and AV25-Ta; and the brown cluster (4) included AV25-P samples and a single AV25-S sample.

Meanwhile, the number of migrants per generation ($N_e m$) ranked from 0.7 to 2.7 (Figure 3b) and was negatively correlated with the F_{ST} ($r = 0.93$, *p*-value < 0.001). The width and color intensity of the green lines were proportional to the F_{ST} score in the sub-figure (a), and proportional to number of migrants per generation ($N_e m$) in sub-figure (b). The thinnest lines in the sub-figure corresponded to $N_e m$ values below one. $N_e m$ values above one indicated mixtures.

4. Discussion

Cultivated oat genotypes, whether they are for cereal or fodder, are considered autogamous, self-pollinating plants [1]. Therefore, the mixture of genotypes in the original AV25-C population, uncovered by SNP analysis, show that the forage variety had become contaminated and was a composite variety with multi-lines of different inbred oat lines. Phenotypic analysis showed that the most variable feature of this population was phenological, including flowering and harvest time. The genotypes AV-P and AV-Ta selected from this population showed that contaminant plant types were part of the composite AV variety and genotyping showed that this could be related to other varieties grown in Colombia, such as Cajica and Cayuse, which are both late flowering and maturing. Although the AV variety was developed in southern Colombia (Obonuco station in Nariño), purification by mass selection (MS) was successful in both locations in central Colombia (Tibaitatá station in Cundinamarca and Sutacón farm in Paipa, Boyacá), producing two sub-populations that were uniform in SNP fingerprint and in phenological characteristics important for having a uniform forage crop.

The results of genetic structure analysis coincided with field observations, since the original multi-line oat cultivar AV25-C presented plants of different phenotypes (early, intermediate and late). The purification of AV25-S and AV5-T was achieved to a certain extent as shown by the admixture levels and PCA/dendogram results and given that they all presented intermediate phenology and no longer included early and late plants. With the structure $K = 4$ value, it was still possible to find a common genetic profile of the oat varieties Cayuse and Cajicá with the early and late mass selections of AV25-P and AV25-Ta. However, clear differentiation between the AV25-T, AV25-S and AV25-C selections was not possible, showing that the derived mass selections for the central Colombian sites was indeed derived from the original composite variety from southern Colombia.

The rapid advances in the development of large-scale genotyping platforms with the consequent decrease in processing costs have made genotyping with SNP chips an attractive and practical approach to rapidly characterize genomes and populations [6]. In the short term, these resources have opened the door to mass selection (MS) and the evaluation of populations. Such processes can also lead to genomic selection (GS) programs that use genome-wide molecular markers to predict the genetic merit of complex and quantitative traits in the improvement plant populations [37]. However, these GS breeding programs are often beyond the capacity of developing country national programs, such as the one at AGROSAVIA. For example, our budget only allowed one 96-well plate to be analyzed limiting the sampling for off-types, although providing room for evaluation of mass selections for the new purified varieties of AV25-S and AV25-T is useful for central Colombia. Therefore, in our study we emphasized the use of SNP arrays to evaluate varietal purity in MS derived sub-populations from a forage oat originally from southern Colombia.

The same SNP chip we used here validated SNPs on 1100 genotypes from 6 recombinant inbred line mapping populations [7]. Here we used fewer samples but had almost as many validated SNP markers as that study with 4950 loci called in our study and 4975 loci called in that previous study. This was better than some previous testing on pools of various oat cultivars and lines, which only provided 3500 polymorphic Mendelian loci [38].

We evaluated fewer genotypes but found more SNPs to be successfully called on the variety AV25-C and the AV25-S and AV25-T subpopulations derived from mass selections. In addition, our ability to capture outlier SNP markers in our oat population further endorsed the implementation of indirect polygenic selection strategies for the species in addition to the uses of SNP chips for multi-or pure-line varietal certification.

The application of SNP technology is already a routine practice in the regulatory processes of varietal identity and protection of plant breeders' rights. SNP chips have been used to strengthen the production of pure seed in other countries. For instance, the 4004 loci SoySNP6K chip (Lee et al., 2016) was used in 858 *Glycine max* varieties by the Argentine National Seed Commission to regulate varietal purity. Similar scenarios have been reported for other grain species: wheat (*Triticum aestivum* L., Akhunov et al., 2009; [9], barley (*Hordeum vulgare* L., [11]), rice (*Oriza sativa* L., [18]) and quinoa (*Chenopodium quinua* Willd, [39]).

Oat selections in Colombia have been classified as early, intermediate and late due to the long growing period available in the tropical environment where no winter or summer limits physiological growth. Here we had all three types of oats ranging from the earliest AV25-P to the latest AV25-Ta. Early oats are characterized by <100 days to time of forage harvest (Z7.1 stage). Intermediate oats are harvestable for forage at 135 days on average, and late oats exceed 150 days to be harvested for forage [3]. Days to harvest affects dry matter production and nutritional value of forage oats [30].

In addition to yield differences, the forage oat varieties showed differences in growth habit, plant height, stems per plant, and leaves per plant. The use of oats for livestock feed is very common in some countries, especially where processing of oatmeal and rolled oats for human consumption is limited. In some cases, oats are a dual function crop providing a boiled porridge food for humans and a fodder for animals, such as milk cows and sheep. Oats used for forage purposes in Colombia are grown in the same way as the cereal grain that is grown for its seed, but fields of livestock oats are opened up for pasture directly by animals, often sheep or cattle. On the other hand, the forage oats can be harvested for hay to supplement dairy cows or for use by the large horse breeding industry.

In Colombia, varietal purity regulations follow an agreement of the Andean Community of Nations (adopted by decision 345 and regulatory decree 533 in 1993–1994) called the Common Regime for Protection of Plant Variety and Breeders' Rights. The decree assigned the Colombian Agricultural Institute (ICA) as the national authority for the management of variety regulations. Likewise, the ICA standardized and implemented phenotypic evaluation techniques for the identification of cultivars (Valencia et al., 2010). This technology follows UPOV rules that a new plant variety must be different (D) from any other well-known variety, sufficiently uniform (U) and stable (S), according to Breeder's Rights (Community Plant Variety Office, 2015).

Conventional tests for DUS as outlined above, especially for multi-lines like the variety used here, make use of morphological characteristics, such as descriptors, and are based on UPOV's or national protocols. The reliance on phenotyping alone is problematics, as it is well known that morphological traits are highly influenced by the environment, which makes the predictions unreliable [9,25]. The morphological testing should be complemented using molecular markers, as a biotechnology tool for varietal identity. In the specific case of the application of SNPs in varietal quality control, the markers are a complementary tool to the use of morphological descriptors [7] and can give both within variety and between variety diversity, as found in this study. Its application in variety development programs among row crops has been successful in corn [40], chickpea (*Cicer arietinum*, [6]), pigeon pea (*Cajanus cajan*, [41]) and canola (*Brassica napus*, [42]). Here we show the success of marker fingerprinting in the first study of forage oats in Colombia.

5. Conclusions

Overall, our results showed that SNP marker analysis confirmed that visual selection for phenology was reliable. SNP genotyping allowed the identification of genetic differences existing within the original multi-line oat variety AV25. The information on the mass selected sub-populations AV25-S and AV25-T will facilitate registration [19,38] and will support plant breeders' rights through better fingerprinting [42]. Moving forward, other cultivars might require the implementation of procedures for the purification of single plant derived lines. The AV25 variety is now a candidate for wider release across more departments of Colombia. Previous use of the forage variety was limited to the southern department of Nariño and had become contaminated with off types. However, now, it has been multiplied in two central highland locations (Susacón and Tibaitatá) for distribution within other highland regions (>2000 masl) in Boyacá and Cundinamarca departments. With a better understanding of the population structure for the variety and its sub-populations we can disseminate the cultivar more widely with greater uniformity. Varietal purity in oats is especially important due to the delayed production of late-maturing genotypes compared to the mid-season intermediate maturing selections made here. Apart from the varietal implications of this study, the results of this research also represent an opportunity to generate new breeding techniques for oats, which might leverage GS for complex polygenic selection (Heffner et al., 2010) in an improvement program for forage oats, which would help accelerate a growing dairy industry. For this breeding to be successful, it would be necessary to broaden the base population for local oats with novel genomic and phenotypic resources while developing machine learning predictors [43] to boost prediction of key yield and adaptive traits (Jannink et al., 2010). The oat SNP chip can continue to be used for varietal certification of other forage oat cultivars. Similar procedures would be useful in wheat and barley.

Supplementary Materials: The following supporting information can be downloaded at: https://www.mdpi.com/article/10.3390/agronomy12071710/s1. Table S1: Variation in development for *Avena sativa* phenological stages in days after planting (dap) for forage oat selections grown in highland Colombia (>2000 masl), including, AV25-P (early), AV25-C/AV25-S/AV25-T (all intermediate) and AV25-Ta (late) with growth periods as defined by the Zadoks et al. [30] scale. Figure S1: neighbor joining (NJ) dendogram for genetic clustering among oat genotypes. Dots are colored following same coding, as shown in Figure 2. Figure S2: Unsupervised genetic clustering of 88 individual plants from seven forage oat cultivars. Each individual is represented by a single vertical line divided into K colors, where this is the number of groups; length of the colored segment represents the estimated proportion of sample membership to a particular group. Bars in $K = 4$ are colored following Figure 2: orange and blue dots shown as blue bars for purified genotypes AV25-S (29 samples) and AV25-T (30 samples) with some of the AV25-C genotypes (13 samples); the bars in brown were AV25-P and a single AV25-S sample; and purple bars included commercial varieties and AV25-Ta.

Author Contributions: Conceptualization, L.F.C.-D. and D.B.-G.; methodology, L.F.C.-D., D.B.-G. and J.C.-S.; formal analysis, L.F.C.-D., D.B.-G., D.R.T.-C. and A.J.C.; investigation, L.F.C.-D. and M.W.B.; resources, L.F.C.-D.; data curation, D.B.-G., D.R.T.-C. and A.J.C.; writing—original draft preparation, L.F.C.-D. and M.W.B.; writing—review and editing, A.J.C. and M.W.B.; visualization, D.R.T.-C. and A.J.C.; supervision, L.F.C.-D. and M.W.B.; funding acquisition, L.F.C.-D. and M.W.B. All authors have read and agreed to the published version of the manuscript.

Funding: Agrosavia-Tibaitatá Station and Tennessee State University—Nashville and their sponsors the Ministry of Agriculture and Rural Development (MADR) and United States Department of Agriculture (USDA) for support to this work. USDA funding included Evans Allen (TEN-X07) and AFRI-Plant Breeding (New Crops Conference). Additional funding to Agrosavia was from the Fund for Milk and Meat of MADR.

Data Availability Statement: Data is available in the Supplementary Materials.

Acknowledgments: We appreciate help form Rodrigo Martinez Sarmiento of AGROSAVIA for experimental design, and Jhon Alexander Berdugo Cely for preliminary data analysis.

Conflicts of Interest: The authors declare no conflict of interest.

References

1. Yan, H.; Martin, S.L.; Bekele, W.A.; Latta, R.G.; Diederichsen, A.; Peng, Y.; Tinker, N.A. Genome size variation in the genus *Avena*. *Genome* **2016**, *59*, 209–220. [CrossRef] [PubMed]
2. Arango-Quispe, S.J.S.; Viera-Valencia, M.; Gómez Bravo, C.A. Agronomic Evaluation and Nutritive Value of Oat (*Avena Sativa*) under Rainfall Restriction in the Andes. *Cienc. Tecnol. Agropecu.* **2022**, *23*. [CrossRef]
3. Campuzano-Duque, L.F.; Castro-Rincón, E.; Torres-Cuesta, D.; Castillo-Sierra, J.; Nieto-Sierra, D.; Portillo-Lopez, P.A. Altoandina: Nueva variedad forrajera para la zona andina en Colombia. *Agron. Mesoam.* **2020**, *31*, 581–591. [CrossRef]
4. McCouch, S.R.; Wright, M.H.; Tung, C.W.; Maron, L.G.; McNally, K.L.; Fitzgerald, M.; Singh, N.; DeClerck, G.; Agosto-Perez, F.; Korniliev, P.; et al. Open access resources for genome-wide association mapping in rice. *Nat. Commun.* **2016**, *7*, 10532. [CrossRef]
5. Poland, J.; Endelman, J.; Dawson, J.; Rutkoski, J.; Wu, S.; Manes, Y.; Dreisigacker, S.; Crossa, J.; Sánchez-Villeda, H.; Sorrells, M.; et al. Genomic selection in wheat breeding using genotyping-by-sequencing. *Plant Genome J.* **2012**, *5*, 103–113. [CrossRef]
6. Tinker, N.A.; Chao, S.; Lazo, G.R.; Oliver, R.E.; Huang, Y.F.; Poland, J.A.; Jellen, E.; Maughan, P.J.; Kilian, A.; Jackson, E.W. A SNP genotyping array for hexaploid oat. *Plant Genome* **2014**, *7*, 1–8. [CrossRef]
7. You, Q.; Yang, X.; Peng, Z.; Xu, L.; Wang, J. Development and applications of a high throughput genotyping tool for polyploid crops: Single nucleotide polymorphism (SNP) Array. *Front. Plant Sci.* **2018**, *9*, 104. [CrossRef]
8. Lu, F.; Lipka, A.E.; Glaubitz, J.; Elshire, R.; Cherney, J.H.; Casler, M.D.; Buckler, E.S.; Costich, D.E. Switchgrass Genomic Diversity, Ploidy, and Evolution: Novel Insights from a Network-Based SNP Discovery Protocol. *PLoS Genet.* **2013**, *9*, e1003215. [CrossRef] [PubMed]
9. Tinker, N.A.; Kilian, A.; Wight, C.P.; Heller-Uszynska, K.; Wenzl, P.; Rines, H.W.; Bjørnstad, A.; Howarth, C.J.; Jannink, J.-L.; Anderson, J.M.; et al. New DArT markers for oat provide enhanced map coverage and global germplasm characterization. *BMC Genom.* **2009**, *10*, 39. [CrossRef]
10. Korir, P.C.; Zhang, J.; Wu, K.; Zhao, T.; Gai, J. Association mapping combined with linkage analysis for aluminum tolerance among soybean cultivars released in Yellow and Changjiang River Valleys in China. *Theor. Appl. Genet.* **2013**, *126*, 1659–1675. [CrossRef]
11. Bayer, M.M.; Rapazote-Flores, P.; Ganal, M.; Hedley, P.; Macaulay, M.; Plieske, J.; Ramsay, L.; Russell, J.; Shaw, P.D.; Thomas, W.; et al. Development and evaluation of a barley 50k iSelect SNP array. *Front. Plant Sci.* **2017**, *8*, 1792. [CrossRef] [PubMed]

12. Bekele, W.A.; Wight, C.P.; Chao, S.; Howarth, C.; Tinker, N.A. Haplotype based genotyping-by-sequencing in oat genome research. *Plant Biotechnol. J.* **2018**, *16*, 1452–1463. [CrossRef] [PubMed]

13. Waters, D.; Bundock, P.; Henry, R. Genotyping by allele-specific PCR. In *Plant Genotyping II: SNP Technology*; Henry, R., Ed.; CABI: Oxfordshire, UK, 2008; pp. 88–97.

14. Chen, H.; Xie, W.; He, H.; Yu, H.; Chen, W.; Li, J.; Yu, R.; Yao, Y.; Zhang, W.; He, Y.; et al. A high-density SNP genotyping array for rice biology and molecular breeding. *Mol. Plant* **2014**, *7*, 541–553. [CrossRef]

15. Felcher, K.J.; Coombs, J.J.; Massa, A.N.; Hansey, C.N.; Hamilton, J.P.; Veilleux, R.E.; Buell, C.R.; Douches, D.S. Integration of two diploid potato linkage maps with the potato genome sequence. *PLoS ONE* **2012**, *7*, e36347. [CrossRef]

16. Ganal, M.W.; Durstewitz, G.; Polley, A.; Bérard, A.; Buckler, E.S.; Charcosset, A.; Clarke, J.D.; Graner, E.-M.; Hansen, M.; Joets, J.; et al. A large maize (*Zea mays* L.) SNP genotyping array: Development and germplasm genotyping, and genetic mapping to compare with the B73 reference genome. *PLoS ONE* **2011**, *6*, e28334. [CrossRef] [PubMed]

17. Clarke, W.E.; Higgins, E.E.; Plieske, J.; Wieseke, R.; Sidebottom, C.; Khedikar, Y.; Batley, J.; Edwards, D.; Meng, J.; Li, R.; et al. A high-density SNP genotyping array for *Brassica napus* and its ancestral diploid species based on optimised selection of single-locus markers in the allotetraploid genome. *Theor. Appl. Genet.* **2016**, *129*, 1887–1899. [CrossRef]

18. Yu, H.; Xie, W.; Li, J.; Zhou, F.; Zhang, Q. A whole-genome SNP array (RICE6K) for genomic breeding in rice. *Plant Biotechnol. J.* **2013**, *12*, 28–37. [CrossRef]

19. Hulse-Kemp, A.M.; Lemm, J.; Plieske, J.; Ashrafi, H.; Buyyarapu, R.; Fang, D.D.; Frelichowski, J.; Giband, M.; Hague, S.; Hinze, L.L.; et al. Development of a 63K SNP Array for Cotton and High-Density Mapping of Intraspecific and Interspecific Populations of *Gossypium* spp. *Genes | Genomes | Genet.* **2015**, *5*, 1187–1209. [CrossRef]

20. Unterseer, S.; Bauer, E.; Haberer, G.; Seidel, M.; Knaak, C.; Ouzunova, M.; Meitinger, T.; Strom, T.M.; Fries, R.; Pausch, H.; et al. A powerful tool for genome analysis in maize: Development and evaluation of the high density 600 k SNP genotyping array. *BMC Genom.* **2014**, *15*, 823. [CrossRef]

21. Wang, S.; Wong, D.; Forrest, K.; Allen, A.; Chao, S.; Huang, B.E.; Maccaferri, M.; Salvi, S.; Milner, S.G.; Cattivelli, L.; et al. Characterization of polyploid wheat genomic diversity using a high-density 90,000 single nucleotide polymorphism array. *Plant Biotechnol. J.* **2014**, *12*, 787–796. [CrossRef]

22. Cortés, A.J.; Blair, M.W. Genotyping by Sequencing and Genome—Environment Associations in Wild Common Bean Predict Widespread Divergent Adaptation to Drought. *Front. Plant Sci.* **2018**, *9*, 128. [CrossRef] [PubMed]

23. Blair, M.W.; Cortes, A.J.; Farmer, A.D.; Huang, W.; Ambachew, D.; Penmetsa, R.V.; Carrasquilla-Garcia, N.; Assefa, T.; Cannon, S.B. Uneven recombination rate and linkage disequilibrium across a reference SNP map for common bean (*Phaseolus vulgaris* L.). *PLoS ONE* **2018**, *13*, e0189597. [CrossRef] [PubMed]

24. Bianco, L.; Cestaro, A.; Linsmith, G.; Muranty, H.; Denance, C.; Théron, A.; Troggio, M. Development and validation of the Axiom Apple 480k SNP genotyping array. *Plant J.* **2016**, *86*, 62–74. [CrossRef] [PubMed]

25. Aitken, K.; Farmer, A.; Berkman, P.; Muller, C.; Wei, X.; Demano, E.; Jackson, P.; Magwire, M.; Dietrich, B.; Kota, R. Generation of a 345K sugarcane SNP chip. *Proc. Int. Soc. Sugar Cane Technol.* **2016**, *29*, 1165–1172. Available online: http://www.issct.org/pdf/proceedings/2016/Molecular-Biology-papers/document-4.pdf (accessed on 12 July 2022).

26. Bassil, N.V.; Davis, T.M.; Zhang, H.; Ficklin, S.; Mittmann, M.; Webster, T.; Mahoney, L.L.; Wood, D.; Alperin, E.S.; Rosyara, U.R.; et al. Development and preliminary evaluation of a 90K Axiom SNP array for the allo-octoploid cultivated strawberry *Fragaria x ananassa*. *BCM Genom.* **2015**, *16*, 155. Available online: https://bmcgenomics.biomedcentral.com/articles/10.1186/s12864-015-1 310-1 (accessed on 12 July 2022). [CrossRef]

27. Clevenger, J.; Chu, Y.; Chavarro, C.; Agarwal, G.; Bertioli, D.J.; Leal-Bertioli, S.C.M.; Pandey, M.K.; Vaughn, J.; Abernathy, B.; Barkley, N.A.; et al. Genome-wide SNP Genotyping Resolves Signatures of Selection and Tetrasomic Recombination in Peanut. *Mol. Plant* **2017**, *10*, 309–322. [CrossRef]

28. Pandey, M.K.; Agarwal, G.; Kale, S.M.; Clevenger, J.; Nayak, S.N.; Sriswathi, M.; Chitikineni, A.; Chavarro, C.; Chen, X.; Upadhyaya, H.D.; et al. Development and evaluation of a high-density genotyping "axiom_arachis" array with 58 K SNPs for accelerating genetics and breeding in groundnut. *Sci. Rep.* **2017**, *7*, 40577. [CrossRef]

29. Oliver, R.E.; Tinker, N.A.; Lazo, G.R.; Chao, S.; Jellen, E.N.; Carson, M.L.; Rines, H.W.; Obert, D.E.; Lutz, J.D.; Shackelford, I.; et al. SNP discovery and chromosome anchoring provide the first physically anchored hexaploid oat map and reveal synteny with model species. *PLoS ONE* **2013**, *8*, e58068. [CrossRef]

30. Zadoks, J.C.; Chang, T.T.; Konzak, C.F. A decimal code for the growth stages of cereals. *Weed Res.* **1974**, *14*, 415–421. [CrossRef]

31. Purcell, S.; Neale, B.; Todd-Brown, K.; Thomas, L.; Ferreira, M.A.R.; Bender, D.; Maller, J.; Sklar, P.; de Bakker, P.I.W.; Daly, M.J.; et al. PLINK: A tool set for whole-genome association and population-based linkage analyses. *Am. J. Hum. Genet.* **2007**, *81*, 559–575. [CrossRef]

32. Alexander, D.H.; Novembre, J.; Lange, K. Fast model-base estimation of ancestry in unrelated individuals. *Genome Res.* **2009**, *19*, 1655–1664. [CrossRef] [PubMed]

33. Beerli, P.; Felsenstein, J. Maximum likelihood estimation of migration rates and effective population numbers in two populations using a coalescent approach. *Genetics* **1999**, *152*, 763–773. [CrossRef] [PubMed]

34. Watterson, G.A. Number of segregating sites in genetic models without recombination. *Theor. Popul. Biol.* **1975**, *7*, 256–276. [CrossRef]

35. Tajima, F. Statistical method for testing the neutral mutation hypothesis by DNA polymorphism. *Genetics* **1989**, *123*, 585–595. [CrossRef]
36. Bradbury, P.J.; Zhang, Z.; Kroon, D.E.; Casstevens, R.M.; Ramdoss, Y.; Buckler, E.S. TASSELL Software for association mapping of complex traits in diverse samples. *Bioinformatics* **2007**, *23*, 2633–2635. [CrossRef]
37. Meuwissen, T.H.E.; Hayes, B.J.; Goddard, M.E. Prediction of total genetic value using genome-wide dense marker maps. *Genetics* **2001**, *157*, 1819–1829. Available online: https://www.ncbi.nlm.nih.gov/pmc/articles/PMC1461589 (accessed on 12 July 2022). [CrossRef]
38. Tumino, G.; Voorrips, R.E.; Rizza, F.; Badeck, F.W.; Morcia, C.; Ghizzoni, R.; Germeier, C.U.; Paulo, M.-J.; Terzi, V.; Smulders, M.J.M. Population structure and genome-wide association analysis for frost tolerance in oat using continuous SNP array signal intensity ratios. *Theor. Appl. Genet.* **2016**, *129*, 1711–1724. [CrossRef]
39. Maughan, P.J.; Smith, S.M.; Rojas-Beltran Elzinga, D.; Raney, J.A.; Jellen, E.N.; Fairbanks, J.D. Single nucleotide polymorphisms identification, characterization, and linkage mapping in *Chenopodium quinua*. *Plant Genome* **2012**, *5*, 1–7. [CrossRef]
40. Winfield, M.O.; Allen, A.M.; Burridge, A.J.; Barker, G.L.A.; Benbow, H.R.; Wilkinson, P.A.; Coghill, J.; Waterfall, C.; Davassi, A.; Scopes, G.; et al. High-density SNP genotyping array for hexaploid wheat and its secondary and tertiary gene pool. *Plant Biotechnol. J.* **2015**, *14*, 1195–1206. [CrossRef]
41. Saxena, R.K.; Varma Penmetsa, R.; Upadhyaya, H.D.; Kumar, A.; Carrasquilla-Garcia, N.; Schlueter, J.A.; Farmer, A.; Whaley, A.M.; Sarma, B.; May, G.D.; et al. Large-scale development of cost-effective single-nucleotide polymorphism marker assays for genetic mapping in Pigeonpea and comparative mapping in legumes. *DNA Res.* **2012**, *19*, 449–461. [CrossRef]
42. Tinker, N.A.; Bekele, W.A.; Hattori, J. Haplotag: Software for Haplotype-Based Genotyping-by-Sequencing Analysis. *Genes | Genomes | Genet.* **2016**, *6*, 857–863. [CrossRef] [PubMed]
43. Cortés, A.J.; López-Hernández, F. Harnessing Crop Wild Diversity for Climate Change Adaptation. *Genes* **2021**, *12*, 783. [CrossRef] [PubMed]

 *agronomy*

Article

Marker-Assisted Backcross Breeding for Improvement of Submergence Tolerance and Grain Yield in the Popular Rice Variety 'Maudamani'

Elssa Pandit [1,2,†], Swapnil Pawar [1,†], Saumya Ranjan Barik [1], Shakti Prakash Mohanty [1], Jitendriya Meher [1] and Sharat Kumar Pradhan [1,*]

[1] Crop Improvement Division, ICAR—National Rice Research Institute, Cuttack 753006, Odisha, India; elsambio@gmail.com (E.P.); swapnilpwr189@gmail.com (S.P.); saumya.bt06@gmail.com (S.R.B.); evenpunk22@gmail.com (S.P.M.); jmehercrri@gmail.com (J.M.)
[2] Department of Biosciences and Biotechnology, Fakir Mohan University, Balasore 754089, Odisha, India
* Correspondence: pradhancrri@gmail.com
† Authors contributed equally.

Citation: Pandit, E.; Pawar, S.; Barik, S.R.; Mohanty, S.P.; Meher, J.; Pradhan, S.K. Marker-Assisted Backcross Breeding for Improvement of Submergence Tolerance and Grain Yield in the Popular Rice Variety 'Maudamani'. *Agronomy* **2021**, *11*, 1263. https://doi.org/10.3390/agronomy11071263

Academic Editors: Roxana Yockteng, Andrés J. Cortés and María Ángeles Castillejo

Received: 19 May 2021
Accepted: 14 June 2021
Published: 22 June 2021

Publisher's Note: MDPI stays neutral with regard to jurisdictional claims in published maps and institutional affiliations.

Abstract: Submergence stress due to flash floods reduces rice yield drastically in sensitive varieties. Maudamani is a high yielding popular rice variety but is highly susceptible to submergence stress. The selection of progenies carrying Sub1 and GW5 (wide-grain) enhanced the submergence stress tolerance and grain yield of theMaudamani variety by following the marker-assisted backcross breeding method. Foreground screening detected 14 BC1F1, 17 BC2F1, and 12 BC3F1 backcross progenies that carried the target QTLs for submergence tolerance and grain width. Background screening was performed in the progenies carrying the target QTL and enhanced the recovery of a recipient parent's genome by upto 96.875% in the BC3 pyramided line. The BC3F1 plant containing the highest recipient parent genome content and the target QTLs was self-pollinated. In BC3F2 generation, the target QTLs the Sub1 and GW5 (wide-grain) alleles and recipient parent's yield component QTL OsSPL14 were tracked for homozygous states in the progenies. Seven pyramided lines showed tolerance to submergence for 14 days and higher grain yield than both the parents. The pyramided lines were similar to the recipient parent for the majority of the studied morphological and quality traits. The pyramided lines are useful as cultivars and can serve as potential donors for transfer of Sub1, OsSPL14, Gn1a, GW5 (wide-grain), and SCM2 QTLs.

Keywords: background selection; foreground selection; gene pyramiding; marker-assisted breeding; submergence tolerance; yield component QTL

1. Introduction

Rice, the golden cereal, sustains life for millions of people around the world. Rice is important not only as a staple food but also for its association with life in India, which is seen in its use as worship material for important ceremonies and rituals since ancient time. Rice provides important compounds, namely carbohydrates, quality proteins, vitamins, specific oils, many minerals, dietary fiber, and a few phyto-compounds thatprovide added health benefits [1]. The crop is very unique in its adaptation and cultivation from very high elevations to below sea level. The crop covers around 160 million hectares of land around the world. The crop is cultivated as a rainfed crop in about 45% of the total rice cultivated area [1]. Rice crop provides livelihood to nearly 4 billion people, which constitutes about 55% of the global population. The crop generates about $206 billion global annual earnings, which is 17% of the total crop value [2]. In recent years, the crop has become highly affected by the adverse effects of climate change. The higher production from rainfed rice cultivation is now challenged by the climate change-related yield-reducing factors in India [3]. About 22 mha of rainfed rice area is cultivated in India, of which 90% is confined to the eastern region of the country [4]. Submergence tolerance and high grain yield along with resistance to major diseases and insect pests should be transferred to a superior variety to ensure stable production.

Although there has been impressive growth in rice production and productivity during recent years, still, the demand for rice continues to be higher, projecting upwards in the future due to the population growth rate being higher than the production increase rate of rice. Again, in the coming years, the production increase must be harvested under the adverse effects of climate change. In addition, the future production needs to be obtained from the least available land with less use of inputs and lower chemical usage and in a more environmentally friendly manner. The breeding strategy in this investigation focuses on simultaneous improvement of stress tolerance and yield in rice so as to fulfill future requirements [4]. Yield potential in a rice variety can be enhanced by pyramiding the suitable yield component QTLs absent in that variety. Among the yield component traits, grain number enhancement is controlled by QTLs Gn1a [5], Ghd7 [6], Ghd8 [7,8], APO1 [9], DEP1 [10,11], DEP2 [12], and DEP3 [13]. An increase in yield is also regulatedby the grain weight and grain dimension QTLs GW2 [14], GS3 [15], GS5 [16], and GW5 [17,18]. Tiller is controlled by the QTLs MOC1 [19], LRK1 [20], EP3 [21], and IPA1 [22,23]. High yield through higher grain filling is controlled by the QTLs GIF1 [11] and FLO [24,25]. Transfer of these yield component QTLs into a variety lacking these QTLs will be a useful way of raising the yield potential further in the variety.

Unpredictable and flash floods during rainy season are now a common occurrence in India, particularly in the eastern region of the country. This is a major cause of yield reduction in susceptible varieties affected by submergence stress. The Maudamani variety produces 7 to 9 t/ha normally but harvests upto 11 t/ha grain yield under favorable conditions. However, total crop failure occurs if the crop is affected by flash flood causing submergence for more than a week. The submergence tolerance QTL, Sub1, confers tolerance to submergence for about two weeks [26,27]. Recently, gene-based markers are available for transfer of Sub1 QTL through marker-assisted breeding. Submergence tolerance has been improved in many high yielding varieties including Swarna using this QTL introgression [28–32]. The transfer of Sub1 QTL from Swarna-Sub1 to Maudamani may contribute negligible undesirable genetic effects from the donor variety as the donor parent is a popular variety [33].

The yield component QTL, GW5, is associated with reduced grain width and the effects of the QTL was consistent under multiple environmental conditions. The loss of a GW5 segment resulted in wide-grain genotypes in most japonica and indica rice [18]. The donor parent, Swarna-Sub1, developed through the marker-assisted backcross breeding approach, carries the Sub1 QTL for submergence tolerance [1]. Additionally, the presence of yield component QTLs Gn1a, GW5, and SCM2 were also confirmed in the parent by the parental line validation study. Additionally, the recipient parent, Maudamani, a high yielding super rice variety, showed the presence of yield component QTLs OsSPL14, Gn1a, and SCM2 via the parental line validation study in this investigation. We report here the successful development of pyramided lines in a Maudamani background carrying Sub1, OsSPL14, and GW5 (wide grain) QTLs in a homozygous state for submergence tolerance and high grain yield.

2. Materials and Methods

2.1. Plant Materials and Breeding Program

The rice variety Swarna-Sub1 carrying Sub1 QTL for submergence tolerance and yield component QTLs SCM2 and GW5 was used as the donor male parent in a hybridization program. The recipient parent, Maudamani, is a high yielding variety of eastern India inbuilt with the yield component QTLs OsSPL14, Gn1a, and SCM2 but shows susceptibility to submergence stress. The recipient parent was crossed with the donor variety, Swarna-Sub1, during a dry season in 2014 as per the scheme depicted for the marker-assisted breeding (Figure 1). The donor and recipient parents were obtained from the gene bank of

National Rice Research Institute (NRRI), Cuttack, India. One true F1 plant was hybridized with the recipient parent during the rainy season in 2014 to generate BC1F1 generation seeds. True hybridity was checked using the direct Sub1 marker, Sub1-A203, and with a co-dominant marker, RM8300, as well. The BC1F1 seeds were grown, and the progenies were screened for the target genes, yield component, and submergence tolerance QTLs by using the established molecular markers (Table 1). The grain size QTL, GW5, is associated with narrow grained rice, and hence, negative selection was performed to obtain the Maudamani grain width. In the background selection, progenies of the BC1F1 generation carrying the two target QTLs were screened using the polymorphic markers. The foreground positive progenies containing the highest genome content of the recurrent parent was hybridized with the recipient parent Maudamani to obtain BC2F1 seeds. During the dry season in 2015, BC2F1 seeds were harvested. Among the derived progenies detected with the target QTLs in BC2F1 generation, those with highest recurrent genome content was again crossed with the recipient parent during the wet season in 2015 to produce BC3F1 seeds. The background analysis of BC3F1 progenies was performed during the dry season in 2016. The BC3F1 plant containing the highest recipient genome content along with two target QTLs wereselfed in the dry season in 2016. BC3F2 progenies were genotyped to search for the presence of homozygosity for the target QTLs and recipient parent's yield QTLs during the wet season in 2016. The seed increase in the pyramided progenies detected with homozygous target genes were increased during the dry season in 2017. Evaluations for agronomic and other traits were performed during the wet seasons in 2017, 2018, and 2019.

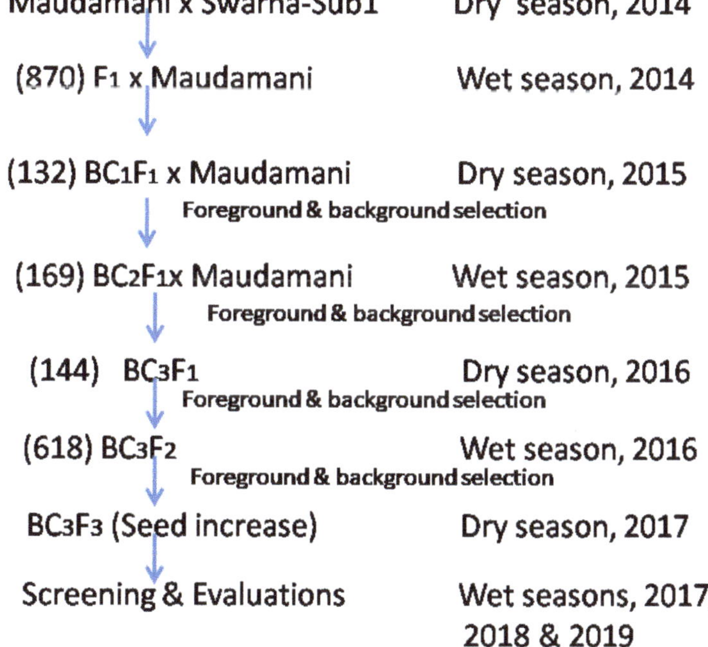

Figure 1. Breeding scheme used for transfer of Sub1 and GW5 (wide-grain) for submergence tolerance and yield component QTLs through marker-assisted backcross breeding into popular variety, 'Maudamani' (Inside the parentheses, number of hybrids/derived progenies were raised in the generation).

Table 1. Molecular markers used for screening of yield component QTLs and submergence tolerance during foreground selection.

Sl No.	Trait	QTL	Chromosome	Position (Mb)	Primer Name	Primer Sequence	Reference
1.	High grain number	Gn1a	1	5.27	Gn1a(F)	5′ TGAGGATGCCGTGGAAGACGA 3′	Ashikari et al. [5]
					Gn1a (R)	5′ TTCGTGTTCGCGCAGGACGT 3′	
2.	Grain width	GW2	2	8.11	GW2(F)	5′ CCAATAAAGATGTCCATTCTGTTA 3′	Song et al. [14]
					GW2 (R)	5′ GCTCTTCCTGTAACACATATTATG 3′	
3.	Grain weight	GW5	5	27.4	GW5 (F)	5′ GCGTCGTCAGAGGTAGA 3′	Weng et al. [18]
					GW5 (R)	5′ GACCTAACCCATCTCATTCCA 3′	
4.	Strong culm	SCM 2	6	27.66	SCM 2 (F)	5′ ATTCAGATCAATAGGTTGAGTGT 3′	Ookawa et al. [33]
					SCM 2 (R)	5′ TGCTATGTATATCCTATCGGTTC 3′	
5.	Wealthy Farmers Panicle	OsSPL14	8	25.27	OsSPL14(F)	5′ CAAGGGTTCCAAGCAGCGTAA 3′	Miura et al. [23]
					OsSPL14(R)	5′ TGCACCTCATCAAGTGAGAC 3′	
6.	Submergence tolerance	Sub1	9	5.9	Sub1-A203(F)	5′ CTTCTTGCTCAACGACAACG 3′	Pradhan et al. [3] Xu et al. [34] Septiningsih et al. [35]
					Sub1-A203(R)	5′ AGGCTCCAGATGTCCATGTC 3′	
					Sub1-BC2(F)	5′ AAAACAATGGTTCCATACGAGAC 3′	
					Sub1-BC2(R)	5′ GCCTATCAATGCGTGCTCTT 3′	
				5.7	RM8300(F)	5′ GCTAGTGCAGGGTTGACACA 3′	
					RM8300(R)	5′ CTCTGGCCGTTTCATGGTAT 3′	

2.2. Genomic DNA Isolation, Polymerase Chain Reaction and Marker Analysis

Genomic DNA was isolated following the standard extraction protocol [36]. APCR reaction was performed following the procedure used in a previous publication [32]. The information regarding chromosome number, position, and sequence of the primers used in the polymerase chain reaction are presented in Table 1. Eight gene-specific and tightly linked markers for the two target QTLs and four recipient QTLs were used in foreground selection (Table 1). These markers information were taken from earlier publications reported for these target traits [3,5,14,18,23,33,35,36]. A total of 644 publicly available SSR markers were used for the study of polymorphism between the two parents. The polymorphic markers detected were used for background selection (Table 2). Agarose gel electrophoresis was used to separate the amplification products obtained from PCR reactions. The images were recorded in a gel documentation system (SynGene, Cambridge, UK). Data analysis and dendrogram construction were performed following the standard publications [37–39]. Graphical Geno Types (GGT) Version 2.0 software was used to construct the genome recovery graph of recipient parent in the pyramided lines based on the SSR marker data [40].

2.3. Screening for Submergence Tolerance

The BC3F4 generation pyramided lines and parents were transplanted in the screening tank of ICAR NRRI, Cuttack, at around 3 weeks' seedling age during the wet seasons in 2018 and 2019. The screening trial was laid out in a randomized complete block design (RBD) with three replications/entries accommodating a population size of 66 plants/entry. The experiment materials were transplanted with a spacing of 15 × 20 cm^2 by providing three rows/entry. Two weeks of complete submergence stress upto 1.5 water depth was maintained in the tank. De-submergence was performed just after completion of the 14-day stress period, and subsequently, regeneration was assessed one week after de-submergence. The data recording and scoring the genotypes were collected following the procedures of earlier publications [1,3].

Table 2. Polymorphic microsatellite markers obtained between the rice varieties Maudamani and Swarna-Sub1.

Chromosome	No. of Markers Tested	No. of Polymorphic Markers	Name of Polymorphic Markers
1	48	4	RM11694, RM10346, RM495, RM594
2	40	4	RM1347, RM263, RM521, RM6374
3	40	5	RM14723, RM426, RM1278, RM570, RM3392
4	40	5	RM1113, RM2416, RM470, RM551, RM335
5	40	5	RM440, RM430, RM334, RM7452, RM122
6	60	3	RM20377, RM469, RM225
7	48	4	RM432, RM336, RM429, RM8007
8	96	6	RM337, RGNMS2900A, AUT22718, RGNMS2866, RGNMS2873, RM23444
9	72	5	RM23668, RM23722, RM257, RM444, RM201
10	40	5	RM6100, RM258, RM590, RM222, RM25181
11	40	6	RM1812, RM167, RM1341, RM206, RM287, RM224
12	40	5	RM235, RM7003, RM20A, RM1337, RM309
Total	644	57	

2.4. Evaluation of the Pyramided Lines for Various Traits

The seedlings of 25-day-old pyramided lines carrying Sub1 and yield QTLs were transplanted along with the parents during the wet seasons in 2017, 2018, and 2019. A plot slot size of 12 m^2 was provided to each entry, with 40 plants per row, at a spacing of 15 × 20 cm^2, and planted in RBD with three replications in the research farm of NRRI, Cuttack. The data for ten plants for morpho-quality traits viz., plant height, panicles/plant, panicle weight (g), number of filled grains, total spikelets, number of primary branches, secondary branches and number of tertiary branches per panicle, grain length, grain breadth, 1000-grain weight, milling (%), head rice recovery (%), and amylose content (%) from each entry and replications were recorded. Plot yield and days to 50% flowering were recorded on a whole plot basis. The standard protocols published for head rice recovery [41] and gel consistency [42] were adopted. Amylose content in the grains of the pyramided and parental lines was estimated following the standard procedures described in an earlier publication [43]. An analysis of the various morpho-quality traits of the pyramided and parental lines were analyzed using SAS 2008, version 9.2 [44]. The Principal Component analysis (PCA) for the pyramided and parental lines was performed by using multivariate analysis (Past Software version 4.03) data of the 15 morphological traits. A scatter plot was generated by using two major components: Principal Component 1 (PC1) and Principal Component 2 (PC2). The Eigen value and percentage of variance were generated by the interaction of a variance–covariance matrix. The interaction between all morphological traits was depicted through biplot graph in the matrix. All of the plots and results of PCA were generated as per a standard procedure following previous publications [45–47].

3. Results

3.1. Development of Improved Lines

3.1.1. Validation of Donor and Recipient Parents for the Target Traits

The target QTLs controlling the traits were validated in the donor and recipient parents before starting hybridization and selection. The presence of submergence tolerance

QTL Sub1 and the yield component QTLs Gn1a and GW5 was confirmed in the parent, Swarna-Sub1 (Figure 1). The recipient parent is a high yielding popular variety. The genetic basis of high yield was checked by validating the presence of QTLs contributing higher yield through Gn1a, OsSPL14, and SCM2 in the recipient parent Maudamani. The yield component QTLs common in both parents were observed to be Gn1a and SCM2 (Figure 2). The gene-based and tightly linked molecular markers for Sub1 QTL and the direct marker for the yield component QTL, GW5, were used for validation and tracking of the target genes in the parental and backcross-derived lines (Table 1). A parental polymorphism survey was performed by using 644 simple sequence repeats markers covering all of the chromosomes. A total of 57 polymorphic markers were detected between the two parents and deployed for background screening (Table 2).

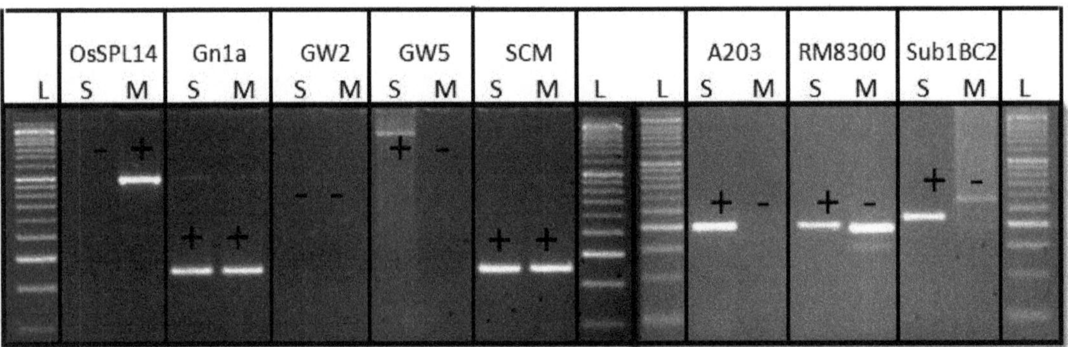

Figure 2. PCR amplification of OsSPL14, Gn1a, GW5 and SCM2 gene specific markers for yield component QTLs and the markers controlling submergence tolerance, Sub1 by deploying Sub1-A203 and Sub1-BC2 in both the parents Maudamani and Swarna-Sub1. S: Swarna-Sub1; M: Maudamani; L—Molecular weight marker (50 bp plus ladder) are the gel lanes.

3.1.2. Marker-Assisted Selection in BC1F1 Generation

Maudamani was hybridized with 'Swarna-Sub1', and 870 F1 seeds were obtained. The hybridity in F1 plants was confirmed by genotyping the hybrid plants using the Sub1 specific marker. One true F1 plant was crossed with the recipient parent, Maudamani, and a total of 132 BC1F1 seeds were generated. The backcross generation was grown, and foreground screening was performed in all of the BC1F1 plants using the two markers for the QTLs, Sub1 and GW5 (Figure 3).

The screening results of the BC1F1 progenies of the cross revealed the presence of Sub1 QTL in 78 derivatives detected by the markers A203 and Sub1-BC2 (200 bp and 240 bp). Screening for the presence of GW5 (wide-grain) gene controlling the grain width and weight identified 14 progenies to carry both traits. Negative selection was performed for GW5 in order to obtain the Maudamani grain width and weight. The other yield component QTL desired, OsSPL14, and validated in the recipient parent, Maudamani, isalso inherited in these progenies. Additionally, the common yield component QTLs present in both parents namely, Gn1a and SCM2, detected from both parents, are expected to be in homozygous states in these progenies. Background screening was performed in the 14 BC1F1 foreground-positive progenies by using 57 SSR markers. Out of these 14 plants, the progeny carrying maximum recipient genome content was selected for the next backcross. The recipient parent's genome content in those 14 progenies varied from 64.58 to 81.25% with an average value of 76.26% (Table 3). The backcross derivatives MSS128 and MSS84showed the highest recurrent genome content of 81.25%. The BC1F1 lines generated from MSS84 and MSS128 were backcrossed with the recipient parent, Maudamani, to obtain BC2F1 seeds.

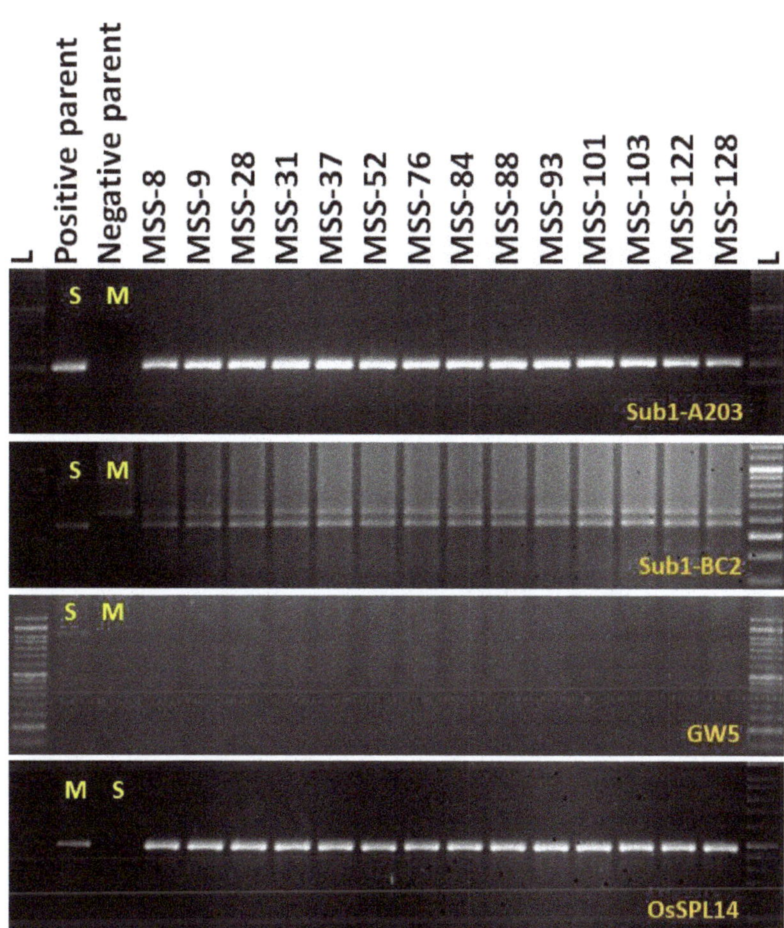

Figure 3. PCR amplification of markers for submergence tolerance, Sub1 deploying Sub1-A203 and Sub1-BC2 along with yield QTLs GW5 and OsSPL14 in BC1F1 progenies. L: Molecular weight marker (50 bp plus ladder) and lanes on the top of the gel indicate BC1F1 progenies.

Table 3. Genotyping of backcross progenies for two QTLs and recovery of recipient parent's genome in the foreground positive backcross progenies.

Generation	No. of Plants Scored	No. of Progenies Heterozygotes for 2 Target QTLs	Expected % of Recurrent Parent Genome to Selected Backcross Plants	Average Recipient Parent Genome Content (%) in the Backcross Progenies	Estimated Maximum % Genome Recovery of Recurrent Parent to Selected Backcross Progenies
BC_1F_1	132	14	75.0	76.26	81.25
BC_2F_1	169	17	87.5	87.74	91.67
BC_3F_1	144	12	93.25	94.88	96.87

3.1.3. Marker-Assisted Selection in BC2F1 Generation

One hundred and sixty-nine BC2F1 plants were grown in the field for selection. The target QTLs were tracked by foreground selection using gene-specific and linked markers.

The genotyping results of 169 BC2F1 progenies showed 93 positive progenies for Sub1 QTL. These 93 positive progenies were checked for the presence/absence of the GW5 (wide-grain) gene using gene-specific markers. Seventeen plants with the desired QTLs were identified for further background selection (Figure 4). The other yield QTLs OsSPL14 desired from Maudamani, Gn1a and SCM2, from both parents, is also present in those 17 progenies. Background screening for recovery of the recipient parent genome in those 17 identified plants containing the target QTLs ranged from 82.29 to 91.67% with an average of 87.74% (Table 3). The plant MSS 128-102 showing 91.67% of the Maudamani genome content was used for the next BC3 back crossing.

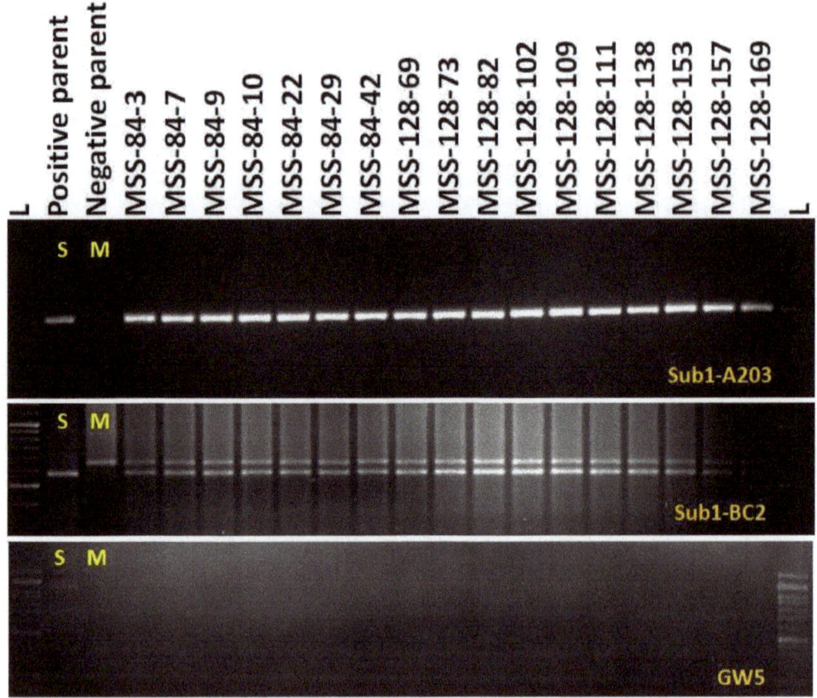

Figure 4. PCR amplification of gene specific markers for submergence tolerance, Sub1 using Sub1-A203 and Sub1-BC2 along with yield component QTLs GW5 and OsSPL14 in BC2F1 progenies. L—Molecular weight marker (50 bp plus ladder) and Lanes on the top of the gel indicate BC2F1 progenies.

3.1.4. Marker-Assisted Selection in BC3F1 and BC3F2 Generations

The BC3F1 seeds were generated by crossing BC2F1 plant no. MSS 128-102 and the recurrent parent 'Maudamani'. A total of 144 BC3F1 seeds were generated and raised for molecular screening by foreground and background selections. The genotyping results for the target QTL, Sub1, were positive in 87 progenies. Those 87 Sub1 carrier plants were genotyped for checking the presence/absence of the GW5 QTL. This analysis identified 12 plants positive for GW5 and further genotyped for background screening (Figure 5). The background analysis using 57 SSR markers in these 12 plants detected 92.7 to 96.875% recurrent parent's genome recovery, with an average of 94.88% (Table 3). The highest recurrent genome containing plant MSS 128-102-97 was selfed, and 31.5g seeds were produced for further evaluation in BC3F2 generation. Around one third of the selfed seeds were raised, and 618 BC3F2 plants were subjected to foreground screening, of which seven plants were identified to be homozygous. Additionally, the yield QTL being inherited from the Maudamani parent was checked for its homozygous state. In addition, the inheritances of yield component QTLs Gn1a and SCM2 from both parents to BC3F2 progenies were also validated through foreground analysis using gene-specific markers. The genotyping results of 618 BC3F2 progenies detected seven plants containing three QTLs, namely Sub1, GW5 (wide-grain), and OsSPL14, along with the yield component QTLs Gn1a and SCM2 in homozygous condition (Figure 6). The seed increase in these seven pyramided lines was assessed for evaluation in BC3F4 generation for various morphological and quality traits. Cluster analysis with agro-morphologic and quality traits showed distinct clusters of Swarna-Sub1, the donor line forming one group (Figure 7A).Additionally, a dendrogram was generated by using the alleles detected using the SSR markers, which grouped the developed pyramided and parental lines into two main groups (Figure 7B). Eight genotypes were accommodated in cluster I along with the recipient parent 'Maudamani', while the donor parent for submergence tolerance and yield component QTL remained in cluster II. The backcross derived lines in the cluster I were found to form different subclusters based on the 15 agro-morphologic traits studied but were similarto the recipient parent 'Maudamani' for the majority of the studied morphological and quality traits. The pyramided lines MSS 128-102-97-117, MSS128-102-97-601, MSS128-102-97-613, and MSS128-102-97-617 were almost similar in terms of genome recovery among themselves and with recipient parent 'Maudamani' (Figure 7C).

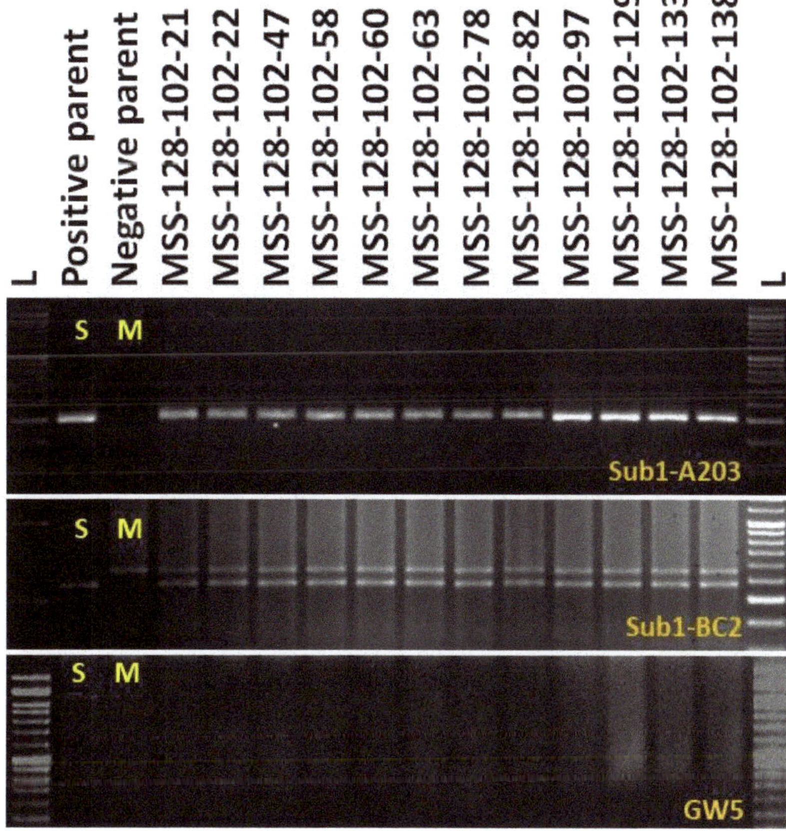

Figure 5. PCR amplification of markers for submergence tolerance, Sub1 deploying Sub1-A203 and Sub1-BC2 along with yield component QTLs Gw5 and OsSPL14 in BC3F1 progenies. L: Molecular weight marker and lanes on the top of the gel indicate BC3F1 progenies.

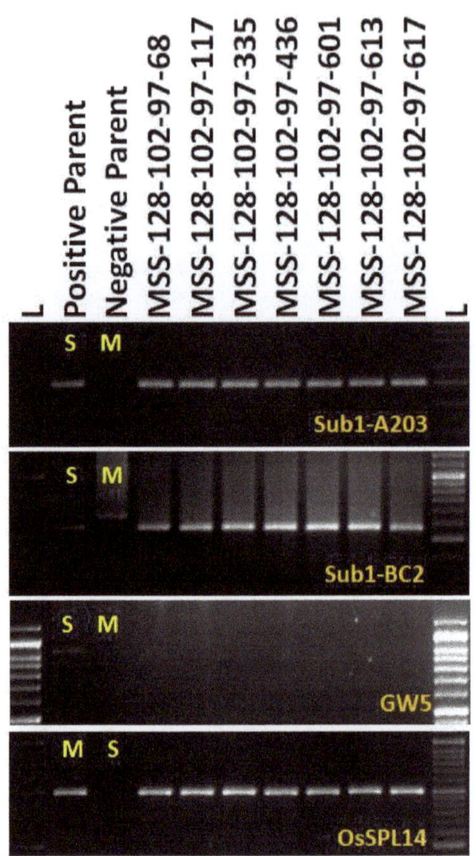

Figure 6. PCR amplification of gene specific markers for submergence tolerance, Sub1 deploying Sub1-A203 and Sub1-BC2 along with yield component QTLs GW5 and OsSPL14 in BC3F2 progenies. L—Molecular weight marker (50 bp plus ladder) and lanes on the top of the gel indicate BC3F2 progenies.

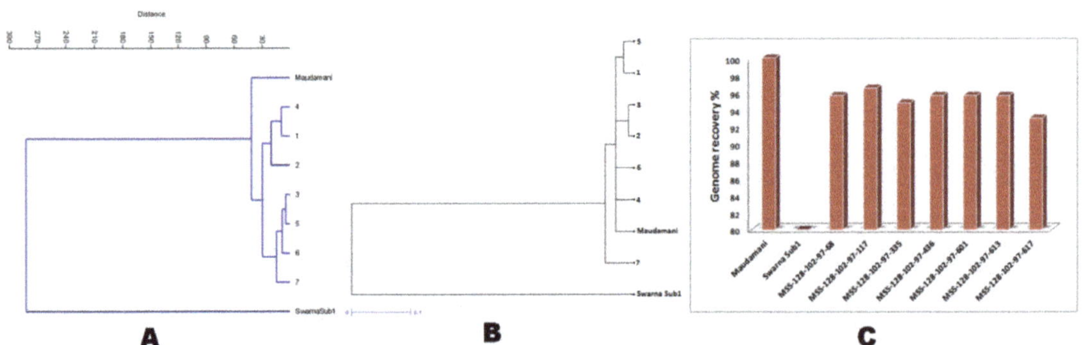

Figure 7. Seven pyramided lines along with parents in (**A**) dendrogram showing relatedness based on 16 morphologic and quality traits; (**B**) Dendrogram showing the genetic relationship between lines based on 57 microsatellite markers and (**C**) % contribution of recurrent genome in the pyramided lines. The numbers indicate the pyramided lines, 1: MSS128-102-97-68; 2: MSS128-102-97-117; 3: MSS128-102-97-335; 4: MSS128-102-97-436; 5: MSS128-102-97-601; 6: MSS128-102-97-613; 7: MSS128-102-97-617.

3.2. Analysis of Recipient Genome Recovery on the Carrier Chromosomes in the Pyramided Lines

The background analysis for recipient genome recovery and genetic drag linked to the donor segments were assessed using 57 background and 8 foreground markers. The markers were carefully selected for all of the chromosomes to obtain maximum coverage in background screening. The foreground analysis detected seven BC3F2 pyramided lines for the presence of homozygous target QTLs in the progenies. The Sub1 carrier chromosome 9 showed linkage drag of the donor fragment on both sides of the marker A203 and Sub1-BC2 in all seven NILs (Figure 8). The GW5 (wide-grain) present on the chromosome 5 showed no drag of the donor segment in all pyramided lines except MSS128-102-97-613 and MSS128-102-97-335, where a drag was noticed in between RM7452 and RM440 (Figure 8).

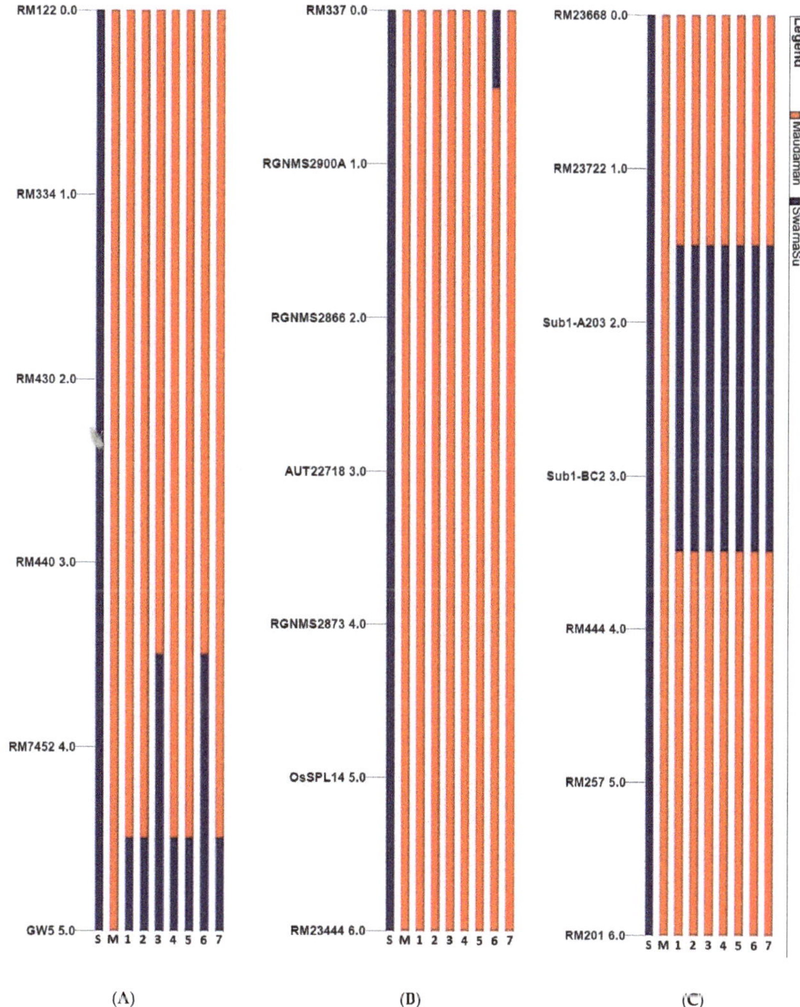

Figure 8. Analyses of QTLs stacking in carrier chromosomes associated with submergence tolerance and yield component QTLs in 7 pyramided lines (**A**) GW5 (wide-grain) yield component QTL present on chromosome 5 (**B**) OsSPL14 yield component QTL present on chromosome 8 and (**C**) Sub1 QTL on chromosome 9 in the BC3F3 progenies of Maudamani/Swarna-Sub1. The numbers indicate the pyramided lines, 1: MSS128-102-97-68; 2: MSS128-102-97-117; 3: MSS128-102-97-335; 4: MSS128-102-97-436; 5: MSS128-102-97-601; 6: MSS128-102-97-613; 7: MSS128-102-97-617.

3.3. Evaluation of the Pyramided Lines for Submergence Tolerance

Nine genotypes including seven BC3F4 pyramided lines carrying target QTLs were evaluated under the controlled submergence screening tank for confirmation of the submergence tolerance trait in the pyramided lines. The test genotypes were exposed to two weeks of submergence stress. After one week of de-submergence, all seven pyramided lines showed regeneration ability from 85 to 95% while the donor parent 'Swarna-Sub1' showed regeneration of 95% (Figure 9). No regeneration was found in the sensitive parent 'Maudamani'. The pyramided lines MSS128-102-97-117 and MSS128-102-97-436 had similar regeneration abilitiesto that of the Swarna-Sub1 parent. Pyramided lines viz., MSS128-102-97-68, MSS128-102-97-335, MSS128-102-97-601, MSS128-102-97-617, and MSS128-102-97-613 showed regeneration abilities of 90%.

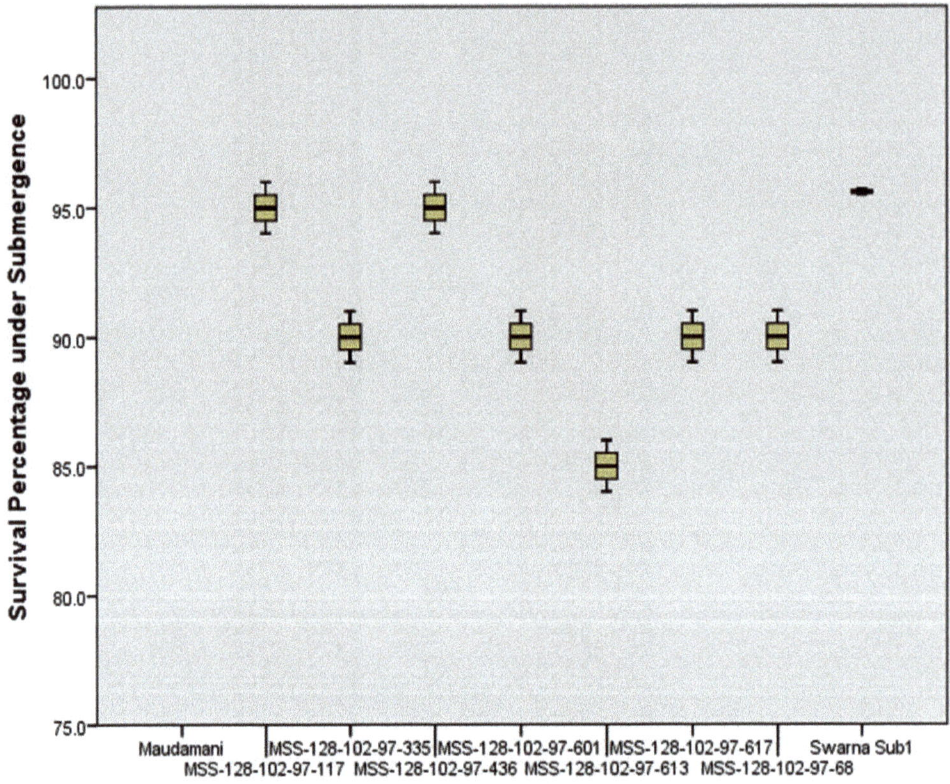

Figure 9. Percent plant regenerated in the pyramided lines carrying Sub1 QTL along with the parents under control screening facility after one week of de-submergence from 14 days of submergence stress.

3.4. Evaluation of Pyramided Lines for Agro-Morphologic, Yield Components and Grain Quality Traits

The pyramided lines carrying submergence tolerance and yield component QTLs in the background of the Maudamani variety were evaluated for various traits during the wet seasons in 2017, 2018, and 2019. The pyramided lines were compared with both the popular rice varieties Maudamani and Swarna-Sub1. The recipient parent 'Maudamani' produced a pooled mean grain yield of 8.69 t/ha. The pyramided lines MSS128-102-97-436, MSS128-102-97-117, MSS128-102-97-617, MSS128-102-97-601, MSS128-102-97-68, and MSS128-102-97-335 produced more yield than the recipient parent Maudamani (Table 4). The pyramided lines MSS128-102-97-117 and MSS128-102-97-436 produced >9 t/ha grain yield, showing an advantage of >5% over the recipient popular variety Maudamani. However, the agro-morphologic traits of all of the pyramided lines were not similar tothat of the parent, Maudamani. The target morphologic traits controlled by the yield component QTLs viz., no. of primary branches, secondary branches and tertiary branches per panicle, and panicle weight finally influencing grain yield in the pyramided lines were observed to be almost similar within each subcluster (Table 4; Figure 10). Much of the grain quality and the cooking characters of the recipient parent such as milling (%), head rice recovery (%), kernel length (mm), kernel breadth (mm), kernel length after cooking, gel consistency, amylose content (%), and alkali spreading value were retained in a few pyramided lines (Table 4). The placement pattern of the parents and the pyramided lines in the quadrants of the genotype-by-trait biplot diagram constructed based on 15 agro-morphologic, yield, and component traits over three years showed similarity among the pyramided lines (Figure 11). The pyramided lines were found in the first and second quadrants along with the recipient parent 'Maudamani'. The pyramided lines closer to each other were almost similar in grain yield, grain quality, and the other studied parameters (Figure 11). These genotypes closer to Maudamani are good candidates for further evaluation and release as cultivars in various parts of the country. The variation observed for the first principal component was61.7%, while 15.1% was explained for the second component.

Figure 10. Panicle photographs of 2 parents and 7 pyramided lines evaluated in BC3F4 generation during wet season, 2018. M: Maudamani, S: Swarna-Sub1 and the numbers in the figure indicate the pyramided lines, 1: MSS128-102-97-68; 2: MSS128-102-97-117; 3: MSS128-102-97-335; 4: MSS128-102-97-436; 5: MSS128-102-97-601; 6: MSS128-102-97-613; 7: MSS128-102-97-617.

Table 4. Agro-morphologic and grain quality parameters of the pyramided lines along with parents pooled over 3 seasons under field evaluation.

Serial Number	Pyramided and Parental Lines	Plan Height (cm)	Days to 50% Flowering	Panicles/ Plant	Grains/ Panicle	Total Spikelets/ Panicle	No. of Primary Branches/ Panicle	No.of Secondary Branches/ Panicle	No.of Tertiary Branches/ Panicle	1000-seed weight (g)	Grain Length (mm)	Grain Breadth (mm)	Milling (%)	Head Rice Recovery (%)	Amylose Content (%)	Plot Yield (t/ha)
1	MSS 128-102-97-68	102 [c]	108 [de]	10.73 [b]	305 [a]	370 [abc]	12 [c]	78 [ab]	7 [c]	20.31 [a]	5.17 [a]	2.61 [abc]	68.4 [a]	64.47 [a]	24.23 [ab]	8.747 [c]
2	MSS 128-102-97-117	104 [abc]	110 [b]	11.43 [b]	316 [a]	388 [ab]	14 [ab]	84 [ab]	9 [a]	21.25 [a]	5.24 [a]	2.47 [c]	68.6 [a]	64.3 [a]	24.23 [ab]	9.195 [a]
3	MSS 128-102-97-335	104 [bc]	108 [e]	11.6 [b]	294 [a]	354 [c]	12 [c]	76 [ab]	8 [bc]	21.02 [ab]	5.24 [a]	2.31 [d]	67.7 [a]	64.5 [a]	24.13 [ab]	8.664 [c]
4	MSS 128-102-97-436	105 [abc]	108 [de]	11.17 [b]	287 [a]	369 [abc]	13 [bc]	84 [ab]	8 [ab]	21.31 [a]	5.17 [a]	2.69 [ab]	68.1 [a]	63.86 [a]	24.42 [ab]	9.125 [ab]
5	MSS 128-102-97-601	103 [c]	109 [cd]	11.43 [b]	297 [a]	353 [c]	12 [c]	75 [ab]	7 [c]	20.93 [ab]	5.17 [a]	2.55 [bc]	67.5 [a]	64.43 [a]	24.26 [ab]	8.783 [c]
6	MSS 128-102-97-613	103 [bc]	110 [b]	11.47 [b]	303 [a]	357 [bc]	12 [c]	76 [b]	7 [c]	20.75 [ab]	5.32 [a]	2.66 [ab]	67.5 [a]	63.53 [a]	24.19 [ab]	8.314 [d]
7	MSS 128-102-97-617	105 [abc]	110 [bc]	10.57 [b]	306 [a]	359 [bc]	13 [bc]	83 [ab]	8 [ab]	20.93 [ab]	5.3 [a]	2.6 [abc]	68.1 [a]	63.67 [a]	24.24 [ab]	8.965 [b]
8	Swarna-Sub1	107 [ab]	113 [a]	13.93 [a]	137 [b]	176 [d]	9 [d]	21 [c]	0 [d]	20.01 [b]	5.35 [a]	2.27 [d]	68.5 [a]	63.73 [a]	24.97 [a]	6.129 [e]
9	Maudamani (recipient)	108 [a]	107 [e]	8.86 [c]	312 [a]	394 [a]	14 [a]	87 [a]	9 [a]	21.52 [a]	5.33 [a]	2.73 [a]	68.9 [a]	64.63 [a]	23.5 [b]	8.69 [c]
	LSD$_{5\%}$	7.42	4.35	2.76	27.8	29.2	1.36	7.61	0.89	1.83	0.73	0.154	7.24	7.62	2.187	0.348
	CV%	3.26	0.95	10.58	9.82	9.78	10.3	10.3	9.38	4.61	6.36	8.196	6.68	9.32	6.833	11.634
	Standard Error (SE)	1.114	0.362	0.387	9.106	9.931	0.411	2.592	0.231	0.346	0.112	0.582	1.373	1.910	0.509	0.325
	Mean	104.67	109.22	11.243	285.6	347.44	12.41	74.47	7.037	21.007	5.254	2.542	68.14	64.126	24.23	8.51
	Heritability (h^2)%	69.961	78.441	74.019	80.56	82.325	83.78	86.93	93.94	76.63	76.505	73.525	81.73	68.903	82.48	78.317

Note: Means with the same letter are not significantly different at 5% level (α value 0.05).

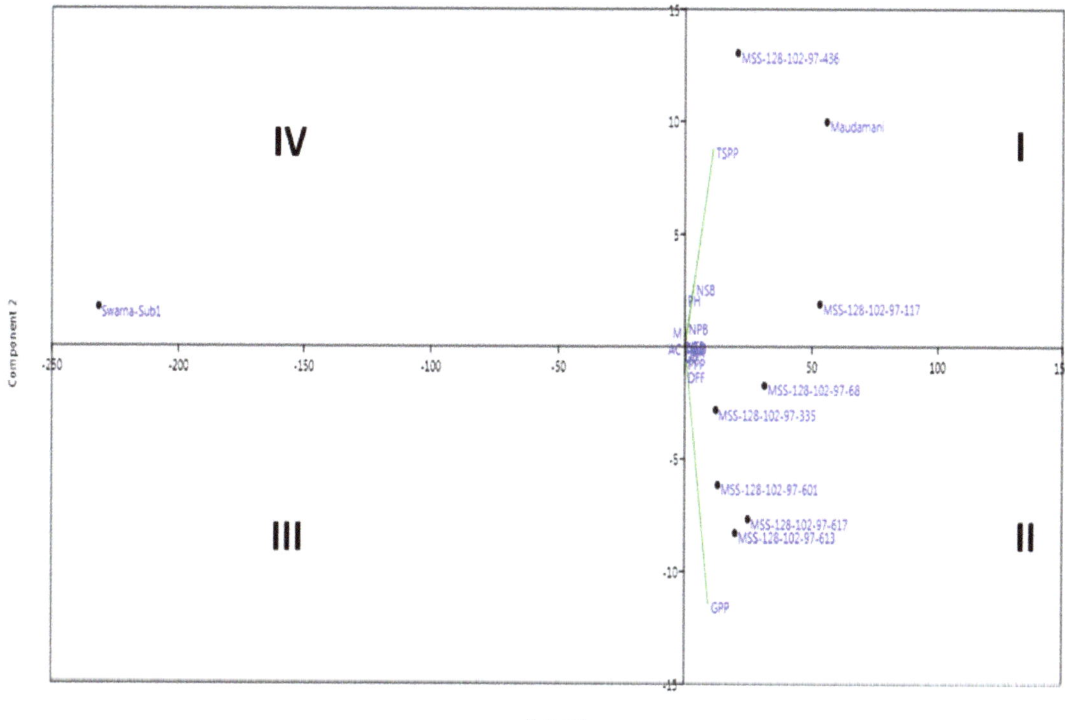

Figure 11. Genotype-trait biplot diagram of 7 pyramided lines carrying Sub1 and GW5 (wide-grain) allele along with the parents for the first two principal components. DFF—Days to 50% flowering; PH—Plant height; PN—Panicles/plant; TW—1000-grain weight; Milling (%); HRR—Head rice recovery (%); AC—Amylose content (%); NPB—Number of primary branches/panicle; NSB—Number of secondary branches/Panicle; NTB—number of tertiary branches/Panicles; GL—Grain length; GB—Grain breadth, and YLD—plot yield.

4. Discussion

The inclusion of marker-aided selection increases the accuracy in transfer of the target genes/QTLs into a recipient variety in a backcross breeding program. In the present marker-assisted breeding program, we successfully developed pyramided lines showing submergence tolerance and yield component QTL in the high yielding background of Maudamani without altering the main features of the recipient variety. Here, the QTLs were transferred simultaneously into the popular variety. In addition, it was possible to reduce the breeding duration for developing a variety compared to the classical backcross breeding approach. In this breeding program, three backcrosses and one selfing generation were utilized to transfer the target QTLs into the popular variety. The desired traits that were lacking in the popular variety could be improved in less time and with more precision. Such examples of variety development by precise transfer of genes and with a shorterduration through marker-assisted breeding are available in rice crop [29,30,48–51].

Previous reports of many successful gene transfer and pyramiding cases have been published in rice crop [47–60]. This study of QTL pyramiding for submergence tolerance and yield component QTLs is clearly different from earlier gene pyramiding work. Earlier gene stacking publications on bacterial blight resistance with submergence tolerance transfer into rice varieties, namely improved Lalat and improved Tapaswini, have been published [58,59]. However, here, yield improvement and submergence tolerance through gene pyramiding is a typical example of gene stacking. Other publications using MAB breeding are mostly for the development of cultivars through pyramiding of resistance genes for insects and diseases in rice [48,49,52–58].

In this investigation, the presence of the OsSPL14, Gn1a, and SCM2 yield QTLs were detected in the popular variety Maudamani while Gn1a, SCM2, and GW5 were confirmed in the variety Swarna-Sub1. These two varieties are good sources of yield component QTLs in indica rice. The yield component QTL OsSPL14 showed negative regulation for tiller/panicle number per plant in rice. In our study, OsSPL14 is present in a homozygous state in Maudamani and produces low to moderate tiller numbers (average, 8.86/plant). However, Swarna-Sub1 lacks this QTL and produces high tiller numbers (average, 13.93/plant). As expected, the pyramided lines also showed similar trends in tiller no./plant to that of the recipient parent, Maudamani, but different from that of Swarna-Sub1 (i.e., not a higher tiller number). The current pyramided lines will serve as potential sources of QTLs containing Sub1+ OsSPL14+ Gn1a + gw5 + SCM2 and may be suitable as cultivars. The research results of a mapping study of Gn1a QTL indicated the source of the grain-number controlling trait from a japonica variety, Habataki [5]. The presence of SCM2 provides non-lodging to the rice culm, and the japonica variety Habataki was the source for the QTL [33]. Similarly, the donor line for high panicle branching was from the ST-12 variety. The yield QTL GW5 is responsible for grain width and weight [15,61,62]. A 1212 bp deletion is responsible for enhanced grain width and weight, whereas the presence of it reduces the grain width, thereby making it a narrow grain [18]. Now, the pyramided lines in a Maudamani background carrying five yield QTLs along with submergence tolerance genes are much better and potential sources for transfer to indica rice rather than from different japonica varieties.

Work on gene pyramiding for the transfer of various traits in rice has been published earlier [48,52–55,60]. By using this precision breeding for transfer of target traits, pyramids containing Sub1+ OsSPL14+ Gn1a + gw5 + SCM2 QTLs along with recipient parents' genome of >95% in the pyramided lines was possible. The undesirable drag expected from the donor genome may come from theselection of additional unlinked loci in backcross generations [51]. In our investigation, such effects were detected in the elite pyramided lines while transferring the Sub1 QTL into the Maudamani background. The graphical representation of genotyping data as seen in the diagram constructed for the pyramided lines showed the linkage drag on the chromosome carrying the target QTLs (Figure 5). This region was previouslyreported to be a recombination hotspot [18]. However, no linkage drag was observed on the chromosome 8 carrying OsSPL14 as the QTL was inherited from the recipient parent and was not from the donor parent. Less linkage drag from donor parents was also reported by earlier researchers assessingmarker-assisted breeding in rice using more background markers [49,54,55]. Here, the donor parent was a popular variety, and hence, the drag may not show any undesirable effects in the developed pyramided lines (Figure 5). Similar findings in other publications suggest the use of an improved variety as the donor results in less or no undesirable drag compared to the wild and landraces source [49–51].

A few elite pyramided lines hadimportant features similar to the recipient parent though variation was seen among the pyramided lines. The dendrogram drawn based on the studied traits indicated grouping of the pyramided and parental lines into main three clusters with similarity within the clusters (Figure 7A). All of the pyramided lines and recipient parents were observed in quadrants I and II in the biplot diagram drawn based on the 15 morpho-quality traits, indicating minor variations among the lines (Figure 11). An evaluation of the pyramided lines for yield and quality traits showed higher yieldsin pyramided lines MSS128-102-97-436, MSS128-102-97-117, MSS128-102-97-617, MSS128-102-97-601, MSS128-102-97-68, and MSS128-102-97-335 than the recipient parent (Table 4). The transfer of traits and achieving similar or better yield in the pyramided lines were also reported earlier in a few gene-pyramiding publications [48,49,52–57].

The biplot diagram places the six pyramided lines closer to each other, while the donor parent is quite far away and placed in a separate quadrant. This shows resemblance among the pyramided and recipient lines and no undesirable drag from the donor parent during transfer of the target genes into the pyramided lines. The performance of a few pyramided lines was better than the recipient parent in yield, quality, and morphological traits (Table 4). The analysis of background genotyping results showed higher recovery of the recipient parent's genome in a few pyramided lines than the expected value in various backcross generations. Again, it revealed that transfer of Sub1 and the yield component QTLs into one genetic background may not show antagonistic effects for yield and other traits [1,48,54,56].

5. Conclusions

The pyramided lines MSS128-102-97-436, MSS128-102-97-117, MSS128-102-97-617, MSS128-102-97-601, MSS128-102-97-68, and MSS128-102-97-335 showed higher yield and submergence tolerance than the recipient parent Maudamani. The higher yield obtained in the pyramided lines might be due to an accumulation of additional yield QTLs, and no yield penalty happened due to these QTLs. In addition, the elite pyramided lines in the background of the popular variety 'Maudamani' may serve as potential donors of QTLs possessing Sub1+ OsSPL14+ gw5 + SCM2 in future breeding programs. Moreover, a few promising pyramided lines may be released as cultivars for flood-prone target regions in the country. Much of the grain quality and thecooking characteristics of the recipient parents such as milling %, head rice recovery %, kernel length (mm), kernel breadth (mm), kernel length after cooking, gel consistency, amylose content (%), and alkali spreading value were retained in few pyramided lines. The quality features of the popular rice variety Maudamani remained unchanged along with high grain yield. This study established the application of marker-assisted selection for transferring abiotic stresses tolerance and enhancing yield in rice.

Author Contributions: S.K.P. conceptualized the work, provided the resources, procured the funding, and supervised the work. S.K.P. wrote the original draft, E.P. and S.K.P. reviewedand edited the manuscript. S.K.P. and J.M. planned, executed, and supervised the work. E.P., S.P., S.R.B. and S.P.M. performed the investigation. Formal analysis andsoftware work were performed by E.P. and S.R.B., E.P., S.P. and S.P.M. performed the data curation. S.P. and S.P.M. validated the work. E.P., S.P. and S.R.B. visualized the work. All authors have read and agreed to the published version of the manuscript.

Funding: No externally aided fund was availed for this research work. However, the institute's internal funding was used for this investigation.

Institutional Review Board Statement: The authors declare that this study complies with the current laws of the countries in which the experiments were performed.

Informed Consent Statement: Not applicable.

Data Availability Statement: The data generated or analyzed in this study are included in this article.

Acknowledgments: The authors are highly grateful to the Head, Crop Improvement Division and Director, ICAR-NRRI, Cuttack for encouraging the team and for providing all the necessary facilities.

Conflicts of Interest: The authors declare that there is no competing interest and that the article is submitted without any commercial or economic interest that could be generated as a potential conflict of interest.

Abbreviations

BC1F1	Backcross generation 1
BC2F1	Backcross generation 2
BC3F1	Backcross generation 3
GC	Gel consistency
QTL	Quantitative trait loci
MSS	Maudamani Swarna-Sub1
Sub1	Submergence tolerance
RBD	Randomized block design

References

1. Pradhan, S.K.; Pandit, E.; Pawar, S.; Baksh, S.Y.; Mukherjee, A.K.; Mohanty, S.P. Development of flash-flood tolerant and durable bacterial blight resistant versions of mega rice variety 'Swarna' through marker-assisted backcross breeding. *Sci. Rep.* **2019**, *9*, 12810. [CrossRef] [PubMed]
2. Food and Agriculture Organization. Food and Agriculture Organization of the United Nations. *Rice Mark. Monit.* **2017**, *20*, 1–38.
3. Pradhan, S.K.; Barik, S.R.; Sahoo, J.; Pandit, E.; Nayak, D.K.; Pani, D.R.; Anandana, A. Comparison of Sub1 markers and their combinations for submergence tolerance and analysis of adaptation strategies of rice in rainfed lowland ecology. *Comptes Rendus Biol.* **2015**, *338*, 650–659. [CrossRef]
4. Pradhan, S.K.; Chakraborti, M.; Chakraborty, K.; Behera, L.; Meher, J.; Subudhi, H.N.; Mishra, S.K.; Pandit, E.; Reddy, J.N. Genetic Improvement of Rainfed Shallow-lowland Rice for Higher Yield and Climate Resilience. In *Rice Research for Enhancing Productivity, Profitability and Climate Resilience*; Pathak, H., Nayak, A.K., Jena, M., Singh, O.N., Samal, P., Sharma, S.G., Eds.; ICAR-National Rice Research Institute: Cuttack, India, 2018; pp. 107–121. Available online: https://icar-nrri.in/wp-content/uploads/2019/02/Rice_Research_book_nrri.pdf (accessed on 19 May 2021).
5. Ashikari, M.; Sakakibara, H.; Lin, S.; Yamamoto, T.; Takashi, T.; Nishimura, A.; Angeles, E.R.; Qian, Q.; Kitano, H.; Matsuoka, M. Cytokinin oxidase regulates rice grain production. *Science* **2005**, *309*, 741–745. [CrossRef]
6. Ikeda, K.; Ito, M.; Nagasawa, N.; Kyozuka, J.; Nagato, Y. Rice *ABERRANT PANICLE ORGANIZATION 1*, encoding an F-box protein, regulates meristem fate. *Plant J.* **2007**, *51*, 1030–1040. [CrossRef]
7. Wang, E.; Wang, J.; Zhu, X.; Hao, W.; Wang, L.; Li, Q.; Zhang, L.; He, W.; Lu, B.; Lin, H.; et al. Controlofricegrainfillingandyieldbyagenewithapotentialsignatureofdomestication. *Nat. Genet.* **2008**, *40*, 1370–1374. [CrossRef]
8. Huang, X.; Qian, Q.; Liu, Z.; Sun, H.; He, S.; Luo, D.; Xia, G.; Chu, C.; Li, J.; Fu, X. Natural variation at the *DEP1* locus enhances grain yield in rice. *Nat. Genet.* **2009**, *41*, 494–497. [CrossRef]
9. Xue, W.; Xing, Y.; Weng, X.; Zhao, Y.; Tang, W.; Wang, L.; Zhou, H.; Yu, S.; Xu, C.; Li, X.; et al. Natural variation in Ghd7is an important regulator of heading date and yield potential in rice. *Nat. Genet.* **2008**, *40*, 761–767. [CrossRef] [PubMed]
10. Wei, X.; Xu, J.; Guo, H.; Jiang, L.; Chen, S.; Yu, C.; Zhou, Z.; Hu, P.; Zhai, H.; Wan, J. DTH8 suppresses flowering in rice, influencing plant height and yield potential simultaneously. *Plant Physiol.* **2010**, *153*, 1747–1758. [CrossRef] [PubMed]
11. Yan, W.H.; Wang, P.; Chen, H.X.; Zhou, H.J.; Li, Q.P.; Wang, C.R.; Ding, Z.H.; Zhang, Y.S.; Yu, S.B.; Xing, Y.Z.; et al. A major QTL, Ghd8, plays pleiotropic roles in regulating grain productivity, plant height, and heading date in rice. *Mol. Plant* **2011**, *4*, 319–330. [CrossRef]
12. Yuki, A.; Keiko, M.; Tsuyu, A.; Izumi, K.; Masahiro, Y.; Hidemi, K.; Yukimoto, I. The SMALL AND ROUND SEED1 (SRS1/DEP2) gene is involved in the regulation of seed size in rice. *Genes Genet. Syst.* **2010**, *85*, 327–339.
13. Qiao, Y.L.; Piao, R.H.; Shi, J.X.; Lee, S.I.; Jiang, W.Z.; Kim, B.K.; Lee, J.; Han, L.; Ma, W.; Koh, H.J. Fine mapping and candidate gene analysis of *dense and erect panicle 3, DEP3*, which confers high grain yield in rice (*Oryza sativa* L.). *Theor. Appl. Genet.* **2011**, *122*, 1439–1449. [CrossRef] [PubMed]
14. Song, X.-J.; Huang, W.; Shi, M.; Zhu, M.-Z.; Lin, H.-X. A QTL for rice grain width and weight encodes a previously unknown RING-type E3 ubiquitin ligase. *Nat. Genet.* **2007**, *39*, 623–630. [CrossRef] [PubMed]
15. Fan, C.; Xing, Y.; Mao, H.; Lu, T.; Han, B.; Xu, C.; Li, X.; Zhang, Q. GS3, a major QTL for grain length and weight and minor QTL for grain width and thickness in rice, encodes a putative transmembrane protein. *Theor. Appl. Genet.* **2006**, *112*, 1164–1171. [CrossRef] [PubMed]
16. Li, Y.; Fan, C.; Xing, Y.; Jiang, Y.; Luo, L.; Sun, L.; Shao, D.; Xu, C.; Li, X.; Xiao, J.; et al. Natural variation in *GS5* plays an important role in regulating grain size and yield in rice. *Nat. Genet.* **2011**, *43*, 1266–1269. [CrossRef] [PubMed]
17. Shomura, A.; Izawa, T.; Ebana, K.; Ebitani, T.; Kanegae, H.; Konishi, S.; Yano, M. Deletion in a gene associated with grain size increased yields during rice domestication. *Nat. Genet.* **2008**, *40*, 1023–1028. [CrossRef]
18. Weng, J.; Gu, S.; Wan, X.; Gao, H.; Guo, T.; Su, N.; Lei, C.; Zhang, X.; Cheng, Z.; Guo, X.; et al. Isolation and initial characterization of *GW5*, a major QTL associated with rice grain width and weight. *Cell Res.* **2008**, *18*, 1199–1209. [CrossRef]
19. Li, X.; Qian, Q.; Fu, Z.; Wang, Y.; Xiong, G.; Zeng, D.; Wang, X.; Liu, X.; Teng, S.; Hiroshi, F.; et al. Control of tillering in rice. *Nature* **2003**, *422*, 618–621. [CrossRef]
20. Zha, X.; Luo, X.; Qian, X.; He, G.; Yang, M.; Li, Y.; Yang, J. Over-expression of the rice LRK1 gene improves quantitative yield components. *Plant Biotechnol. J.* **2009**, *7*, 611–620. [CrossRef]

21. Piao, R.; Jiang, W.; Ham, T.-H.; Choi, M.-S.; Qiao, Y.; Chu, S.-H.; Park, J.-H.; Woo, M.-O.; Jin, Z.; An, G.; et al. Map-based cloning of the *ERECT PANICLE 3* gene in rice. *Theor. Appl. Genet.* **2009**, *119*, 1497–1506. [CrossRef]

22. Jiao, Y.; Wang, Y.; Xue, D.; Wang, J.; Yan, M.; Liu, G.; Dong, G.; Zeng, D.; Lu, Z.; Zhu, X.; et al. Regulation of *OsSPL14* by OsmiR156 defines ideal plant architecture in rice. *Nat. Genet.* **2010**, *42*, 541–544. [CrossRef] [PubMed]

23. Miura, K.; Ikeda, M.; Matsubara, A.; Song, X.-J.; Ito, M.; Asano, K.; Matsuoka, M.; Kitano, H.; Ashikari, M. *OsSPL14* promotes panicle branching and higher grain productivity in rice. *Nat. Genet.* **2010**, *42*, 545–549. [CrossRef]

24. Qiao, Y.; Lee, S.-I.; Piao, R.; Jiang, W.; Ham, T.-H.; Chin, J.-H.; Piao, Z.; Han, L.; Kang, S.-Y.; Koh, H.-J. Fine mapping and candidate gene analysis of the floury endosperm gene, *FLO(a)*, in rice. *Mol. Cells* **2010**, *29*, 167–174. [CrossRef] [PubMed]

25. She, K.-C.; Kusano, H.; Koizumi, K.; Yamakawa, H.; Hakata, M.; Imamura, T.; Fukuda, M.; Naito, N.; Tsurumaki, Y.; Yaeshima, M.; et al. A novel factor *FLOURY ENDOSPERM2* is involved in regulation of rice grain size and starch quality. *Plant Cell* **2010**, *22*, 3280–3294. [CrossRef] [PubMed]

26. Xu, K.; Mackill, D.J. A major locus for submergence tolerance mapped on rice chromosome 9. *Mol. Breed.* **1996**, *2*, 219–224. [CrossRef]

27. Chen, M.; Presting, G.; Barbazuk, W.B.; Goicoechea, J.L.; Blackmon, B.; Fang, G.; Kim, H.; Frisch, D.; Yu, Y.; Sun, S.; et al. An integrated physical and genetic map of the rice genome. *Plant Cell* **2002**, *14*, 537–545. [CrossRef]

28. Neeraja, C.N.; Maghirang-Rodriguez, R.; Pamplona, A.; Heuer, S.; Collard, B.C.; Septiningsih, E.M.; Vergara, G.; Sanchez, D.; Xu, K.; Ismail, A.M.; et al. A marker-assisted backcross approach for developing submergence-tolerant rice cultivars. *Theor. Appl. Genet.* **2007**, *115*, 767–776. [CrossRef]

29. Iftekharuddaula, K.M.; Newaz, M.A.; Salam, M.A.; Ahmed, H.U.; Mahbub, M.A.A.; Septiningsih, E.M.; Collard, B.C.Y.; Sanchez, D.L.; Pamplona, A.M.; Mackill, D.J. Rapid and high-precision marker assisted backcrossing to introgress the *SUB1* QTL into BR11, the rainfed lowland rice mega variety of Bangladesh. *Euphytica* **2011**, *178*, 83–97. [CrossRef]

30. Manivong, P.; Korinsak, S.; Siangliw, J.L.; Vanavichit, A.; Toojinda, T. Marker-assisted selection to improve submergence tolerance, blast resistance and strong fragrance in glutinous rice. *Thai J. Genet.* **2014**, *7*, 110–122.

31. Khush, G.S.; Mackill, D.J.; Sidhu, G.S. *Breeding Rice for Resistance to Bacterial Leaf Blight*; IRRI: Manila, Philippines, 1989; pp. 207–217.

32. Pradhan, S.K.; Pandit, E.; Pawar, S.; Naveenkumar, R.; Barik, S.R.; Mohanty, S.P.; Nayak, D.K.; Ghritlahre, S.K.; Rao, D.S.; Reddy, J.N.; et al. Linkage disequilibrium mapping for grain Fe and Zn enhancing QTLs useful for nutrient dense rice breeding. *BMC Plant Biol.* **2020**, *20*, 57. [CrossRef]

33. Ookawa, T.; Hobo, T.; Yano, M.; Murata, K.; Ando, T.; Miura, H.; Asano, K.; Ochiai, Y.; Ikeda, M.; Nishitani, R.; et al. New approach for rice improvement using a pleiotropic QTL gene for lodging resistance and yield. *Nat. Commun.* **2010**, *1*, 132. [CrossRef]

34. Xu, K.; Xu, X.; Fukao, T.; Canlas, P.; Maghirang-Rodriguez, R.; Heuer, S.; Ismail, A.M.; Bailey-Serres, J.; Ronald, P.C.; Mackill, D.J. *Sub1A* is an ethylene-response-factor-like gene that confers submergence tolerance to rice. *Nature* **2006**, *442*, 705–708. [CrossRef]

35. Septiningsih, E.M.; Pamplona, A.M.; Sanchez, D.L.; Neeraja, C.N.; Vergara, G.V.; Heuer, S.; Ismail, A.M.; Mackill, D.J. Development of submergence tolerant rice cultivars: The Sub1 locus and beyond. *Ann. Bot.* **2009**, *103*, 151–160. [CrossRef]

36. Dellaporta, S.L.; Wood, J.; Hicks, J.B. A plant DNA mini preparation: Version II. *Plant Mol. Biol. Rep.* **1983**, *1*, 19–21. [CrossRef]

37. Pavalíce, A.; Hrda, S.; Flegr, J. Free Tree—freeware program for construction of phylogenetic trees on the basis of distance data and bootstrap/jackknife analysis of the tree robustness. Application in the RAPD analysis of genus Frenkelia. *Folia Biol. (Pragua)* **1999**, *45*, 97–99.

38. Hampl, V.; Pavlicek, A.; Flegr, J. Construction and bootstrap analysis of DNA fingerprinting based phylogenetic trees with the freeware program FreeTree: Application to trichomonad parasites. *Int. J. Syst. Evol. Microbiol.* **2001**, *51*, 731–735. [CrossRef]

39. Page, R.D. TreeView: An application to display phylogenetic trees on personal computers. *Comput. Appl. Biosci.* **1996**, *12*, 357–358. [PubMed]

40. Van Berloo, R. GGT: Software for display of graphical genotypes. *J. Hered.* **1999**, *90*, 328–330. [CrossRef]

41. Tan, Y.F.; Li, J.X.; Yu, S.B.; Xing, Y.Z.; Xu, C.G.; Zhang, Q. The three important traits for cooking and eating quality of rice grains are controlled by a single locus in an elite rice hybrid, Shanyou63. *Theor. Appl. Genet.* **1999**, *99*, 642–648. [CrossRef]

42. Cagampang, G.B.; Perez, C.M.; Juliano, B.O. A gel consistency test for eating quality of rice. *J. Sci. Food Agric.* **1973**, *24*, 1589–1594. [CrossRef]

43. Juliano, B.O. Rice quality screening with the Rapid ViscoAnalyser. In *Applicntions of the Rapid ViscoAnalyser*; Walker, C.E., Hazelton, J.L., Eds.; Newport Scientific: Sydney, Australia, 1996; p. 19.

44. SAS Institute Inc. *Statistical Analysis System*, version 9.2.; SAS Institute Inc.: Cary, NC, USA, 2008.

45. Pandit, E.; Tasleem, S.; Nayak, D.K.; Barik, S.R.; Mohanty, D.P.; Das, S.; Pradhan, S.K. Genome-wide association mapping reveals multiple QTLs governing tolerance response for seedling stage chilling stress in *Indica* rice. *Front. Plant Sci.* **2017**, *8*, 552. [CrossRef]

46. Pandit, E.; Panda, R.K.; Sahoo, A.; Pani, D.R.; Pradhan, S.K. Genetic relationship and structure analyses of root growth angle for improvement of drought avoidance in early and mid-early maturing rice genotypes. *Rice Sci.* **2020**, *27*, 124–132. [CrossRef]

47. Pradhan, S.K.; Pandit, E.; Pawar, S.; Bharati, B.; Chatopadhyay, K.; Singh, S.; Dash, P.; Reddy, J.N. Association mapping reveals multiple QTLs for grain protein content in rice useful for biofortification. *Mol. Genet. Genom.* **2019**, *294*, 963–983. [CrossRef]

48. Pradhan, S.K.; Nayak, D.K.; Mohanty, S.; Behera, L.; Barik, S.R.; Pandit, E.; Lenka, S.; Anandan, A. Pyramiding of three bacterial blight resistance genes for broad-spectrum resistance in deepwater rice variety, Jalmagna. *Rice* **2015**, *8*, 19. [CrossRef]

49. Sundaram, R.M.; Vishnupriya, M.R.; Biradar, S.K.; Laha, G.S.; Reddy, G.A.; Rani, N.S.; Sarma, N.P.; Sonti, R.V. Marker assisted introgression of bacterial blight resistance in Samba Mahsuri, an elite indica rice variety. *Euphytica* **2008**, *160*, 411–422. [CrossRef]

50. Pradhan, S.K.; Nayak, D.K.; Pandit, E.; Behera, L.; Anandan, A.; Mukherjee, A.K.; Lenka, S.; Barik, D.P. Incorporation of bacterial blight resistance genes into lowland rice cultivar through marker assisted backcross breeding. *Phytopathology* **2016**, *6*, 710–718. [CrossRef]

51. Nayak, D.K.; Pandit, E.; Mohanty, S.; Barik, D.P.; Pradhan, S.K. Marker assisted selection in back cross progenies for transfer of bacterial leaf blight resistance genes into a popular lowland rice cultivar. *Oryza* **2015**, *52*, 163–168.

52. Sonti, R.V. Bacterial leaf blight of rice: New insights from molecular genetics. *Curr. Sci.* **1998**, *74*, 206–212.

53. Sanchez, A.C.; Brar, D.S.; Huang, N.; Li, Z.; Khush, G.S. Sequence tagged site markers-assisted selection for three bacterial blight resistance genes in rice. *Crop Sci.* **2000**, *40*, 792–797. [CrossRef]

54. Singh, S.; Sidhu, J.S.; Huang, N.; Vikal, Y.; Li, Z.; Brar, D.S.; Dhaliwal, H.S.; Khush, G.S. Pyramiding three bacterial blight resistance genes (xa-5, xa-13 and Xa-21) using marker-assisted selection into indica rice cultivar PR-106. *Theor. Appl. Genet.* **2001**, *102*, 1011–1015. [CrossRef]

55. Perez, L.M.; Redona, E.D.; Mendioro, M.S.; Vera Cruz, C.M.; Leung, H. Introgression of Xa4, Xa7 and Xa21 for resistance to bacterial blight in thermo-sensitive genetic male sterile rice (*Oryzasativa* L.) for the development of two-line hybrids. *Euphytica* **2008**, *164*, 627–636. [CrossRef]

56. Dokku, P.; Das, K.M.; Rao, G.J.N. Pyramiding of four resistance genes of bacterial blight in Tapaswini, an elite rice cultivar, through marker-assisted selection. *Euphytica* **2013**, *192*, 87–96. [CrossRef]

57. Pradhan, S.K.; Nayak, D.K.; Pandit, E.; Barik, S.R.; Mohanty, S.P.; Anandan, A.; Reddy, J.N. Characterization of morpho-quality traits and validation of bacterial blight resistance in pyramided rice genotypes under various hotspots of India. *Aust. J. Crop. Sci.* **2015**, *9*, 127–134.

58. Das, G.; Rao, G.J.N. Molecular marker assisted gene stacking for biotic and abiotic stress resistance genes in an elite rice cultivar. *Front. Plant Sci.* **2015**, *6*, 698. [CrossRef] [PubMed]

59. Das, G.; Rao, G.J.; Varier, M.; Prakash, A.; Prasad, D. Improved Tapaswini having four BB resistance genes pyramided with six genes/QTLs, resistance/tolerance to biotic and abiotic stresses in rice. *Sci. Rep.* **2018**, *8*, 2413. [CrossRef] [PubMed]

60. Pradhan, S.K.; Barik, S.R.; Nayak, D.K.; Pradhan, A.; Pandit, E.; Nayak, P.; Das, S.R.; Pathak, H. Genetics, Molecular Mechanisms and Deployment of Bacterial Blight Resistance Genes in Rice. *Crit. Rev. Plant Sci.* **2020**, *39*, 360–385. [CrossRef]

61. Mohanty, S.P.; Kumbhakar, S.; Pandit, E.; Barik, S.R.; Mohanty, D.P.; Nayak, D.K.; Singh, N.R.; Pradhan, S.K. Molecular screening of yield component QTLs for strong culm, grain number and grain width using gene specific markers in *indica-tropical japonica* derived rice lines. *Oryza* **2016**, *53*, 136–143.

62. Mohapatra, S.; Pandit, E.; Mohanty, S.P.; Barik, S.R.; Pawar, S.; Nayak, D.K.; Subudhi, H.N.; Das, L.; Pradhan, S.K. Molecular and phenotypic analyses of yield components QTLs in IR64 backcross progenies and popular high yielding rice varieties of India. *Oryza* **2018**, *55*, 271–277. [CrossRef]

Article

The GASA Gene Family in Cacao (*Theobroma cacao*, Malvaceae): Genome Wide Identification and Expression Analysis

Abdullah [1,*], Sahar Faraji [2], Furrukh Mehmood [1], Hafiz Muhammad Talha Malik [3], Ibrar Ahmed [3], Parviz Heidari [4,*] and Peter Poczai [5,6,7,*]

1   Department of Biochemistry, Faculty of Biological Sciences, Quaid-i-Azam University, Islamabad 45320, Pakistan; furrukhmehmood@gmail.com
2   Department of Plant Breeding, Faculty of Crop Sciences, Sari Agricultural Sciences and Natural Resources University (SANRU), Sari 4818166996, Iran; sahar.faraji@rocketmail.com
3   Alpha Genomics Private Limited, Islamabad 45710, Pakistan; hafiz.talhamalik@gmail.com (H.M.T.M.); iaqureshi_qau@yahoo.com (I.A.)
4   Faculty of Agriculture, Shahrood University of Technology, Shahrood 3619995161, Iran
5   Finnish Museum of Natural History, University of Helsinki, P.O. Box 7, FI-00014 Helsinki, Finland
6   Faculty of Biological and Environmental Sciences, University of Helsinki, P.O. Box 65, FI-00065 Helsinki, Finland
7   Institute of Advanced Studies Kőszeg (iASK), P.O. Box 4, H-9731 Kőszeg, Hungary
*   Correspondence: abd.ullah@bs.qau.edu.pk (A.); heidarip@shahroodut.ac.ir (P.H.); peter.poczai@helsinki.fi (P.P.)

Citation: Abdullah; Faraji, S.; Mehmood, F.; Malik, H.M.T.; Ahmed, I.; Heidari, P.; Poczai, P. The GASA Gene Family in Cacao (*Theobroma cacao*, Malvaceae): Genome Wide Identification and Expression Analysis. *Agronomy* **2021**, *11*, 1425. https://doi.org/10.3390/agronomy11071425

Academic Editors: Roxana Yockteng, Andrés J. Cortés and María Ángeles Castillejo

Received: 30 May 2021
Accepted: 14 July 2021
Published: 16 July 2021

Publisher's Note: MDPI stays neutral with regard to jurisdictional claims in published maps and institutional affiliations.

Abstract: The gibberellic acid-stimulated *Arabidopsis* (*GASA/GAST*) gene family is widely distributed in plants and involved in various physiological and biological processes. These genes also provide resistance to abiotic and biotic stresses, including antimicrobial, antiviral, and antifungal. We are interested in characterizing the *GASA* gene family and determining its role in various physiological and biological process in *Theobroma cacao*. Here, we report 17 *tcGASA* genes distributed on six chromosomes in *T. cacao*. The gene structure, promoter region, protein structure and biochemical properties, expression, and phylogenetics of all *tcGASAs* were analyzed. Phylogenetic analyses divided tcGASA proteins into five groups. Among 17 *tcGASA* genes, nine segmentally duplicating genes were identified which formed four pairs and cluster together in phylogenetic tree. Differential expression analyses revealed that most of the *tcGASA* genes showed elevated expression in the seeds (cacao food), implying their role in seed development. The differential expression of *tcGASAs* was recorded between the tolerant and susceptible cultivars of cacao, which indicating their possible role as fungal resistant. Our findings provide new insight into the function, evolution, and regulatory system of the *GASA* family genes in *T. cacao* and may suggest new target genes for development of fungi-resistant cacao varieties in breeding programs.

Keywords: gibberellic acid-stimulated *Arabidopsis* (GASA); gene expression; phylogenetics; *Phytophthora megakarya*; abiotic stresses; biotic stresses; *Theobroma cacao*; malvaceae

1. Introduction

Theobroma cacao L. belongs to the family Malvaceae [1]. This is an economically important tree and grows in up to 50 countries located in the humid tropics [2]. *Theobroma cacao* L. seeds are enclosed in pods and are used for chocolate production, confectionery, and cosmetics [3]. This plant is adapted to high humidity areas, and is therefore predisposed to various fungal diseases [4,5]. Pod rot, or black rod, is caused by the *Phytophthora* species of fungus (*P. megakarya*, *P. palmivora*, and *P. capsici*), leading to 20–30% loss in yield and 10% death of trees [4]. The elucidation of the whole genome is helping to understand the genetic bases of biotic and abiotic stresses [2]. The availability of the high-quality chromosome-level genome assembly of *Theobroma cacao* [2,6] provides quality resources for the characterization of various gene families to elucidate the role of different gene families in cacao development. However, to the best of our knowledge, few gene families such as WRKY [7], sucrose synthase [8], Stearoyl-acyl carrier protein desaturase [9], sucrose transporter [10], and NAC [11] have been elucidated in *Theobroma cacao*.

The gibberellic acid-stimulated *Arabidopsis* (*GASA/GAST*) gene family is widely distributed in plants and performs various functions [12,13]. *GAST1* was the first gene identified among the *GASA* family's genes in tomato [14]. GASA proteins are comprised of three domains, including a signal peptide of up to 18–29 amino acids at the N-terminal, hydrophilic and high variable regions of up to 7–31 amino acids in the center, and a conserved domain at the C-terminal of up to 60 amino acids which mostly includes 12 cysteine residues [15–17]. The C-terminal domain is the characteristics of all identified GASAs [12,15,16,18]. These cysteine-rich peptides play a vital role in various plant processes. Their roles have been stated in organ development [19], lateral root development [20], stem growth [21], cell division [22,23], fruit ripening and development [24], flowering time [12,18,21], seed development [13,22], and bud dormancy [25]. The detailed expression analyses of various tissues of Arabidopsis, tomato, rice, soybean, and apple showed the tissue-specific expression of the *GASA* genes [12,15,18]. For example, *GASA* genes in tomato, including *Solyc11g011210*, *Solyc12g089300*, and *Solyc01g111075*, showed high expression at the fruit-ripening stage, whereas *Solyc03g113910*, *Solyc01g111075*, *Solyc06g069790*, and *Solyc12g042500* showed high expression level at the flowering stage [15]. Moreover, some studies also showed the contrast effect of GASA proteins. For instance, *AtGASA4* promotes flowering [22], while *AtGASA5* induces the opposite effect [21].

The *GASA* genes also play crucial roles in response to various biotic, abiotic, and hormone-related stresses. For instance, a member of the *GASA* gene family, *GmSN1*, overexpressed and enhances virus resistance in Arabidopsis and soybean [26]. Similarly, a high expression level of *CcGASA4* was reported in citrus leaves after infection with *Citrus tristeza* virus [27]. The antimicrobial properties of various proteins of the GASA family have also been reported [28–30]. The antifungal activity of the GASA proteins has been reported in almost all tissues of potato, including root, tubers, leaves, stem, stolon, axillary bud, and flowers [28–30]. Similarly, the antifungal activity of GASA members has been found in Arabidopsis, tomato, Alfalfa, and Jujuba [16,31–33]. The GASAs also showed resistance to various abiotic stresses, such as salt and drought [34]. The induction of *GASA4* and *GASA6* has been reported in Arabidopsis by growth hormones such as auxin, brassinosteroids (BR), gibberellic acid (GA), and cytokinin. In contrast, repression has been stated by stress hormones including salicylic acid (SA), abscisic acid (ABA), and jasmonic acid (JA) [31]. The expression pattern and evolutionary relationships of *GASA* genes were studied in Arabidopsis [16], apple [12], common wheat [35], grapevine [13], soybean [18], and potato [36].

Here, we are interested in characterizing the *GASA* gene family and providing data about the expression analyses of this gene family specifically regarding fungus-related diseases. To the best of our knowledge, we are the first to provide data of *GASAs* related to their distribution in genome, chemical properties, subcellular localization, and cis-regulatory elements of promoter regions in cacao. We also explored the roles of *GASA* genes in various abiotic and biotic stresses. This helps us to identify *GASAs* that show high expressions against infections of fungus *Phytophthora megakarya*.

2. Materials and Methods

2.1. Identification of GASA Genes in the Genome of Theobroma cacao and Analyses for Conserved GASA Domain

We retrieved the GASA family's protein sequences from The Arabidopsis Information Resource (TAIR10) database (ftp://ftp.arabidopsis.org). We used them as a query in BLAST for the identification of *GASA* genes in the *Theobroma cacao* genome, with an expected value of E^{-10}. The *GASA* genes were identified in the latest version of the *Theobroma cacao* genome (*Theobroma cacao* Belizian Criollo B97-61/B2) [6] and retrieved protein sequences, coding DNA sequences (CDS), genomics, and promoter sequence (1500 bp upstream of gene). The retrieved protein sequences were further analyzed for the presence of the GASA domain using the CDD database, available online: https://www.ncbi.nlm.nih.gov/Structure/cdd/wrpsb.cgi (accessed on 24 November 2020), of the National Center for Biotechnology Information (NCBI). All the sequences that showed the presence of the conserved GASA domain were selected for further analyses, whereas all the proteins with an absent/truncated domain were discarded.

2.2. Chromosome Mapping and Characterization of Physiochemical Properties

The position of each gene, including chromosome number and position on chromosome, was noted. All the genes were renamed according to the location of the chromosome and their position, as shown in Table 1. The MapChart software [37] through Ensemble was used to show the position of each *tcGASA* gene along the position of the chromosome. Various physiochemical properties, including length of the protein, molecular weight (MW), isoelectric point (PI), instability index, and the grand average of hydropathy (GRAVY), were determined using the ExPASy tool [38]. The subcellular localization of *GASA* genes was also predicted using the BUSCA webserver [39].

Table 1. List of the identified *GASA* genes and their characteristics in *Theobroma cacao*.

Gene Locus ID	Gene ID	Location	Length (Amino Acid)	MW (kDa)	PI	Instability Index	GRAVY	Subcellular Localization
Tc01v2_t001590	*tcGASA01*	1: 830,757–831,466	105	11.18	8.93	Unstable	−0.023	Extracellular space
Tc01v2_t005890	*tcGASA02*	1: 3,268,900–3,269,687	107	11.44	8.98	Unstable	−0.057	Extracellular space
Tc02v2_t009150	*tcGASA03*	2: 5,635,674–5,636,363	90	9.90	9.01	Stable	−0.054	Extracellular space
Tc02v2_t021110	*tcGASA04*	2: 30,717,916–30,719,013	106	11.91	9.21	Unstable	−0.195	Extracellular space
Tc04v2_t000240	*tcGASA05*	4: 270,172–272,304	320	33.92	9.83	Unstable	−0.750	Extracellular space
Tc04v2_t006890	*tcGASA06*	4: 15,324,400–15,325,648	106	11.83	9.32	Unstable	−0.343	Extracellular space
Tc04v2_t009440	*tcGASA07*	4: 19,321,294–19,322,037	95	10.37	9.26	Unstable	−0.072	Extracellular space
Tc04v2_t016520	*tcGASA08*	4: 26,297,870–26,298,743	123	13.42	8.27	Unstable	−0.026	Plasma membrane
Tc04v2_t016530	*tcGASA09*	4: 26,301,782–26,302,647	95	10.62	8.88	Unstable	−0.251	Extracellular space
Tc05v2_t012230	*tcGASA10*	5: 25,218,365–25,220,141	114	12.76	9.46	Unstable	0.327	Extracellular space
Tc08v2_t003650	*tcGASA11*	8: 1,941,175–1,942,036	102	10.92	6.67	Stable	−0.031	Extracellular space
Tc08v2_t003660	*tcGASA12*	8: 1,943,144–1,944,201	100	10.57	8.66	Unstable	0.004	Extracellular space
Tc08v2_t007890	*tcGASA13*	8: 4,555,655–4,556,418	88	9.64	9.32	Stable	−0.062	Extracellular space
Tc08v2_t014670	*tcGASA14*	8: 15,650,553–15,651,438	102	10.98	8.72	Unstable	−0.164	Extracellular space
Tc08v2_t014680	*tcGASA15*	8: 15,651,664–15,652,892	92	10.01	8.90	Stable	−0.148	Extracellular space
Tc08v2_t014690	*tcGASA16*	8: 15,653,408–15,654,378	110	12.13	9.64	Unstable	−0.263	Extracellular space
Tc09v2_t020100	*tcGASA17*	9: 30,243,952–30,245,432	112	12.32	9.46	Unstable	−0.252	Extracellular space

2.3. Gene Structure and Promoter Region Analyses

We analyzed CDS sequences for exons–introns within all *tcGASA* genes using the Gene Structure Display Server, available online: http://gsds.cbi.pku.edu.cn (accessed on 26 November 2020). PlantCare [40] was used to study cis-regulatory elements in the 1500 bp promoter region.

2.4. Prediction of Post-Translational Modifications of GASA Proteins

The phosphorylation site of the GASA proteins was predicted by the NetPhos 3.1 server [41] with a potential value >0.5. N-glycosylation sites were predicted using the NetNGlyc 1.0 server [42] with default parameters.

2.5. Phylogenetic and Conserved Motif Analyses

The phylogenetic relationship of the *GASA* genes of *T. cacao* was inferred with *GASA* genes of six other species, including Arabidopsis *thaliana*, *Gossypium raimondii*, *Vitis vinifera*,

Oryza sativa, Brachypodium distachyon, and *Zea mays.* A similar approach, which was given for the identification of *GASA* genes in *Theobroma cacao,* was employed for the identification of the genes of all other species. Clustal Omega [43] was used for multiple alignment of the protein sequences of all species. The unrooted neighbor-joining tree was drawn using MEGA X [44] and visualization of the tree was improved by using an interactive tree of life (iTOL) [45]. The conserved motif distribution into GASA proteins was performed using MEME v5.3.0 server [46]. We searched for a maximum number of five motifs with a minimum width of motif 6 and a maximum width of motif 30.

2.6. Gene Duplications and Estimation of Ka/Ks Values

The identity of >85% in nucleotide sequences of genes is considered a sign of duplication [47]. Hence, we aligned DNA coding sequences using Clustal Omega [43], and the extent of the identity of the genes with each other was determined using Geneious R8.1 [48]. Gene duplication events, as compared to other species, were determined using the MCScan v0.8 program [49] through the Plant Genome Duplication Database.

The selection pressure on *GASA* genes was determined by calculating non-synonymous (Ka) and synonymous (Ks) substitutions and the ratio of non-synonymous to synonymous substitutions (Ka/Ks) using DnaSP v6 software [50]. The time divergence and duplication were assessed by a synonymous mutation rate of λ substitutions per synonymous site per year as T = (Ks/2λ) (λ = 6.5 $\times$ 10^{-9}) $\times$ 10^{-6} [51]. The synteny relationships of *GASA* genes among the orthologous pairs of *T. cacao-O. sativa, T. cacao-A. thaliana, T. cacao-Zea mays, T. cacao-B. distachyon, T. cacao-G. raimondii,* and *T. cacao-Vitis vinifera* at both gene and chromosome levels were visualized using Circos software [52].

2.7. Three-Dimensional Protein Modeling and Molecular Docking

We used iterative template-based fragment assembly simulations in I-TASSER [53] to build three-dimensional protein structures of GASAs after selection of best models by the 3D-refine program [54]. We also used P2Rank in the PrankWeb software [55] and the CASTp tool [56] to analyze the refined structure of GASA proteins to predict protein pockets and cavities. Finally, PyMOL [57] was used to visualize results.

2.8. In Silico Expression Analysis of GASA Genes through RNA-seq Data

The publicly available RNA-seq data related to the cacao genome were employed for expression assays to measure *GASA* family members in multiple tissues and during various biotic and abiotic stimuli exposure. The RNA-seq data of cacao inoculated with *Phytophthora megakarya* for 0 h (0 h), 6 h, 24 h, and 72 h in susceptible cultivar Nanay (NA32) and fungal resistant cultivar Scavina (SCA6) were downloaded from GEO DataSets under accession number GSE116041 [58]. The comparison of susceptible cultivar NA32 and fungal resistant cultivar SCA6 was performed to determine differentially expressed genes within both cultivars and to identify those genes that are specifically induced in fungal resistant cultivar SCA6. These data were log2 transformed to generate heatmaps via the TBtools package [59]. Furthermore, the expression levels of *GASA* genes for tissue specific expression and under multiple abiotic stresses, including cold, osmotic, salt, drought, UV, wounding, and heat, have been detected in the *Arabidopsis* orthologous genes for *tcGASAs* (SAMEA5755003 and PRJEB33339).

3. Results
3.1. Identification of GASA Genes and Their Distributions on Chromosomes within Genomes

We detected 17 *GASA* genes in the genome of *T. cacao* distributed on six chromosomes out of ten. These genes were named from *tcGASA1* to *tcGASA17* based on their distribution on chromosomes starting from chromosome 1. When two or more genes were present on the same chromosome, then the gene present at the start of a chromosome was named first (Table 1 and Figure 1). Six genes were distributed on chromosome 8 and five genes were distributed on chromosome 4. Chromosome 1 and chromosome 2 each contained two genes, whereas chromosome 5 and chromosome 9 contained one gene each. These data showed the unequal distribution of *tcGASA* genes within the cacao genome. The location of each gene on the chromosome is mentioned in Table 1, as well as the start and end. Moreover, the sequences of genes, proteins, coding regions, and promoter regions are provided in Table S1.

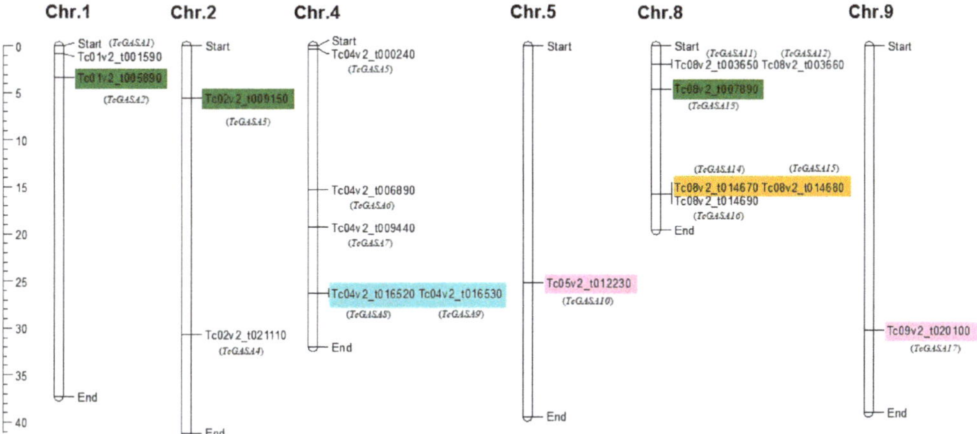

Figure 1. Location of *tcGASA* genes on the chromosome in *Theobroma cacao*. Each gene is shown with a gene identification number along with the number given to each gene in the current study. The pairs of segmentally duplicated genes are shown with same color.

3.2. Protein Length, Molecular Weight, and Isoelectric Point of tcGASA Proteins

In the current study, tcGASAs were characterized based on their physiochemical properties (Table 1). The identified tcGASA proteins were low molecular weight proteins ranging in length from 88 (tcGASA13) to 320 (tcGASA05) amino acids with molecular weight (MW) ranging from 9.64 to 33.92 kDa. Except for tcGASA05 and tcGASA08, with molecular weights of 13.42 kDa and 33.92 kDa, respectively, the MW of all other tcGASAs was found to be less than 13 kDa. The isoelectric point also showed similarities among tcGASA and suggested the alkaline nature of the proteins. Except for tcGASA11, which has an isoelectric point of 6.67, all other tcGASA proteins have isoelectric points of more than 8, ranging from 8.27 to 9.83.

3.3. Analyses of Instability Index, GRAVY, and Subcellular Localization of tcGASA Proteins

The instability index provides information about the stable and non-stable features of proteins in the various biochemical processes. The instability index indicated 4 stable tcGASAs including tcGASA03, tcGASA11, tcGASA13, and tcGASA15, as well as 13 unstable tcGASAs (Table 1). The positive value of GRAVY indicates its hydrophobic nature, whereas the negative value indicates the hydrophilic nature of proteins. The GRAVY value was recorded as negative for sixteen tcGASA proteins ranging from −0.75 to −0.023, but as positive (0.004) for tcGASA12 (Table 1). Hence, the data indicate the hydrophilic nature of most tcGASA proteins. Subcellular localization provides information about the function of proteins. Based on BUSCA, we predicted extracellular localization of tcGASA proteins, except for tcGASA08, which localized in the plasma membrane (Table 1).

3.4. tcGASA Proteins 3D Structure Analyses and Post-Translational Modifications

The predicted 3D structure of all tcGASA proteins showed that these proteins contain β sheets, α helices, random coils, and extended strands. The random coils were the most abundant and were more extensive than α helices, while the β sheets were the least

(Figure 2). In addition, the active sites of GASA proteins were predicted in their structure. The proline, cysteine, lysine, serine, and threonine amino acids were more predicted, as the binding sites in all candidate GASA proteins in cacao (Figure 2). GASAs were diverse based on predicted 3D structure and pocket sites, indicating that they could have various functions. In the present study, the post-translational modifications of tcGASAs were predicted in terms of phosphorylation and glycosylation (Figure 3, Table S2). We predicted a total of 224 potential phosphorylation events on amino acids serine, threonine, and tyrosine within tcGASA proteins. Most of the phosphorylation events were predicted related to serine (92) followed by threonine (86) and then by tyrosine (46). Among tcGASA proteins, most of the phosphorylation sites (57 sites) were predicted in tcGASA05, whereas in other proteins, phosphorylation events ranged from 9 to 14 sites. Three tcGASA, including tcGASA10, tcGASA15, and tcGASA17, were also identified with a potential glycosylation site (Table S2).

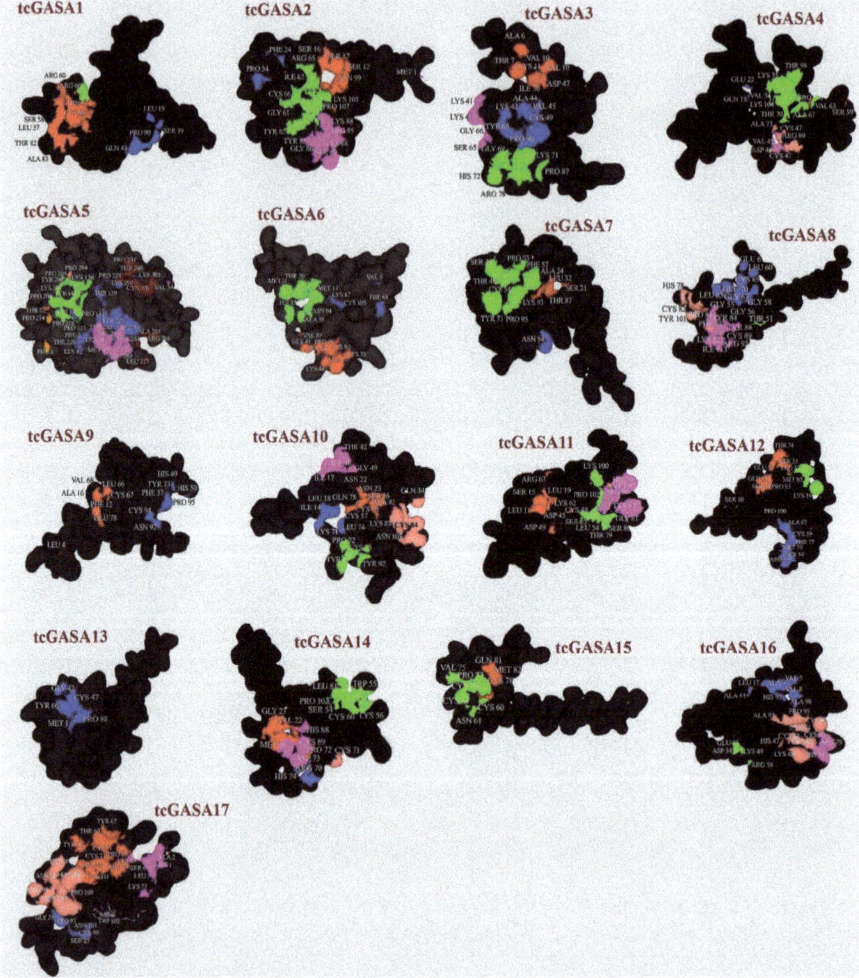

Figure 2. Distribution of ligand-binding sites in predicted 3D structure of tcGASA proteins. The distribution of the major protein pocket sites in ligand regions are highlighted in the predicted 3D structure of tcGASA proteins.

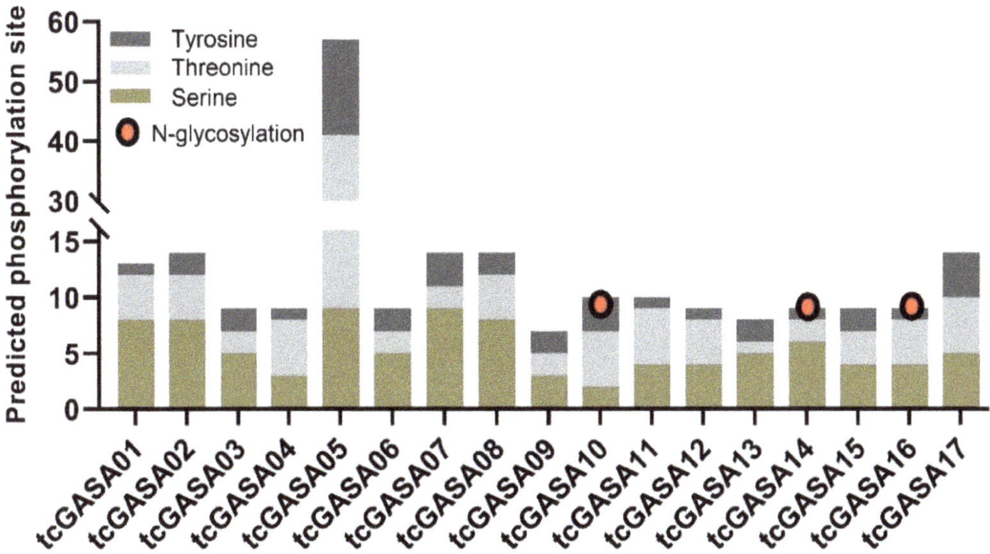

Figure 3. Prediction of post translational modifications based on phosphorylation and glycosylation site of tcGASA proteins. The red circle indicates the predicted glycosylation sites, and one glycosylation site was predicted each in tcGASA10, tcGASA14, and tcGASA16.

3.5. Phylogenetic Analyses of tcGASA Proteins

The phylogenetic inference of ninety-nine GASA-protein sequences from seven species (mentioned in methodology), including 17 sequences of tcGASA, resolved into five groups (Figure 4). The GASA proteins of cacao showed high variations and were distributed into all five groups. Group I include only one sequence tcGASA16 (Tc08v2_t014690). Group II was confined to five protein sequences including tcGASA01 (Gene ID: Tc01v2_t001590), tcGASA04 (Tc02v2_t021110), tcGASA05 (Tc04v2_t000240), tcGASA11 (Tc08v2_t003650), and tcGASA12 (Tc08v2_t003660). Group III comprised two sequences—tcGASA14 (Tc08v2_t014670) and tcGASA15 (Tc08v2_t014680). Group IV included tcGASA02 (Tc01v2_t005890), tcGASA03 (Tc02v2_t009150), and tcGASA13 (Tc08v2_t007890). Group V consisted of six protein sequences, including tcGASA06 (Tc04v2_t006890), tcGASA07 (Tc04v2_t009440), tcGASA08 (Tc04v2_t016520), tcGASA09 (Tc04v2_t016530), tcGASA10 (Tc05v2_t012230), and tcGASA17 (Tc09v2_t020100). The tcGASA04 (Tc02v2_t021110) shared a node with one of the protein sequences of *Vitis vinifera*, whereas the other 16 proteins showed close relationships with the sequences of another species of family Malvaceae (*Gossypium raimondii*).

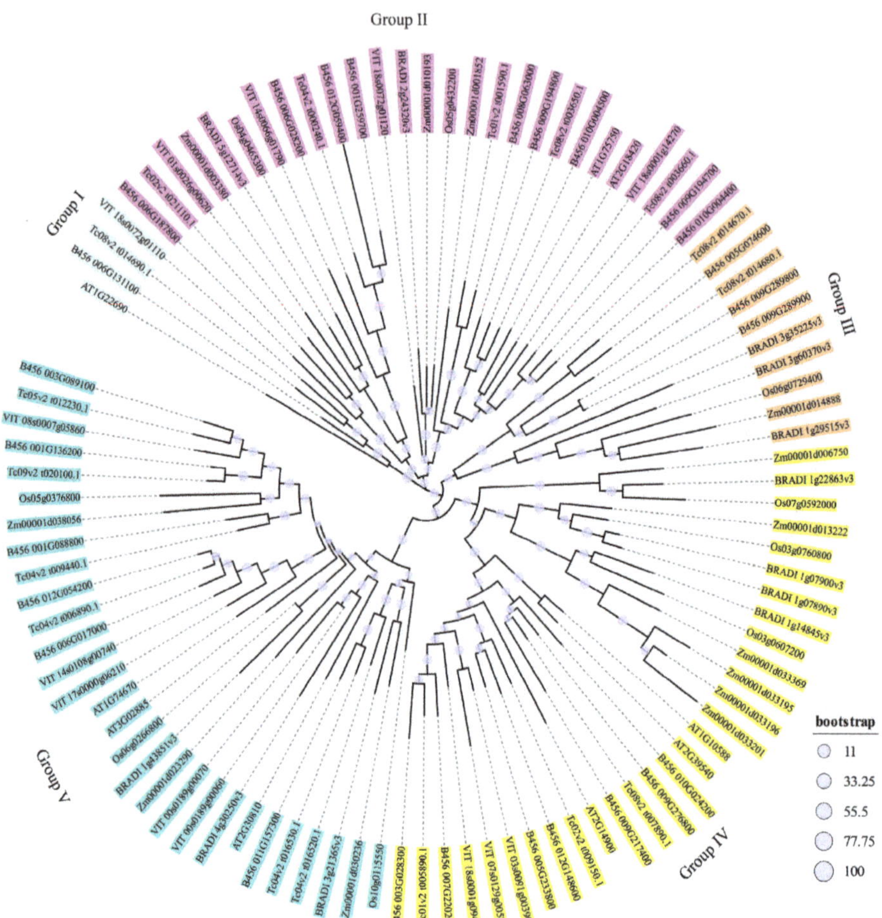

Figure 4. The phylogenetic inference of tcGASA proteins along with protein sequences of six other species. Each sequence is presented with its gene number. The start of each sequence contained code for the species as: Tc: *Theobroma cacao*; AT: *Arabidopsis thaliana*; B456: *Gossypium raimondii*; VIT: *Vitis vinifera*; Os: *Oryza sativa*; BRADI: *Brachypodium distachyon*; Zm: *Zea mays*.

3.6. Gain and Loss of Intron(s) and Conserved Motifs

We also determined numbers of introns–exons within the genes and motifs in protein sequences. We drew a separate phylogeny of the *tcGASA* genes of *T. cacao* to find out the extent to which genes that cluster together are similar in term of introns numbers and pattern, and in terms of protein number and pattern. Each phylogenetic group was shown with different colors for clarity (Figure 5a). The analyses of genomic sequences showed the absence of intron in one gene, presence of one intron in five genes, two introns in eleven genes, and three introns in one gene (Figure 5b). Genes clustered together showed differences in their number and in intron–exon distributions (Figure 5a,b). Five motifs were revealed in tcGASA. Four motifs (1–4) were distributed in all tcGASAs, whereas a fifth motif was limited to Tc04v2_t016520 (tcGASA08), Tc05v2_t012230 (tcGASA10), and Tc09v2_t020100 (tcGASA17) (Figure 5c). The motifs of proteins that clustered together within the phylogenetic tree presented similarities in the distribution of motifs to some extent (Figure 5a,c).

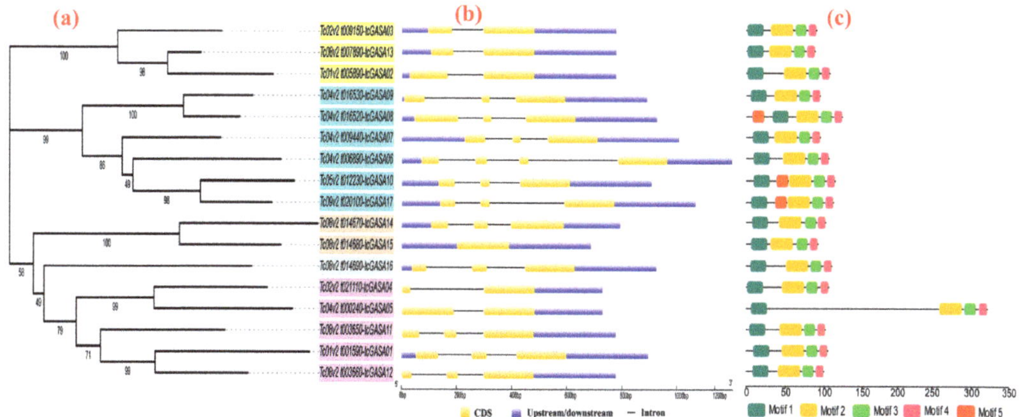

Figure 5. The phylogenetic analyses in comparison to introns–exons in genes and motifs in the protein sequence of tcGASAs. Phylogeny tree of tcGASAs (**a**), introns–exons distribution in *tcGASA* genes (**b**), and conserved motifs distribution in tcGASA proteins (**c**).

3.7. Duplications, Divergence, and Synteny among GASA Genes

We analyzed the paralogous relationships among the *GASA* genes within cacao and analyzed the orthologous relationships of *GASA* genes by comparing them with the other six species (mentioned in methodology). Segmental duplications were found in nine *GASA* genes that were paired into four groups as: *tcGASA02-tcGASA03-tcGASA13*; *tcGASA08-tcGASA09*; *tcGASA10-tcGASA17*; and *tcGASA14-tcGASA15* (Table 2) (as shown with similar colors in Figure 1). The analyses of synonymous and non-synonymous substitutions revealed that high purifying selection pressure exists on these genes after duplication. The analyses of divergence time indicated that the event of duplication of these pairs occurred 50 MYA to 204 MYA (Table 2). The synteny analyses with *GASAs* of other species showed high resemblance and identified orthologous genes among *T. cacao* and compared species (Figure 6). The 17 *GASA* genes in cacao showed the syntenic relationship with 7 and 11 ortholog genes in the Arabidopsis and *Gossypium raimondii*, respectively (Figure 6a,b). Moreover, *tcGASAs* had the syntenic relationship with 7, 5, 4, and 7 *GASA* genes from *Vitis vinifera*, *Oryza sativa*, *Brachypodium distachyon*, and *Zea mays*, respectively (Figure 6c–f). Interestingly, *Os05g0432200*, as a rice-GASA, and a GASA gene of Brachypodium, BRADI.2g24320v3, showed most syntenic blocks with tcGASAs.

Table 2. Gene duplications, synonymous and non-synonymous substitutions and time of divergence.

Gene 1	Gene 2	Duplication Type	K_a	K_s	K_a/K_s	*p*-Value	Divergence Time (MYA)
tcGASA02	*tcGASA03*	Segmental	0.2679	2.6607	0.1007	1.04×10^{-6}	204.67
tcGASA02	*tcGASA13*	Segmental	0.1708	2.2495	0.0759	2.52×10^{-8}	173.04
tcGASA08	*tcGASA09*	Segmental	0.1515	0.6514	0.2325	1.30×10^{-6}	50.11
tcGASA10	*tcGASA17*	Segmental	0.283	1.0283	0.2752	3.28×10^{-5}	79.1
tcGASA14	*tcGASA15*	Segmental	0.3234	0.7971	0.4058	1.46×10^{-8}	61.32

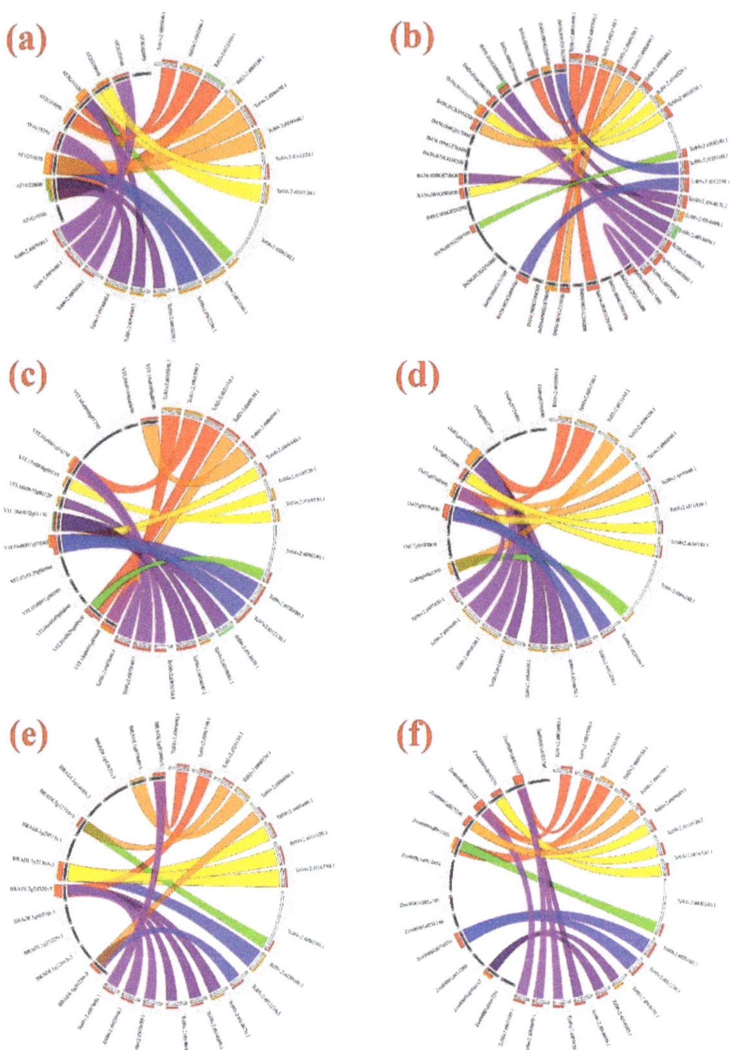

Figure 6. Synteny analysis of *GASA* genes. The syntenic blocks of cacao *GASA* genes are compared with six other species, including *Arabidopsis thaliana* (**a**), *Gossypium raimondii* (**b**), *Vitis vinifera* (**c**), *Oryza sativa* (**d**), *Brachypodium distachyon* (**e**), and *Zea mays* (**f**).

3.8. Promoter Regions Analysis

The analyses of cis-regulatory elements in promoter regions revealed the presence of binding sites for key transcription factors related to light-responsive elements (48.88%), hormone-responsive elements (25.40%), stress-related elements (19.06%), growth-response elements (5.40%), and DNA- and protein-related binding sites (1.26) (Figure 7a). Regulatory sides were found for various hormones such as auxin, salicylic acid, abscisic acid, gibberellin, and methyl jasmonate (MeJA) (Figure 7b). Similarly, regulatory elements were identified for drought, elicitor, anaerobic induction, low temperature, and plant defense/stress (Figure 7c). The complete detail of each element, along with sequence and function is provided in Table S3.

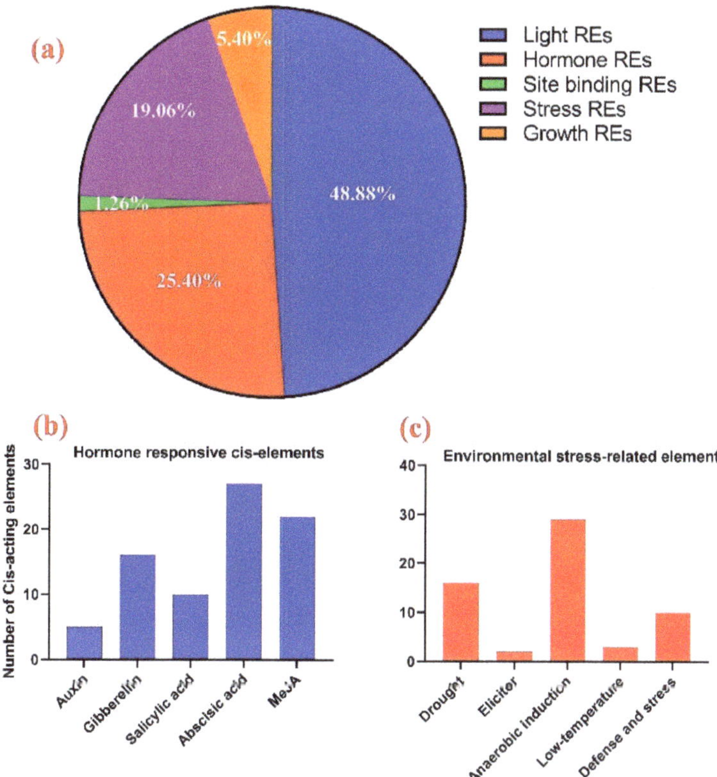

Figure 7. Distribution of cis-regulatory elements into promoter regions of *tcGASA* genes. Classification of identified regulatory elements based on function, such as light, hormone, growth, stress, and binding domain excluding the TATA box and CAAT box (**a**). Distribution of different types of hormone-related cis-regulatory elements (**b**). Distribution of cis-regulatory elements related to various types of environmental stresses (**c**).

3.9. In-Silico Tissue-Specific Expression of tcGASA Genes

We evaluated the expression of orthologous *tcGASAs* in various tissues to evaluate their role in the functions of *T. cacao* (Figure 8a). Differential expression was noted for *tcGASA* genes in various tissues of cacao. Five genes, including *tcGASA*02, *tcGASA*03, *tcGASA*08, *tcGASA*09, and *tcGASA*13, showed high expression in the beans (food part of cacao). In addition, *tcGASA*16 showed a high expression in leaves and entire seedlings. However, *tcGASA*02 and *tcGASA*03 were significantly downregulated in leaves compared to beans. In-silico expression results showed that *tcGASA*12 and *tcGASA*17 were less induced in pistil tissues.

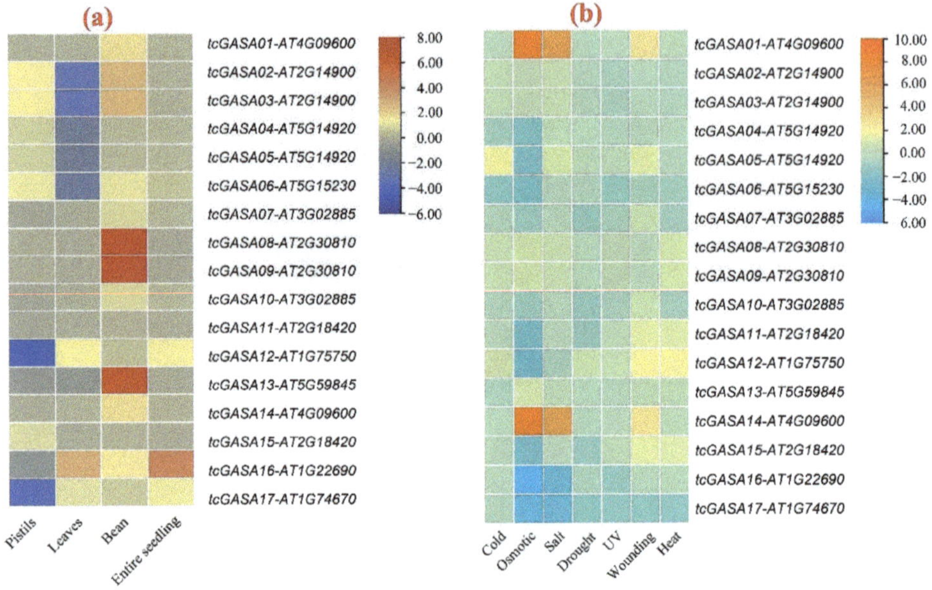

Figure 8. Expression analyses of orthologous *tcGASA* genes in Arabidopsis. Expression profile in different types of tissues (**a**), and in response to various abiotic stresses (**b**).

3.10. Expression Analyses of tcGASAs in Abiotic and Biotic Stresses

We explicated the role of *tcGASAs* under various abiotic stress by analyzing orthologous genes of Arabidopsis (Figure 8b). The orthologous genes of tcGASA01, *tcGASA05*, *tcGASA12*, and *tcGASA14* showed an upregulation in response to wound healing and the orthologs of *tcGASA05* were more expressed in response to cold stress. The expression of orthologous *GASA* genes in Arabidopsis was less induced in response to drought and UV stresses. The *tcGASA01* and *tcGASA14* were highly upregulated in response to osmotic pressure and salt stress, while *tcGASA016* and *tcGASA17* were downregulated.

3.11. Expression Analyses of tcGASAs in Biotic Stress (P. megakarya)

The role of *tcGASAs* against *P. megakarya* was accessed using RNA-seq data of cacao inoculated with *P. megakarya* for 0 h, 6 h, 24 h, and 72 h in susceptible cultivar Nanay (NA32) and fungal resistant cultivar Scavina (SCA6) (Figure 9a,b). In susceptible cultivar, *tcGASA12* and *tcGASA13* as early responses were upregulated after 6 h of inoculation while *tcGASA12* was more expressed after 72 h (Figure 9a). The expression patterns of *tcGASAs* were different in fungal resistant cultivar Scavina, whereas most genes were induced after 72 h of incubation of *P. megakarya* (Figure 9b). Moreover, *tcGASA03*, *tcGASA05*, and *tcGASA13* were more upregulated in response to fungi treatment after 72 h of incubation in resistant cultivar Scavina (Figure 9b). Four genes, *tcGASA01*, *tcGASA08*, *tcGASA09*, *tcGASA15*, were not expressed. Probably, they are induced at specific conditions or at a specific step of growth and development. The analyses showed differential expression of *tcGASAs* expression under biotic stress of fungus *P. megakarya* in SCA6 and NA32 (Figure 10). The *tcGASA03*, *tcGASA05*, *tcGASA06*, *tcGASA16*, and *tcGASA17* showed high expression after 24 h of inoculation and *tcGASA03*, *tcGASA04*, and *tcGASA13* more expressed after 72 h in SCA6. This showed their possible role for fungal resistance. The *tcGASA05*, *tcGASA12*, and *tcGASA14* also showed high expression in SCA6 but these genes also showed high expression in NA32. Hence, these may not be involved in resistance to fungus.

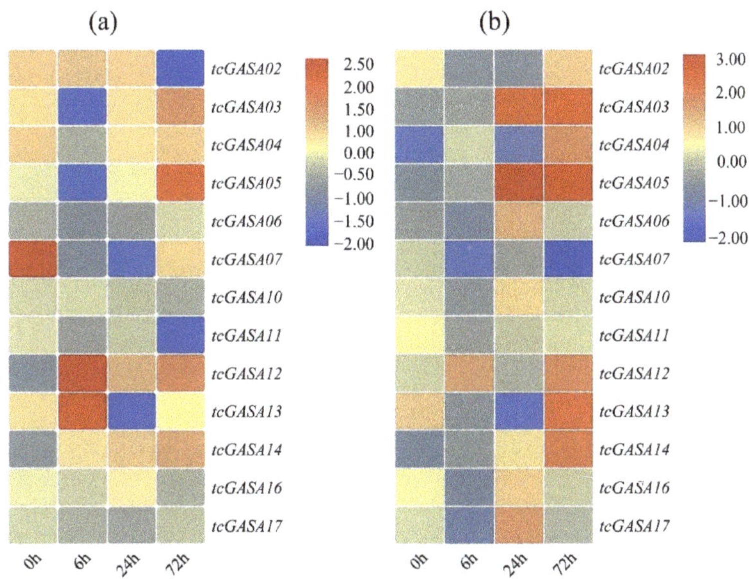

Figure 9. Expression analyses of *tcGASA* genes in cacao plants under inoculation with *P. megakarya* after 0 h, 6 h, 24 h, and 72 h. Nanay (NA-32) susceptible cultivar (**a**), Scavina (SCA6) tolerant cultivar (**b**). The expression analysis is presented after log2 transformation. The strong appearance of the blue color is linked to downregulation of genes whereas the strong red color indicates upregulation of genes

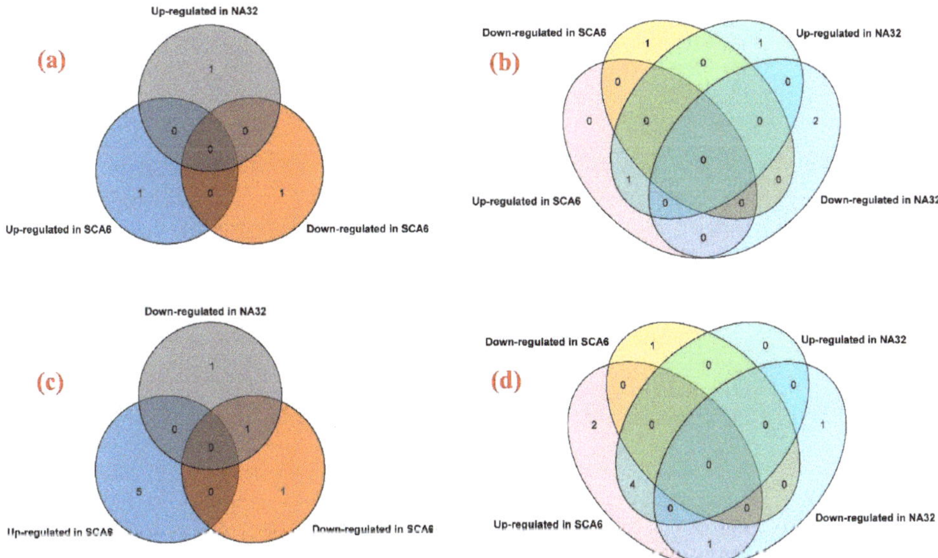

Figure 10. Venn diagram to represent differential expression in susceptible cultivar Nanay (32) and tolerant Cultivar (SCA6) of cacao under inoculation with *P. megakarya* after 0 h (**a**), 6 h (**b**), 24 h (**c**), 72 h (**d**). The downregulated genes are missing after 0 h (**a**) and up-regulated genes are missing after 24 h (**c**). Thus, the data are not shown.

4. Discussion

In the current study, we identified 17 tcGASAs and analyzed their genomic distributions and chemical properties. RNAseq data analyses explicated their possible roles in bean development (food of cacao), various abiotic stresses, and the biotic stress of *P. megakarya*.

Cacao is an economically important plant, and its seeds are used for chocolate, and it is also the main source of income for 40–50 million farmers [2,3]. Despite this importance, up to the best of our knowledge, five gene families, *WRKY* [7], sucrose synthase [8], Stearoyl-acyl carrier protein desaturase [9], sucrose transporter [10], and *NAC* [11] are studied in cacao. The role of the *NAC* family was not explained in abiotic and biotic stresses. The *GASA* family's role was reported in plant development, function regulation, and biotic and abiotic stresses, as mentioned in the introduction (*vide infra*). Cacao production is also affected by various biotic and abiotic stresses. In the current study, we focused on the GASA gene family in cacao. We identified 17 GASA genes, which were unequally distributed on six chromosomes among ten. The GASA proteins have low molecular weight with conserved GASA and cysteine domains. Our findings are in agreement with previous studies that revealed that *GASA* genes mostly exist in lower numbers, have low molecular weight, and are unequally distributed on chromosomes within genomes as reported for rice, 9 *GASA* genes [15], 14 genes in Grapevine [13], 15 genes in Arabidopsis [12], and 19 genes in tomato [15]. However, a somewhat high number of GASA have also reported, such as in apple, 26 genes [12], and 37 genes in soybean [18]. The number of introns were variable in tcGASA within genes that cluster together in phylogeny. The loss and gain of introns occurs in the course of evolution within protein-coding genes of the plants and is also reported within the GASA of other plant species [13,18,60–62].

The tcGASAs were predicted to be alkaline, hydrophilic, and mostly unstable proteins. However, we also detected four stable proteins: tcGASA03, tcGASA11, tcGASA13, and tcGASA15. The stability demonstrates the lifetime of proteins related to cellular enzymatic reactions [62]. Hence, these four proteins may play extensive roles in various enzymatic activities. The data of subcellular localization also provide insight into the function of proteins [63]. Apart from tcGASA08 (which localizes to the plasma membrane), other proteins locate in the extracellular space. Extracellular localization of GASA proteins in several plants has also been reported previously [15,18,21]. Localization of GASA proteins in the plasma membrane, cytoplasm, and the nucleus has also been described. The variations in subcellular localization may occur due to various factors, such as protein–protein interaction and post-translational modifications [36,64]. The prediction of 3D structure and pocket site of proteins can provide valuable information about protein function based on ligand-binding sites [65,66]. In the present study, the cysteine, proline, lysine, leucine, serine, and threonine were frequently predicted as the key binding residues in the structure of GASA proteins, in which proline, serine, and leucine are known as the amino acid residues associated with responses to environmental stimuli [65,67].

Post-translational modifications are processes of chemical modifications of proteins and they produce diversity in structure and function, including subcellular localization, protein–protein interaction, and regulating enzyme activity by allosteric phenomena [36,68,69]. We predicted phosphorylation sites in all tcGASA, ranging in number from 7 to 57. The phosphorylation of proteins is also vital for cell signaling, regulation of various protein mechanisms, and as a substrate for various kinases [70–72]. It could be of interest to further study the function and structure of tcGASA proteins using tandem mass spectrometry (MS/MS), and CRISPR/Cas9 genome editing along with transcriptomic analysis of transgenic lines. Similarly, glycosylation is also an abundant and varied modification that plays an essential role in the biological and physiological functions of a living organism [73]. We detected glycosylation sites on the N-terminal of tcGASA10, tcGASA14, and tcGASA16. These tcGASAs may play significant roles in plant function and regulations.

We performed the phylogenetic analyses of tcGASAs with GASA of 6 other species. We also included *Gossypium raimondii*, a closely related species from the plant family Malvaceae. The phylogenetic analyses distributed GASA proteins into 5 groups and the tcGASAs were also distributed into all five groups. Tc08v2_t01469 (tcGASA16) was the only protein which grouped with Arabidopsis, while other tcGASAs showed sister relationships with GASA of *Gossypium*. The cacao belongs to the basal group, whereas *Gossypium* belongs to the crown group of the family Malvaceae [74,75], but the close phylogenetic relationships of proteins of these two species support their family-level relationship. The motifs and exon–intron analyses showed variations within proteins that cluster together in phylogeny. This shows that GASA of some groups may have evolved during evolution, which led to variations in motifs and introns in some groups. The same was observed in GASA of other species and has also been reported for other gene families [13,18,76,77]. *GASAs* in a phylogenetic group showed different expression patterns, indicating that *GASAs* are controlled by various regulatory systems. These findings also show that some other processes are related to the function of proteins instead of their close phylogenetic relationships. A similar observation was reported in *Nicotiana tabacum* L. [72]. However, some studies also proposed that the closely related proteins on a phylogenetic tree have a similar function [78,79].

The tandem and segmental duplicated genes play an important role in evolution, domestication, functional regulation, and biotic and abiotic stresses [80–84]. We determined segmental duplication of nine genes that form four groups as: *tcGASA02-tcGASA03-tcGASA13, tcGASA08-tcGASA09, tcGASA14-tcGASA15,* and *tcGASA10-tcGASA17*. These gene pairs were present in the same group within the phylogenetic tree. Similar results were reported in genome-wide analyses of *GASAs* in soybean [18]. A previous study also proposed that segmentally duplicated genes also showed similar functions and stable expressions. [18,85]. In the current study, each pair of segmentally duplicated genes did not agree with the previous finding. In biotic stress of fungus (*P. megakarya*), *tcGASA08-tcGASA09* and *tcGASA10-tcGASA17* showed similar expression that *tcGASA08* and *tcGASA09* both did not express while *tcGASA10-tcGASA17* both are less regulated/down-regulated genes. The other two pairs *tcGASA02-tcGASA03-tcGASA13* and *tcGASA14-tcGASA15* showed a different expression, as some genes were found down-regulated while some genes were found up-regulated (Figure 9). These findings suggest that segmentally duplicated pairs may also perform different functions, and functional analyses of each segmentally duplicated gene can provide authentic information about their roles in various physiological and biochemical processes. Moreover, our study, along with the previous report [18], suggests that genes within the same groups have more chances of segmental duplication events among them. The Ka/Ks < 1 indicates that purifying selection pressure exists on the *GASA* genes after duplication, as reported previously [13]. We also observed mostly purifying selection pressure on *tcGASAs*, including duplicating genes (Table S1).

The cis-acting regulatory elements involved in transcription of regulation genes are induced through independent signal transduction pathways under biotic and abiotic stresses [72,86]. We observed several key cis-regulating elements in response to light, hormones, stresses, and growth in the promoter site of *tcGASAs*. The cis-regulating elements for drought, anerobic induction, low temperature, and plant defense were also evident. The existence of diverse cis-regulating elements in promoter regions indicates their roles in the regulation of the *tcGASAs* and different pathways of cacao. Further study such as using the CRISPR/Cas genome-editing system and T-DNA can shed light on the roles of these cis-regulating elements.

Different strategies are used to develop cacao cultivars that produce food in high quantities and are resistant to abiotic and biotic stresses [4]. Here, we studied the role of *GASA* genes in cacao development and in protection against biotic and abiotic stresses. Tissue-specific expression was observed for *tcGASAs*, which showed their role in the development and functional regulation of cacao. Up to eight tcGASAs were highly expressed in the cacao bean, which is the food part used for making chocolate. This high expression may reveal the conserved function of these genes in the development of the bean and cacao flavor. Gene expression analysis of *tcGASA* genes in cacao beans in disease resistance (bulk) and disease susceptibility (fine flavor cacao) could be of interest [87,88]. The role of *tcGASAs* was also stated in the development of grapevine seeds [13]. Further data based on cloning can provide new insight into their roles in bean development. Expression analyses of the orthologous genes of Arabidopsis also indicated the role of *GASA* genes in various abiotic stresses, including drought, which significantly affects the growth of cacao [89]. Hence, these genes may also be important to produce drought-resistant cultivars. Expression analyses also provide insight into the gene's function in response to biotic stresses [72]. The black rod disease of genus *Phytophthora* caused up to 20–25% loss (700,000 metric tons) to the world cacao production annually. In some regions, the *Phytophthora* caused losses of 30–90% of the crops [90]. Here, we explored the function of *tcGASAs* based on RNAseq data against black rod causing pathogen *P. megakarya* and observed highly expressed *tcGASA* genes in plants inoculated at 24 h and 72 h in the tolerant cultivar SCA6 as compared to susceptible cultivar NA32. These data indicate that *tcGASAs* respond to fungus. Hence, the complete characterization of these upregulated genes can provide target genes for the development of resistant cultivars to the disease of genus *Phytophthora* to enhance the production of cacao for the welfare of not only farmers involved in cacao cultivation, but also for the welfare of all humanity to provide high-quality delicious chocolate with quality nutrition.

In conclusion, our study provides new insight into the identification, characterization, and expression of the *GASA* genes in the *Theobroma cacao* plant. Expression analyses revealed the role of the *GASA* genes in seed development. Our findings reveal that tcGASAs are diverse based on their structure and regulatory systems, indicating that they are involved in various cellular pathways related to development and stress responses. Furthermore, our result indicates that the *GASA* genes could be related to resistance against the fungus *Phytophthora megakarya*, which causes significant losses to cacao production each year. Our study may be helpful for the generation of cultivars that are resistant to the fungus of the genus *Phytophthora*. The present work, as an in-silico study, revealed many aspects of structural, regulatory systems, and expression of *GASA* gene family members in cacao. However, it is necessary to evaluate the expression of these genes in different tissues of sensitive and fungus-resistant clones of cacao. We also suggest using new techniques such as CRISPR/Cas genome-editing to determine the function and interactions of cacao-GASAs.

Supplementary Materials: The following are available online at https://www.mdpi.com/article/10.3390/agronomy11071425/s1, Table S1: The complete detail of each sequence of Theobroma cacao analyzed in current study including protein sequences, coding sequences, genomic sequences, and promoter regions, Table S2. Phosphorylation and glycosylation sites in GASA proteins, Table S3. Cis-regulatory elements in promoter regions of *tcGASA*.

Author Contributions: Manuscript drafting: A.; Data analyses: A.; P.H.; H.M.T.M.; S.F.; Data curation: A.; Data interpretation: A.; P.H.; F.M.; P.P.; Conceptualization: A.; P.P.; I.A.; P.H.; Review and editing of first draft: P.H.; P.P.; I.A.; Supervision: P.P.; P.H. All authors have read and agreed to the published version of the manuscript.

Funding: This research received no external funding while the APC was funded by University of Helsinki.

Institutional Review Board Statement: Not applicable.

Informed Consent Statement: Not applicable.

Data Availability Statement: The public data set is analyzed in the manuscript. All the analyzed data are available in the main manuscript or as Supplementary Materials.

Conflicts of Interest: The authors declare no competing financial interests. Author Ibrar Ahmed and Hafiz Muhammad Talha Malik were employed by the company Alpha Genomics Private limited. The remaining authors declare that the research was conducted in the absence of any commercial or financial relationships that could be construed as a potential conflict of interest.

References

1. Purseglove, J.W. *Tropical Crops: Dicotyledons 1 and 2*; Longmans, Green and Co. Ltd.: London, UK, 1968.
2. Motamayor, J.C.; Mockaitis, K.; Schmutz, J.; Haiminen, N.; Livingstone, D.; Cornejo, O.; Findley, S.D.; Zheng, P.; Utro, F.; Royaert, S.; et al. The genome sequence of the most widely cultivated cacao type and its use to identify candidate genes regulating pod color. *Genome Biol.* **2013**, *14*, 1–25. [CrossRef]
3. Litz, R. Theobroma Cacao. In *Biotechnology of Fruit and Nut Crops*; CABI: Wallingford, UK, 2005; pp. 639–669.
4. Bridgemohan, P.; Mohammed, M. The Ecophysiology of Abiotic and Biotic Stress on the Pollination and Fertilization of Cacao (*Theobroma cacao* L.; formerly Sterculiaceae family). In *Abiotic and Biotic Stress in Plants*; IntechOpen: London, UK, 2019. [CrossRef]
5. McElroy, M.S.; Navarro, A.J.R.; Mustiga, G.; Stack, C.; Gezan, S.; Peña, G.; Sarabia, W.; Saquicela, D.; Sotomayor, I.; Douglas, G.M.; et al. Prediction of Cacao (Theobroma cacao) Resistance to Moniliophthora spp. Diseases via Genome-Wide Association Analysis and Genomic Selection. *Front. Plant Sci.* **2018**, *9*, 343. [CrossRef]
6. Argout, X.; Martin, G.; Droc, G.; Fouet, O.; Labadie, K.; Rivals, E.; Aury, J.M.; Lanaud, C. The cacao Criollo genome v2.0: An improved version of the genome for genetic and functional genomic studies. *BMC Genom.* **2017**, *18*, 1–9. [CrossRef]
7. Dayanne, S.M.D.A.; Oliveira Jordão Do Amaral, D.; Del-Bem, L.E.; Bronze Dos Santos, E.; Santana Silva Raner, J.; Peres Gramacho, K.; Vincentz, M.; Micheli, F. Genome-wide identification and characterization of cacao WRKY transcription factors and analysis of their expression in response to witches' broom disease. *PLoS ONE* **2017**, *12*, e0187346.
8. Li, F.; Hao, C.; Yan, L.; Wu, B.; Qin, X.; Lai, J.; Song, Y. Gene structure, phylogeny and expression profile of the sucrose synthase gene family in cacao (*Theobroma cacao* L.). *J. Genet.* **2015**, *94*, 461–472. [CrossRef]
9. Zhang, Y.; Maximova, S.N.; Guiltinan, M.J. Characterization of a stearoyl-acyl carrier protein desaturase gene family from chocolate tree, *Theobroma cacao* L. *Front. Plant Sci.* **2015**, *6*, 239. [CrossRef] [PubMed]
10. Li, F.; Wu, B.; Qin, X.; Yan, L.; Hao, C.; Tan, L.; Lai, J. Molecular cloning and expression analysis of the sucrose transporter gene family from *Theobroma cacao* L. *Gene* **2014**, *546*, 336–341. [CrossRef]
11. Shen, S.; Zhang, Q.; Shi, Y.; Sun, Z.; Zhang, Q.; Hou, S.; Wu, R.; Jiang, L.; Zhao, X.; Guo, Y. Genome-wide analysis of the NAC domain transcription factor gene family in Theobroma cacao. *Genes* **2020**, *11*, 35. [CrossRef] [PubMed]
12. Fan, S.; Zhang, D.; Zhang, L.; Gao, C.; Xin, M.; Tahir, M.M.; Li, Y.; Ma, J.; Han, M. Comprehensive analysis of GASA family members in the Malus domestica genome: Identification, characterization, and their expressions in response to apple flower induction. *BMC Genom.* **2017**, *18*, 827. [CrossRef] [PubMed]
13. Ahmad, B.; Yao, J.; Zhang, S.; Li, X.; Zhang, X.; Yadav, V.; Wang, X. Genome-Wide Characterization and Expression Profiling of GASA Genes during Different Stages of Seed Development in Grapevine (*Vitis vinifera* L.) Predict Their Involvement in Seed Development. *Int. J. Mol. Sci.* **2020**, *21*, 1088. [CrossRef] [PubMed]
14. Shi, L.; Gast, R.T.; Gopalraj, M.; Olszewski, N.E. Characterization of a shoot-specific, GA3- and ABA-Regulated gene from tomato. *Plant J.* **1992**, *2*, 153–159.
15. Rezaee, S.; Ahmadizadeh, M.; Heidari, P. Genome-wide characterization, expression profiling, and post-transcriptional study of GASA gene family. *Gene Rep.* **2020**, *20*, 100795. [CrossRef]
16. Zhang, S.; Wang, X. Expression pattern of GASA, downstream genes of DELLA, in Arabidopsis. *Chin. Sci. Bull.* **2008**, *53*, 3839–3846. [CrossRef]
17. Silverstein, K.A.T.; Moskal, W.A.; Wu, H.C.; Underwood, B.A.; Graham, M.A.; Town, C.D.; VandenBosch, K.A. Small cysteine-rich peptides resembling antimicrobial peptides have been under-predicted in plants. *Plant J.* **2007**, *51*, 262–280. [CrossRef] [PubMed]
18. Ahmad, M.Z.; Sana, A.; Jamil, A.; Nasir, J.A.; Ahmed, S.; Hameed, M.U. A genome-wide approach to the comprehensive analysis of GASA gene family in Glycine max. *Plant Mol. Biol.* **2019**, *100*, 607–620. [CrossRef]
19. de la Fuente, J.I.; Amaya, I.; Castillejo, C.; Sánchez-Sevilla, J.F.; Quesada, M.A.; Botella, M.A.; Valpuesta, V. The strawberry gene FaGAST affects plant growth through inhibition of cell elongation. *J. Exp. Bot.* **2006**, *57*, 2401–2411. [CrossRef]
20. Zimmermann, R.; Sakai, H.; Hochholdinger, F. The Gibberellic Acid Stimulated-Like gene family in maize and its role in lateral root development. *Plant Physiol.* **2010**, *152*, 356–365. [CrossRef] [PubMed]
21. Zhang, S.; Yang, C.; Peng, J.; Sun, S.; Wang, X. GASA5, a regulator of flowering time and stem growth in Arabidopsis thaliana. *Plant Mol. Biol.* **2009**, *69*, 745–759. [CrossRef] [PubMed]
22. Roxrud, I.; Lid, S.E.; Fletcher, J.C.; Schmidt, E.D.L.; Opsahl-Sorteberg, H.-G. GASA4, one of the 14-member Arabidopsis GASA family of small polypeptides, regulates flowering and seed development. *Plant Cell Physiol.* **2007**, *48*, 471–483. [CrossRef]
23. Nahirñak, V.; Almasia, N.I.; Hopp, H.E.; Vazquez-Rovere, C. Snakin/GASA proteins: Involvement in hormone crosstalk and redox homeostasis. *Plant Signal. Behav.* **2012**, *7*, 1004–1008. [CrossRef]
24. Moyano-Canete, E.; Bellido, M.L.; García-Caparrós, N.; Medina-Puche, L.; Amil-Ruiz, F.; González-Reyes, J.A.; Caballero, J.L.; Munoz-Blanco, J.; Blanco-Portales, R. FaGAST2, a strawberry ripening-related gene, acts together with FaGAST1 to determine cell size of the fruit receptacle. *Plant Cell Physiol.* **2013**, *54*, 218–236. [CrossRef]
25. Yang, Q.; Niu, Q.; Tang, Y.; Ma, Y.; Yan, X.; Li, J.; Tian, J.; Bai, S.; Teng, Y. PpyGAST1 is potentially involved in bud dormancy release by integrating the GA biosynthesis and ABA signaling in 'Suli'pear (Pyrus pyrifolia White Pear Group). *Environ. Exp. Bot.* **2019**, *162*, 302–312. [CrossRef]
26. He, H.; Yang, X.; Xun, H.; Lou, X.; Li, S.; Zhang, Z.; Jiang, L.; Dong, Y.; Wang, S.; Pang, J. Over-expression of GmSN1 enhances virus resistance in Arabidopsis and soybean. *Plant Cell Rep.* **2017**, *36*, 1441–1455. [CrossRef] [PubMed]

27. Wu, T.; Cheng, C.; Zhong, Y.; Lv, Y.; Ma, Y.; Zhong, G. Molecular characterization of the gibberellin-stimulated transcript of GASA4 in Citrus. *Plant Growth Regul.* **2020**, *91*, 1–11. [CrossRef]

28. Kovalskaya, N.; Hammond, R.W. Expression and functional characterization of the plant antimicrobial snakin-1 and defensin recombinant proteins. *Protein Expr. Purif.* **2009**, *63*, 12–17. [CrossRef]

29. Almasia, N.I.; Bazzini, A.A.; Hopp, H.E.; Vazquez-Rovere, C. Overexpression of snakin-1 gene enhances resistance to Rhizoctonia solani and Erwinia carotovora in transgenic potato plants. *Mol. Plant Pathol.* **2008**, *9*, 329–338. [CrossRef] [PubMed]

30. Berrocal-Lobo, M.; Segura, A.; Moreno, M.; López, G.; García-Olmedo, F.; Molina, A. Snakin-2, an antimicrobial peptide from potato whose gene is locally induced by wounding and responds to pathogen infection. *Plant Physiol.* **2002**, *128*, 951–961. [CrossRef]

31. Qu, J.; Kang, S.G.; Hah, C.; Jang, J.-C. Molecular and cellular characterization of GA-Stimulated Transcripts GASA4 and GASA6 in Arabidopsis thaliana. *Plant Sci.* **2016**, *246*, 1–10. [CrossRef] [PubMed]

32. García, A.N.; Ayub, N.D.; Fox, A.R.; Gómez, M.C.; Diéguez, M.J.; Pagano, E.M.; Berini, C.A.; Muschietti, J.P.; Soto, G. Alfalfa snakin-1 prevents fungal colonization and probably coevolved with rhizobia. *BMC Plant Biol.* **2014**, *14*, 248. [CrossRef]

33. Balaji, V.; Smart, C.D. Over-expression of snakin-2 and extensin-like protein genes restricts pathogen invasiveness and enhances tolerance to Clavibacter michiganensis subsp. michiganensis in transgenic tomato (Solanum lycopersicum). *Transgenic Res.* **2012**, *21*, 23–37. [CrossRef] [PubMed]

34. Wang, H.; Wei, T.; Wang, X.; Zhang, L.; Yang, M.; Chen, L.; Song, W.; Wang, C.; Chen, C. Transcriptome Analyses from Mutant Salvia miltiorrhiza Reveals Important Roles for SmGASA4 during Plant Development. *Int. J. Mol. Sci.* **2018**, *19*, 2088. [CrossRef]

35. Cheng, X.; Wang, S.; Xu, D.; Liu, X.; Li, X.; Xiao, W.; Cao, J.; Jiang, H.; Min, X.; Wang, J.; et al. Identification and Analysis of the GASR Gene Family in Common Wheat (*Triticum aestivum* L.) and Characterization of TaGASR34, a Gene Associated With Seed Dormancy and Germination. *Front. Genet.* **2019**, *10*, 1–20. [CrossRef] [PubMed]

36. Nahirñak, V.; Rivarola, M.; De Urreta, M.G.; Paniego, N.; Hopp, H.E.; Almasia, N.I.; Vazquez-Rovere, C. Genome-wide analysis of the Snakin/GASA gene family in Solanum Tuberosum cv. Kennebec. *Am. J. Potato Res.* **2016**, *93*, 172–188. [CrossRef]

37. Voorrips, R.E. MapChart: Software for the graphical presentation of linkage maps and QTLs. *J. Hered.* **2002**, *93*, 77–78. [CrossRef] [PubMed]

38. Gasteiger, E.; Hoogland, C.; Gattiker, A.; Duvaud, S.; Wilkins, M.R.; Appel, R.D.; Bairoch, A. Protein identification and analysis tools on the ExPASy server. In *The Proteomics Protocols Handbook*; Humana Press: Totowa, NJ, USA, 2005; pp. 571–607.

39. Savojardo, C.; Martelli, P.L.; Fariselli, P.; Profiti, G.; Casadio, R. BUSCA: An integrative web server to predict subcellular localization of proteins. *Nucleic Acids Res.* **2018**, *46*, W459–W466. [CrossRef]

40. Lescot, M.; Déhais, P.; Thijs, G.; Marchal, K.; Moreau, Y.; Van de Peer, Y.; Rouzé, P.; Rombauts, S. PlantCARE, a database of plant cis-acting regulatory elements and a portal to tools for in silico analysis of promoter sequences. *Nucleic Acids Res.* **2002**, *30*, 325–327. [CrossRef] [PubMed]

41. Blom, N.; Sicheritz-Pontén, T.; Gupta, R.; Gammeltoft, S.; Brunak, S. Prediction of post-translational glycosylation and phosphorylation of proteins from the amino acid sequence. *Proteomics* **2004**, *4*, 1633–1649. [CrossRef] [PubMed]

42. Gupta, R.; Brunak, S. Prediction of glycosylation across the human proteome and the correlation to protein function—PubMed. *Pac. Symp. Biocomput.* **2002**, *7*, 310–322.

43. Sievers, F.; Wilm, A.; Dineen, D.; Gibson, T.J.; Karplus, K.; Li, W.; Lopez, R.; McWilliam, H.; Remmert, M.; Söding, J.; et al. Fast, scalable generation of high-quality protein multiple sequence alignments using Clustal Omega. *Mol. Syst. Biol.* **2011**, *7*, 539. [CrossRef]

44. Kumar, S.; Stecher, G.; Li, M.; Knyaz, C.; Tamura, K. MEGA X: Molecular evolutionary genetics analysis across computing platforms. *Mol. Biol. Evol.* **2018**, *35*, 1547–1549. [CrossRef]

45. Letunic, I.; Bork, P. Interactive Tree of Life (iTOL) v4: Recent updates and new developments. *Nucleic Acids Res.* **2019**, *47*, W256–W259. [CrossRef] [PubMed]

46. Bailey, T.L.; Boden, M.; Buske, F.A.; Frith, M.; Grant, C.E.; Clementi, L.; Ren, J.; Li, W.W.; Noble, W.S. MEME SUITE: Tools for motif discovery and searching. *Nucleic Acids Res.* **2009**, *37*, W202–W208. [CrossRef] [PubMed]

47. Zheng, L.; Ying, Y.; Wang, L.; Wang, F.; Whelan, J.; Shou, H. Identification of a novel iron regulated basic helix-loop-helix protein involved in Fe homeostasis in Oryza sativa. *BMC Plant Biol.* **2010**, *10*, 166. [CrossRef]

48. Kearse, M.; Moir, R.; Wilson, A.; Stones-Havas, S.; Cheung, M.; Sturrock, S.; Buxton, S.; Cooper, A.; Markowitz, S.; Duran, C.; et al. Geneious Basic: An integrated and extendable desktop software platform for the organization and analysis of sequence data. *Bioinformatics* **2012**, *28*, 1647–1649. [CrossRef]

49. Wang, Y.; Tang, H.; DeBarry, J.D.; Tan, X.; Li, J.; Wang, X.; Lee, T.; Jin, H.; Marler, B.; Guo, H. MCScanX: A toolkit for detection and evolutionary analysis of gene synteny and collinearity. *Nucleic Acids Res.* **2012**, *40*, e49. [CrossRef]

50. Rozas, J.; Ferrer-Mata, A.; Sánchez-DelBarrio, J.C.; Guirao-Rico, S.; Librado, P.; Ramos-Onsins, S.E.; Sánchez-Gracia, A. DnaSP 6: DNA sequence polymorphism analysis of large data sets. *Mol. Biol. Evol.* **2017**, *34*, 3299–3302. [CrossRef]

51. Yang, S.; Zhang, X.; Yue, J.-X.; Tian, D.; Chen, J.-Q. Recent duplications dominate NBS-encoding gene expansion in two woody species. *Mol. Genet. Genom.* **2008**, *280*, 187–198. [CrossRef]

52. Krzywinski, M.; Schein, J.; Birol, I.; Connors, J.; Gascoyne, R.; Horsman, D.; Jones, S.J.; Marra, M.A. Circos: An information aesthetic for comparative genomics. *Genome Res.* **2009**, *19*, 1639–1645. [CrossRef]

53. Yang, J.; Yan, R.; Roy, A.; Xu, D.; Poisson, J.; Zhang, Y. The I-TASSER Suite: Protein structure and function prediction. *Nat. Methods* **2015**, *12*, 7–8. [CrossRef]

54. Bhattacharya, D.; Nowotny, J.; Cao, R.; Cheng, J. 3Drefine: An interactive web server for efficient protein structure refinement. *Nucleic Acids Res.* **2016**, *44*, W406–W409. [CrossRef]

55. Jendele, L.; Krivak, R.; Skoda, P.; Novotny, M.; Hoksza, D. PrankWeb: A web server for ligand binding site prediction and visualization. *Nucleic Acids Res.* **2019**, *47*, W345–W349. [CrossRef] [PubMed]

56. Tian, W.; Chen, C.; Lei, X.; Zhao, J.; Liang, J. CASTp 3.0: Computed atlas of surface topography of proteins. *Nucleic Acids Res.* **2018**, *46*, W363–W367. [CrossRef]

57. DeLano, W.L. Pymol: An open-source molecular graphics tool. *CCP4 Newsl. Protein Crystallogr.* **2002**, *40*, 82–92.

58. Pokou, D.N.; Fister, A.S.; Winters, N.; Tahi, M.; Klotioloma, C.; Sebastian, A.; Marden, J.H.; Maximova, S.N.; Guiltinan, M.J. Resistant and susceptible cacao genotypes exhibit defense gene polymorphism and unique early responses to Phytophthora megakarya inoculation. *Plant Mol. Biol.* **2019**, *99*, 499–516. [CrossRef]

59. Chen, C.; Chen, H.; Zhang, Y.; Thomas, H.R.; Frank, M.H.; He, Y.; Xia, R. TBtools: An Integrative Toolkit Developed for Interactive Analyses of Big Biological Data. *Mol. Plant* **2020**, *13*, 1194–1202. [CrossRef]

60. Ding, Q.; Yang, X.; Pi, Y.; Li, Z.; Xue, J.; Chen, H.; Li, Y.; Wu, H. Genome-wide identification and expression analysis of extensin genes in tomato. *Genomics* **2020**, *112*, 4348–4360. [CrossRef] [PubMed]

61. Ahmadizadeh, M.; Rezaee, S.; Heidari, P. Genome-wide characterization and expression analysis of fatty acid desaturase gene family in Camelina sativa. *Gene Rep.* **2020**, *21*, 100894. [CrossRef]

62. Heidari, P.; Ahmadizadeh, M.; Izanlo, F.; Nussbaumer, T. In silico study of the CESA and CSL gene family in Arabidopsis thaliana and Oryza sativa: Focus on post-translation modifications. *Plant Gene* **2019**, *19*, 100189. [CrossRef]

63. Zhang, S.; Wang, X. Overexpression of GASA5 increases the sensitivity of Arabidopsis to heat stress. *J. Plant Physiol.* **2011**, *168*, 2093–2101. [CrossRef] [PubMed]

64. Wang, L.; Wang, Z.; Xu, Y.; Joo, S.; Kim, S.; Xue, Z.; Xu, Z.; Wang, Z.; Chong, K. OsGSR1 is involved in crosstalk between gibberellins and brassinosteroids in rice. *Plant J.* **2009**, *57*, 498–510. [CrossRef] [PubMed]

65. Heidari, P.; Faraji, S.; Ahmadizadeh, M.; Ahmar, S.; Mora-Poblete, F. New insights into structure and function of TIFY genes in Zea mays and Solanum lycopersicum: A genome-wide comprehensive analysis. *Front. Genet.* **2021**, *12*, 534. [CrossRef]

66. Musavizadeh, Z.; Najafi-Zarrini, H.; Kazemitabar, S.K.; Hashemi, S.H.; Faraji, S.; Barcaccia, G.; Heidari, P. Genome-Wide Analysis of Potassium Channel Genes in Rice: Expression of the OsAKT and OsKAT Genes under Salt Stress. *Genes* **2021**, *12*, 784. [CrossRef]

67. Beauregard, M.; Hefford, M.A. Enhancement of essential amino acid contents in crops by genetic engineering and protein design. *Plant Biotechnol. J.* **2006**, *4*, 561–574. [CrossRef] [PubMed]

68. Webster, D.E.; Thomas, M.C. Post-translational modification of plant-made foreign proteins; glycosylation and beyond. *Biotechnol. Adv.* **2012**, *30*, 410–418. [CrossRef]

69. Duan, G.; Walther, D. The Roles of Post-translational Modifications in the Context of Protein Interaction Networks. *PLoS Comput. Biol.* **2015**, *11*, 1004049. [CrossRef] [PubMed]

70. Silva-Sanchez, C.; Li, H.; Chen, S. Recent advances and challenges in plant phosphoproteomics. *Proteomics* **2015**, *15*, 1127–1141. [CrossRef]

71. Heidari, P.; Mazloomi, F.; Nussbaumer, T.; Barcaccia, G. Insights into the SAM synthetase gene family and its roles in tomato seedlings under abiotic stresses and hormone treatments. *Plants* **2020**, *9*, 586. [CrossRef] [PubMed]

72. Nawaz, Z.; Kakar, K.U.; Ullah, R.; Yu, S.; Zhang, J.; Shu, Q.Y.; Ren, X. liang Genome-wide identification, evolution and expression analysis of cyclic nucleotide-gated channels in tobacco (*Nicotiana tabacum* L.). *Genomics* **2019**, *111*, 142–158. [CrossRef]

73. Yang, S.; Höti, N.; Yang, W.; Liu, Y.; Chen, L.; Li, S.; Zhang, H. Simultaneous analyses of N-linked and O-linked glycans of ovarian cancer cells using solid-phase chemoenzymatic method. *Clin. Proteom.* **2017**, *14*, 3. [CrossRef]

74. Abdullah; Mehmood, F.; Shahzadi, I.; Ali, Z.; Islam, M.; Naeem, M.; Mirza, B.; Lockhart, P.; Ahmed, I.; Waheed, M.T. Correlations among oligonucleotide repeats, nucleotide substitutions and insertion-deletion mutations in chloroplast genomes of plant family Malvaceae. *J. Syst. Evol.* **2020**, *59*, 388–402. [CrossRef]

75. Conover, J.L.; Karimi, N.; Stenz, N.; Ané, C.; Grover, C.E.; Skema, C.; Tate, J.A.; Wolff, K.; Logan, S.A.; Wendel, J.F.; et al. A Malvaceae mystery: A mallow maelstrom of genome multiplications and maybe misleading methods? *J. Integr. Plant Biol.* **2019**, *61*, 12–31. [CrossRef] [PubMed]

76. Chattha, W.S.; Atif, R.M.; Iqbal, M.; Shafqat, W.; Farooq, M.A.; Shakeel, A. Genome-wide identification and evolution of Dof transcription factor family in cultivated and ancestral cotton species. *Genomics* **2020**, *112*, 4155–4170. [CrossRef] [PubMed]

77. Lei, P.; Wei, X.; Gao, R.; Huo, F.; Nie, X.; Tong, W.; Song, W. Genome-wide identification of PYL gene family in wheat: Evolution, expression and 3D structure analysis. *Genomics* **2020**, *113*, 854–866. [CrossRef]

78. Nawaz, Z.; Kakar, K.U.; Saand, M.A.; Shu, Q.Y. Cyclic nucleotide-gated ion channel gene family in rice, identification, characterization and experimental analysis of expression response to plant hormones, biotic and abiotic stresses. *BMC Genom.* **2014**, *15*, 1–18. [CrossRef]

79. Nuruzzaman, M.; Manimekalai, R.; Sharoni, A.M.; Satoh, K.; Kondoh, H.; Ooka, H.; Kikuchi, S. Genome-wide analysis of NAC transcription factor family in rice. *Gene* **2010**, *465*, 30–44. [CrossRef] [PubMed]

80. Liu, C.; Wu, Y.; Liu, Y.; Yang, L.; Dong, R.; Jiang, L.; Liu, P.; Liu, G.; Wang, Z.; Luo, L. Genome-wide analysis of tandem duplicated genes and their contribution to stress resistance in pigeonpea (Cajanus cajan). *Genomics* **2020**, *113*, 728–735. [CrossRef]

81. Ahmad, B.; Azeem, F.; Ali, M.A.; Nawaz, M.A.; Nadeem, H.; Abbas, A.; Batool, R.; Atif, R.M.; Ijaz, U.; Nieves-Cordones, M.; et al. Genome-wide identification and expression analysis of two component system genes in Cicer arietinum. *Genomics* **2020**, *112*, 1371–1383. [CrossRef]
82. Li, Z.; Gao, Z.; Li, R.; Xu, Y.; Kong, Y.; Zhou, G.; Meng, C.; Hu, R. Genome-wide identification and expression profiling of HD-ZIP gene family in Medicago truncatula. *Genomics* **2020**, *112*, 3624–3635. [CrossRef] [PubMed]
83. Piégu, B.; Arensburger, P.; Beauclair, L.; Chabault, M.; Raynaud, E.; Coustham, V.; Brard, S.; Guizard, S.; Burlot, T.; Le Bihan-Duval, E.; et al. Variations in genome size between wild and domesticated lineages of fowls belonging to the Gallus gallus species. *Genomics* **2020**, *112*, 1660–1673. [CrossRef]
84. Song, X.; Li, E.; Song, H.; Du, G.; Li, S.; Zhu, H.; Chen, G.; Zhao, C.; Qiao, L.; Wang, J.; et al. Genome-wide identification and characterization of nonspecific lipid transfer protein (nsLTP) genes in Arachis duranensis. *Genomics* **2020**, *112*, 4332–4341. [CrossRef]
85. Faraji, S.; Filiz, E.; Kazemitabar, S.K.; Vannozzi, A.; Palumbo, F.; Barcaccia, G.; Heidari, P. The AP2/ERF Gene Family in Triticum durum: Genome-Wide Identification and Expression Analysis under Drought and Salinity Stresses. *Genes* **2020**, *11*, 1464. [CrossRef]
86. Ahmadizadeh, M.; Chen, J.-T.; Hasanzadeh, S.; Ahmar, S.; Heidari, P. Insights into the genes involved in the ethylene biosynthesis pathway in Arabidopsis thaliana and Oryza sativa. *J. Genet. Eng. Biotechnol.* **2020**, *18*, 1–20. [CrossRef] [PubMed]
87. De Wever, J.; Tulkens, D.; Verwaeren, J.; Everaert, H.; Rottiers, H.; Dewettinck, K.; Lefever, S.; Messens, K. A combined RNA preservation and extraction protocol for gene expression studies in cacao beans. *Front. Plant Sci.* **2020**, *11*, 992. [CrossRef] [PubMed]
88. De Wever, J.; De Coninck, T.; Everaert, H.; Afoakwa, E.O.; Coppieters, F.; Rottiers, H.; Opoku, S.Y.; Lowor, S.; Dewettinck, K.; Vandesompele, J. Selection and validation of reference genes for accurate RT-qPCR gene expression normalization in cacao beans during fermentation. *Tree Genet. Genomes* **2021**, *17*, 1–14. [CrossRef]
89. Medina, V.; Laliberte, B. *A Review of Research on the Effects of Drought and Temperature Stress and Increased CO_2 on Theobroma cacao L., and the Role of Genetic Diversity to Address Climate Change*; Bioversity: Turrialba, Costa Rica, 2017; ISBN 9789292550745.
90. Adeniyi, D. Diversity of Cacao Pathogens and Impact on Yield and Global Production. In *Theobroma Cacao—Deploying Science for Sustainability of Global Cocoa Economy*; IntechOpen: London, UK, 2019. [CrossRef]

Article

Magnesium transporter Gene Family: Genome-Wide Identification and Characterization in *Theobroma cacao*, *Corchorus capsularis*, and *Gossypium hirsutum* of Family Malvaceae

Parviz Heidari [1],*, Abdullah [2], Sahar Faraji [3] and Peter Poczai [4,5],*

[1] Faculty of Agriculture, Shahrood University of Technology, Shahrood 3619995161, Iran
[2] Department of Biochemistry, Faculty of Biological Sciences, Quaid-i-Azam University, Islamabad 45320, Pakistan; abd.ullah@bs.qau.edu.pk
[3] Department of Plant Breeding, Faculty of Crop Sciences, Sari Agricultural Sciences and Natural Resources University (SANRU), Sari 4818166996, Iran; sahar.faraji@rocketmail.com
[4] Finnish Museum of Natural History, University of Helsinki, P.O. Box 7, 00014 Helsinki, Finland
[5] Faculty of Biological and Environmental Sciences, University of Helsinki, P.O. Box 65, 00065 Helsinki, Finland
* Correspondence: heidarip@shahroodut.ac.ir (P.H.); peter.poczai@helsinki.fi (P.P.); Tel.: +98-912-0734-034 (P.H.)

Citation: Heidari, P.; Abdullah; Faraji, S.; Poczai, P. *Magnesium transporter* Gene Family: Genome-Wide Identification and Characterization in *Theobroma cacao*, *Corchorus capsularis*, and *Gossypium hirsutum* of Family Malvaceae. *Agronomy* 2021, 11, 1651. https://doi.org/10.3390/agronomy11081651

Academic Editors: Roxana Yockteng, Andrés J. Cortés and María Ángeles Castillejo

Received: 26 July 2021
Accepted: 18 August 2021
Published: 19 August 2021

Publisher's Note: MDPI stays neutral with regard to jurisdictional claims in published maps and institutional affiliations.

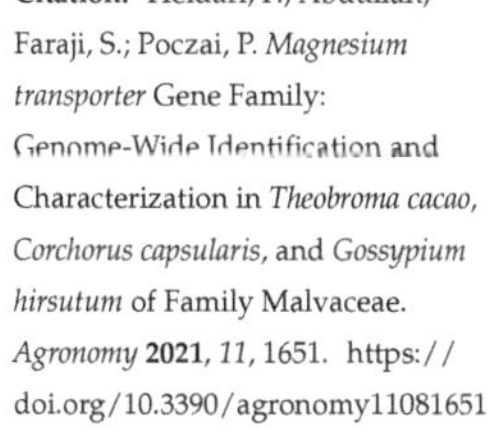

Abstract: Magnesium (Mg) is an element involved in various key cellular processes in plants. Mg transporter (MGT) genes play an important role in magnesium distribution and ionic balance maintenance. Here, MGT family members were identified and characterized in three species of the plant family Malvaceae, *Theobroma cacao*, *Corchorus capsularis*, and *Gossypium hirsutum*, to improve our understanding of their structure, regulatory systems, functions, and possible interactions. We identified 18, 41, and 16 putative non-redundant *MGT* genes from the genome of *T. cacao*, *G. hirsutum*, and *C. capsularis*, respectively, which clustered into three groups the maximum likelihood tree. Several segmental/tandem duplication events were determined between *MGT* genes. *MGTs* appear to have evolved slowly under a purifying selection. Analysis of gene promoter regions showed that *MGTs* have a high potential to respond to biotic/abiotic stresses and hormones. The expression patterns of *MGT* genes revealed a possible role in response to *P. megakarya* fungi in *T. cacao*, whereas *MGT* genes showed differential expression in various tissues and response to several abiotic stresses, including cold, salt, drought, and heat stress in *G. hirsutum*. The co-expression network of MGTs indicated that genes involved in auxin-responsive lipid metabolism, cell wall organization, and photoprotection can interact with MGTs.

Keywords: *magnesium transporter*; comparative analysis; Malvaceae; *Theobroma*; *Gossypium*; *Corchorus*; expression analysis; gene structure; phylogenetic analysis

1. Introduction

Magnesium (Mg) is a critical bimetal that regulates biochemical processes and provides stability to membranes in plants [1,2]. Magnesium acts as a cofactor for polymerase, kinase, and H+-ATPase, which are necessary for synthesizing proteins and nucleic acid and for generating energy [3,4]. It is also required to maintain cation–anion homeostasis in the cell [5]. Various types of adverse effects have been reported in plants during Mg deficiency, including a reduction in photosynthesis, macromolecule synthesis, and plant growth and development [6–8]. Therefore, plants have developed an efficient transport system for absorption, storage, and Mg translocation [2]. The *Mg transporter* (*MGT*) gene family, also known as MRS2 or CorA, has an important role in the aforementioned essential functions [9,10]. Members of the MGT family are defined by two transmembrane domains in which a tripeptide motif GMN (glycine–methionine–asparagine) occurs at the C-terminal domain of the first transmembrane [11,12]. *MGTs* are expressed in root tissues of plants that are more involved in up taking Mg (such as *MGT1* in rice and *MGT6* in *Arabidopsis*), transferring Mg from root to shoot (such as *MGT9* in *Arabidopsis*), homeostasis by maintaining ionic balance (such as *MGT10* in *Arabidopsis*), and accumulation and translocation of Mg within, for instance, the vacuole of the cell (such as *MGT2* and *MGT3* in *Arabidopsis*) [11,13–16]. These genes are also crucial for pollen mitosis and pollen intine formation [15,17,18].

MGT genes also respond to changes in elemental concentration in soil, i.e., *MGT* genes showed high expression due to aluminum (Al) toxicity in acidic soil. In *Arabidopsis* and in maize, Al-tolerant genotypes were observed to have a high capacity for Mg uptake and accumulation [9,19]. The transformation and expression of *Arabidopsis MGT1* in *Nicotiana benthamiana* increased uptake of Mg and reduced toxicity of Al in transgenic lines [20]. In contrast, the knockout of *MGT1* in rice reduced tolerance to salt and was linked to a high content of sodium in shoot tissues [21]. The *MGT* genes are also important for plant adaptation to changing Mg status in soil [13]. Genes of the *MGT* family have been identified and characterized in several plant species, including *Arabidopsis* [22], rice [11], maize [9], pear [23], citrus [24], rapeseed [25], wild sugarcane [26], and tomato [27]. The plant family Malvaceae is one of the largest dicot families with 244 genera and 4225 species [28]. The family includes significant economic plant species, such as cotton (*Gossypium*) and jute (*Corchorus*) are important for fiber, whereas cacao (*Theobroma cacao* L.) is important for chocolate production [29,30]. Besides, the seed of *Theobroma cacao* contained 290 mg/100 g magnesium which is 4–5 time more than pea, corn, white wheat, and rice, and is included among the richest magnesium containing sources [31]. The cotton is considered as a white gold due to its industrial importance and in the world 25 million tons cotton are produce annually of worth of 600 billion dollars [32]. This important plant effected from Magnesium deficiency [33,34] whereas jute is the second most important natural fiber in terms of global consumption after cotton and specifically important in some countries such as Bangladesh [35]. Therefore, the comparative analysis and determining role of the MGT family can provide quality information about the gene structure, variation etc. To date, none of the studies, to the best of our knowledge, focuses on characterization of the *MGT* gene family despite the availability of nuclear genomes for the species of family Malvaceae [36,37] with the advancement of sequencing technologies. Here, we aim to: (i) identify and characterize *MGT* genes within three species of Malvaceae, including *T. cacao*, *C. capsularis*, and *G. hirsutum*, (ii) study evolutionary patterns and phylogenetic relationships, and (iii) determine roles of MGTs in growth and development of Malvaceae species.

2. Materials and Methods

2.1. Identification and Characterization of MGT Genes in T. cacao, C. capsularis, and G. hirsutum

The homologs of *Arabidopsis* MGT proteins, MGT7 (AT5G09690), CorA-like family protein (AT1G29820), MRS2-1 (AT1G16010), MRS2-10 (AT1G80900), NIPA7 (AT4G38730), NIPA1 (AT3G23870), and NIPA5 (AT4G09640), were BLAST with an expected value of E^{-10} in Ensembl Plants [38] for *T. cacao* and *Corchorus capsularis*, and in cotton genome database [39] for *Gossypium hirsutum* to identify *MGT* genes and retrieve protein sequences, coding sequences, genomic sequences, and promoter regions for various analyses following a previous approach [40]. The non-redundant protein sequences were selected based on CDD search (https://www.ncbi.nlm.nih.gov/Structure/cdd/cdd.shtml, accessed on 24 November 2020). The ProtParam program [41] was used to determine molecular weight (MW) and isoelectric points (*pI*), TMHMM Server version 2.0 [42] to predict transmembrane domains, and CELLO2GO [43] to determine the location of MGT proteins in the cell. PlantCARE [44] was used to analyze promoter regions.

2.2. Phylogenetic Inference, Conserved Protein Motifs, and Gene Structure

The protein sequences of *T. cacao*, *C. capsularis*, and *G. hirsutum* aligned in Geneious R8.1 [45] using ClustalW [46] and analyzed in IQ-tree for construction of a maximum likelihood phylogenetic tree under default parameters and 1000 bootstrap replications [47–49].

The Best-fit model JTT+G4 was chosen according to Bayesian Information Criterion using jModelTest 2 [50] and was employed to find relationship among genes. Finally, the iTOL, Integrative Tree of Life version 4 [51], was used for the improvement of tree representation. The MEME (Multiple Em for Motif Elicitation) server [52] was used to identify the conserved protein motifs in MGT. The gene structure of each *MGT* gene was constructed using the Gene Structure Display Server [53].

2.3. Gene Duplications and Synteny Analysis

The *MGT* genes with more than 85% identity in each species were selected as duplicated genes [54]. Then, the location diagram of duplicated genes in cacao and *G. hirsutum* were constructed using TBtools [55]. In addition, the synonymous (Ks) and non-synonymous (Ka) of each duplicated gene pair were calculated by DnaSP version 6 software [56]. Finally, the time of divergence of duplicated genes was determined using the following equation: $T = (Ks/2\lambda) \times 10^{-6}$ [57], where λ is substitutions per synonymous site per year and $\lambda = 6.5 \times 10^{-9}$. Moreover, the synteny relationship diagrams of *MGT* genes among the orthologous pairs of *T. cacao–G. hirsutum* and *T. cacao–C. capsularis* were created using Circos software [58].

2.4. Structure Analyses of MGT Proteins

The transmembrane and three-dimensional structures of the candidates of MGT subgroups MRS2, NIPA, and CorA proteins in *T. cacao*, *G. hirsutum*, and *C. capsularis* were predicted using the Phyre2 server [59], whereas docking analysis was performed to predict the ligand-binding regions (pocket sites) using DeepSite [60] and CASTp [61] tools and finally constructed in PyMOL [62].

2.5. Expression Analysis of TcMGTs and GhMGTs Using RNA-Seq Data

The publicly available RNA-seq data of cacao transcriptome with accession number GSE116041 [63] were used to find a possible differential expression of cacao *MGTs* concerning a fungi disease caused by *Phytophthora megakarya* after inoculation of the different time courses of 0 h, 6 h, 24 h, 48 h, and 72 h in two contrasting cultivars: Nanay (NA-32) as a susceptible cultivar and Scavina (SCA-6) as a fungal-resistant cultivar. Finally, the expression patterns of *TcMGTs* were illustrated in heatmaps based on log2 transformed using TBtools [55]. Furthermore, the expression profile of *GhMGTs* in various tissues (ovule, fiber, anther, bract, filament, leaf, pental, root, sepal, stem, and torus) and in response to various abiotic stresses (cold, heat, salt, and drought) was retrieved from available RNA-seq data of the cotton genome database (https://cottonfgd.org/, accessed on 9 March 2021) under project PRJNA490626 using the gene ID of each gene of the newly assembled genome as query [64]. The expression of *GhMGTs* in various tissues was analyzed and represented as a heatmap based on percentage expression of each gene using TBtools [55], while the data of abiotic stresses were analyzed and represented in heatmap after log2 transformation through TBtools [55].

3. Results

3.1. Sequence and Structure of MGT Genes

Altogether 18, 41, and 16 putative non-redundant MGT genes were identified from the genomes of *T. cacao*, *G. hirsutum*, and *C. capsularis*, respectively. All sequences (genomic, amino acids, coding sequences) of identified MGT genes are shown in Table S1. MGTs were characterized based on their sequences structure (Table S2) and three MGT sub-groups, including MRS2, NIPA, and CorA, were recognized according to the specific domain distribution (Table S2). Our findings revealed that MGTs in the three studied plant species are diverse in sequence length, molecular weight (MW), isoelectric point (pI), and exon number (Table S2). For instance, protein length varied from 321 amino acids (aa) to 632 aa in T. cacao, from 210 aa to 474 aa in *G. hirsutum*, and from 262 aa to 2417 aa in *C. capsularis* (Table 1). In addition, the predicted MW ranged from 32.75 kDa to 70.91 kDa in *T. cacao*, from 32.66 kDa to 53.95 kDa in *G. hirsutum*, and from 29.82 kDa to 268.42 kDa in *C. capsularis* (Table 1). Moreover, the pI of MGTs was between 4.48 and 8.57 in *T. cacao*, between 4.76 and 9.57 in *G. hirsutum*, and between 4.79 and 8.60 in *C. capsularis* (Table 1). Based on pI value, 75% of MGTs in *C. capsularis*, 56% in *T. cacao*, and 49% in *G. hirsutum* were predicted to be acidophilic proteins (Table S2). In addition, the prediction of subcellular localization illustrated that most MGTs are located in the endomembrane or the plasma membrane (Table S2). The exon number of MGT genes varied between 4 and 15 in *T. cacao* and *G. hirsutum*, while the exon number varied between 4 and 21 in *C. capsularis* (Table 1).

Table 1. Summary of MGT properties in the three plant species *Theobroma cacao*, *Gossypium hirsutum*, and *Corchorus capsularis*.

Organism	Gene Number	Gene Length (bp)	Protein Length (aa)	MW (KDa)	pI	Exon Number
T. cacao	18	1212–2632	321–632	32.75–70.91	4.48–8.57	4–15
G. hirsutum	41	633–1425	210–474	32.66–53.95	4.76–9.57	4–15
C. capsularis	16	789–7254	262–2417	29.82–268.42	4.79–8.60	4–21

3.2. Phylogenetic Analysis and Classification of the MGT Gene Family

A phylogenetic tree of MGT proteins was constructed, comprising 18 TcMGT proteins from *T. cacao*, 41 GhMGTs from *G. hirsutum*, and 16 CcMGTs from *C. capsularis*. The MGT proteins clustered into three groups (groups I, II, and III) that group III included three sub-groups, 3a, b, and c (Figure 1). Six MGTs, including a CorA of Jute, namely CcMGT08, a MRS2 protein from cacao, namely TcMGT03, and four MRS2 proteins of cotton, namely GhMGT11, GhMGT12, GhMGT32, and GhMGT33, with similar structure contained 11 exons and were located in group I (Figure 1a,b). In addition, five MGT proteins were located in group II and four CorA proteins, namely TcMGT01, TcMGT16, CcMGT02, and CcMGT16, clustered in group III-a. In addition, 16 MRS2 proteins along with two CorA proteins were located in group III-b and all NIPA-type proteins were located in group III-c (Figure 1a,b). In addition, MGTs were analyzed based on distribution of conserved motifs in their protein sequence. Thirty-five conserved motifs were identified, and MGTs from group III showed more diversity than other groups (Figure 2). Motifs 12, 5, 18, 7, 10, and 2, frequently observed in MGTs and NIPA proteins, illustrated different patterns of conserved motif distribution. Moreover, CorA proteins also showed various conserved motifs (Figure 2).

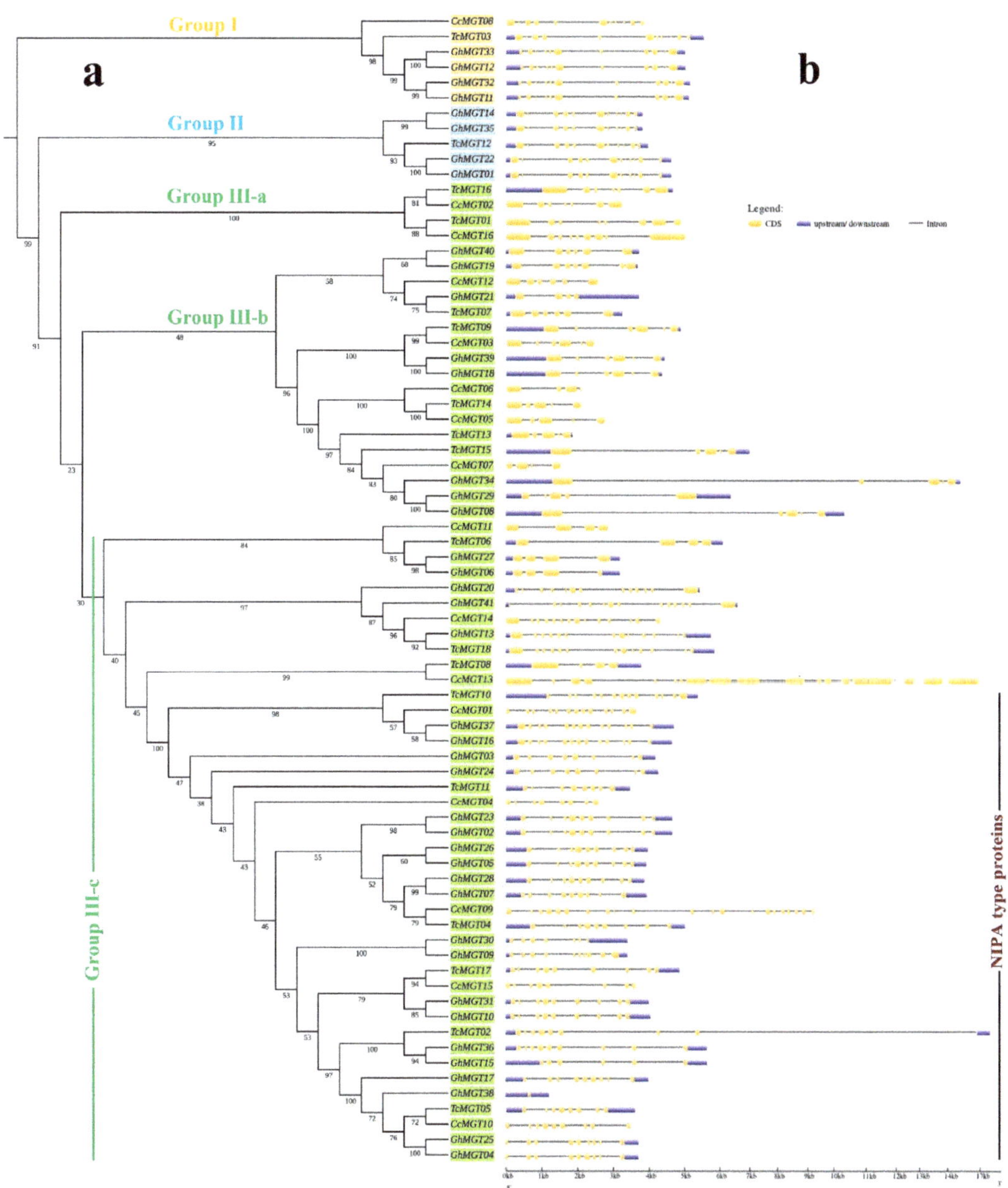

Figure 1. The phylogenetic tree of MGT proteins (**a**) and gene structure of *MGT* genes (**b**) of *Theobroma cacao*, *Gossypium hirsutum*, and *Corchorus capsularis*. The start of each gene name for the species is as follows: Tc: *Theobroma cacao*; Gh: *Gossypium hirsutum*; Cc: *Corchorus capsularis*.

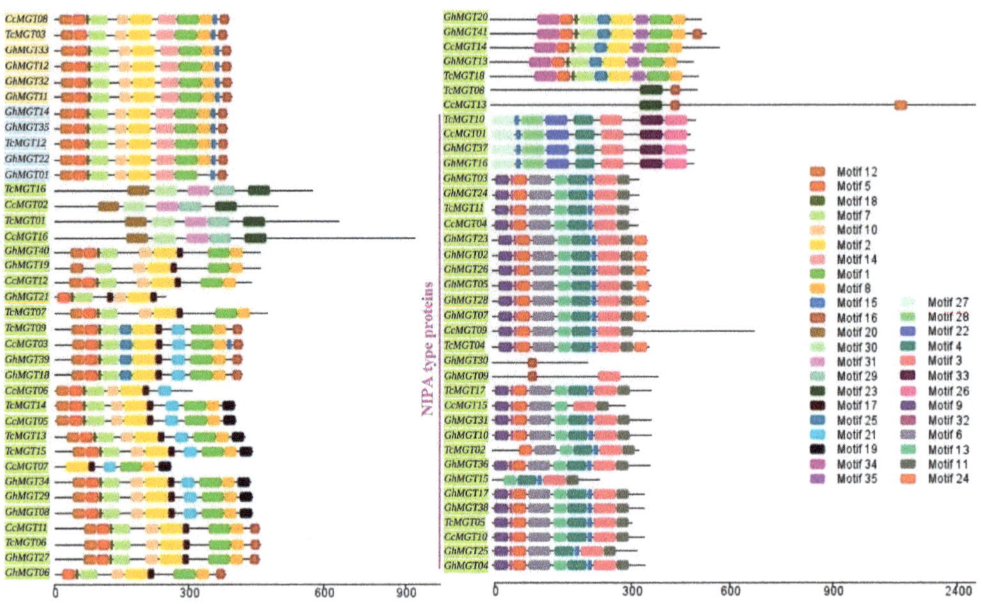

Figure 2. Conserved motif distribution in MGT proteins of *Theobroma cacao*, *Gossypium hirsutum*, and *Corchorus capsularis*.

3.3. Duplication Events and Synteny Analysis

The *TcMGT* genes were mapped onto 5 of 10 chromosomes in the cacao genome (Figure 3a), while *GhMGTs* were distributed over 22 of 26 chromosomes in the *G. hirsutum* genome (Figure 3b). Due to the incomplete physical map of the *C. capsularis* plant, the mapping of *CcMGT* genes is not provided. In the cacao genome, chromosome 2 encompassed the most significant number of *TcMGTs*, with five genes, while in *G. hirsutum* genome, the most significant number of *GhMGTs* were located on chromosomes A04 and D01 (Figure 3). In addition, the duplication events of *MGT* genes in selected plant species were investigated. Eight segmental duplication gene pairs were identified between 12 *TcMGT* genes of cacao (Figure 3a and Table S3). Tandem duplication events seemed to occur on chromosome 6 between three *TcMGT* genes of cacao, including, *TcMGT013*, *TcMGT14*, and *TcMGT15* (Figure 3a). In addition, four segmental duplication events were predicted for the *TcMGT03* gene (Table S3). Notably, a duplication event occurred around 10 MYA between two *CorA*- like genes in cacao, namely *TcMGT01* and *TcMGT16*. In addition, five segmental duplication gene pairs were recognized between *CcMGT* genes, and a triplication event was predicted between *CcMGT05* and *CcMGT06*, *CcMGT07*, and *CcMGT04* (Table S3). According to the *Ka/Ks* ratio, the first duplication was approximately 103 million years ago (MYA) between *CcMGT05* and *CcMGT04*. The most duplication events were observed between *GhMGT* genes in the *G. hirsutum* genome, with 22 segmental duplication gene pairs. Moreover, four *GhMGT* genes, namely *GhMGT02*, *GhMGT05*, *GhMGT23*, and *GhMGT26*, had a common ancestor and probably the first duplication event occurred approximately 68 MYA between *GhMGT23* and *GhMGT26* (Table S3). The intraspecies synteny of *MGT* genes was constructed between *T. cacao* and *G. hirsutum* and between *T. cacao* and *C. capsularis* (Figure 4). The 18 *TcMGT* genes in *T. cacao* illustrated 15 and 16 syntenic block relationships with *MGT* genes in *G. hirsutum* and *C. capsularis*, respectively (Figure 4a,b). Interestingly, TcMGT genes of cacao showed more syntenic relationships with *GhMGTs* of D-genome than A-genome.

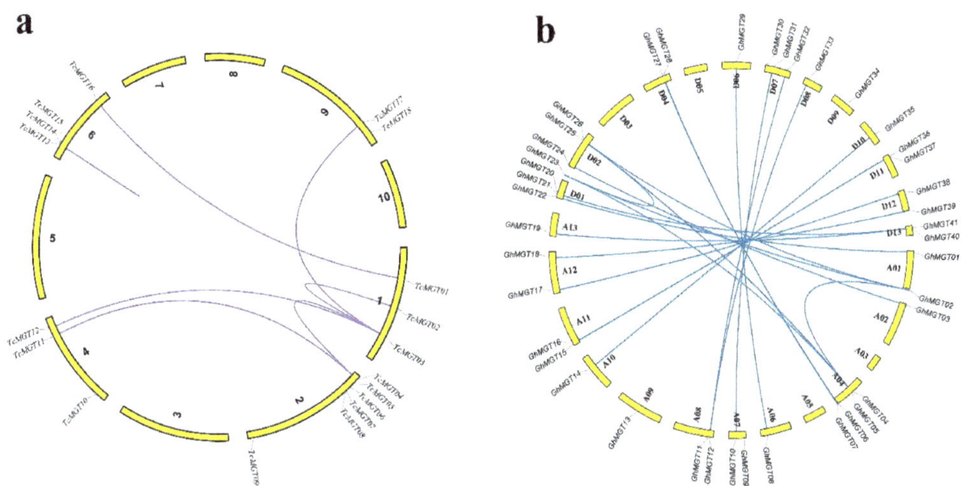

Figure 3. Location of *MGT* genes on the chromosome in *Theobroma cacao* (**a**), and *Gossypium hirsutum* (**b**). The duplicated genes are connected using blue lines.

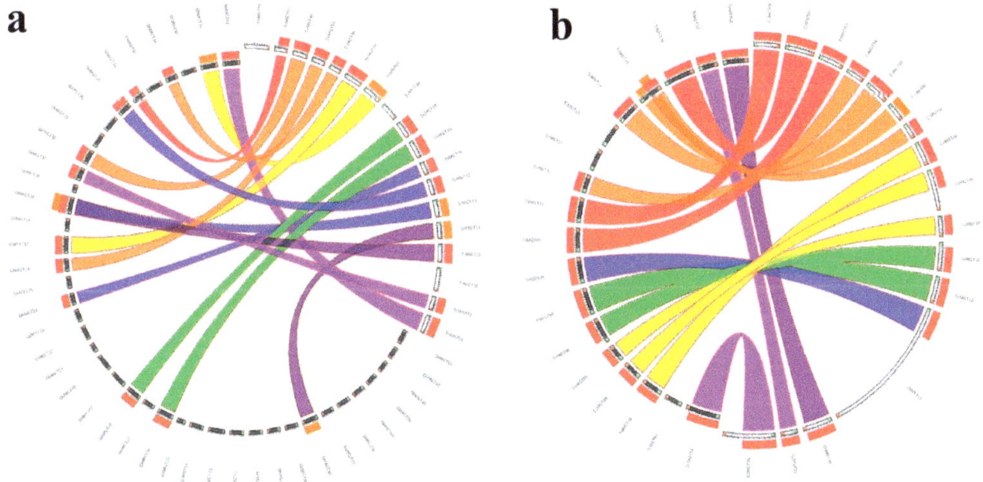

Figure 4. Synteny analysis of *MGT* genes. The syntenic blocks of cacao *MGTs* are constructed with *Gossypium hirsutum* (**a**), and *Corchorus capsularis* (**b**).

3.4. Protein Structure and Docking Analysis

The 3D structures of all candidates of three types of MGTs, namely NIPA, MRS2, and CorA, were predicted in *T. cacao*, *G. hirsutum*, and *C. capsularis* (Figure S1). Nine α-helices were observed in the predicted 3D structure of NIPAs in three studied plants, while fewer α-helices were predicted in the structure of MRS2 and CorA proteins (Figure S1). Furthermore, nine transmembrane helices with eight pores were predicted in the structure of NIPAs in three plants, while in candidate MRS2 proteins two transmembrane helices were observed in all studied plant species. However, both N-terminal and C-terminal of candidate MRS2 proteins from *T. cacao* and *C. capsularis* were predicted in the extracellular part, while in *G. hirsutum*, both N- and C-terminal were observed in the cytoplasmic part.

In the candidate CorA protein, three transmembrane helices were predicted in *G. hirsutum*, and two transmembrane helices were predicted in *T. cacao* and *C. capsularis* (Figure 5).

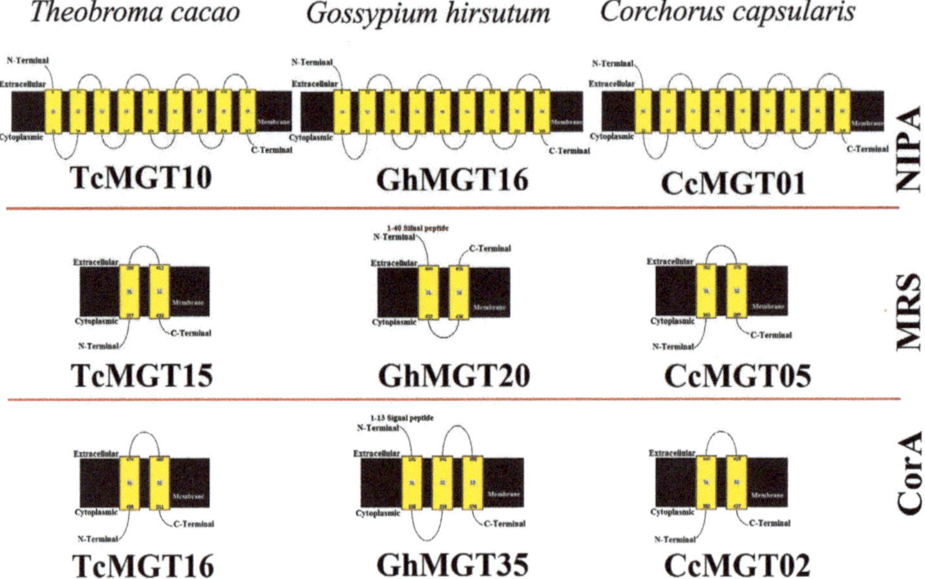

Figure 5. Predicted transmembrane helices in sub-groups of MGTs, namely NIPA, MSR2, and CorA, in *Theobroma cacao*, *Gossypium hirsutum*, and *Corchorus capsularis*.

Moreover, pocket sites of MGT proteins related to the active binding site were predicted in structures of candidate proteins. The results illustrated that sub-groups of MGT proteins are different based on the residues present in predicted pocket sites (Figure 6). Phenylalanine (PHE) amino acid was frequently observed in binding sites of NIPA proteins from *T. cacao* and *G. hirsutum*, while in *C. capsularis* isoleucine (ILE) and lysine (LYS) were more often observed in pocket sites. In candidate MRS2 proteins, proline (PRO), PHE, glutamine (GLN), asparagine (ASN), glutamic acid (GLU), and glycine (GLY) were frequently predicted in *T. cacao* as binding sites, while in *G. hirsutum* tyrosine (TYR), leucine (LEU), GLU, GLY, and GLN were more repeated in pocket sites. In addition, LEU residue was highly observed as a key binding site in candidate CcMGT of *C. capsularis*. In candidate CorA proteins, GLU amino acid was more often predicted in pocket sites of *T. cacao* and *G. hirsutum*, while PHE was frequently observed in pocket sites of candidate CorA protein in *C. capsularis*.

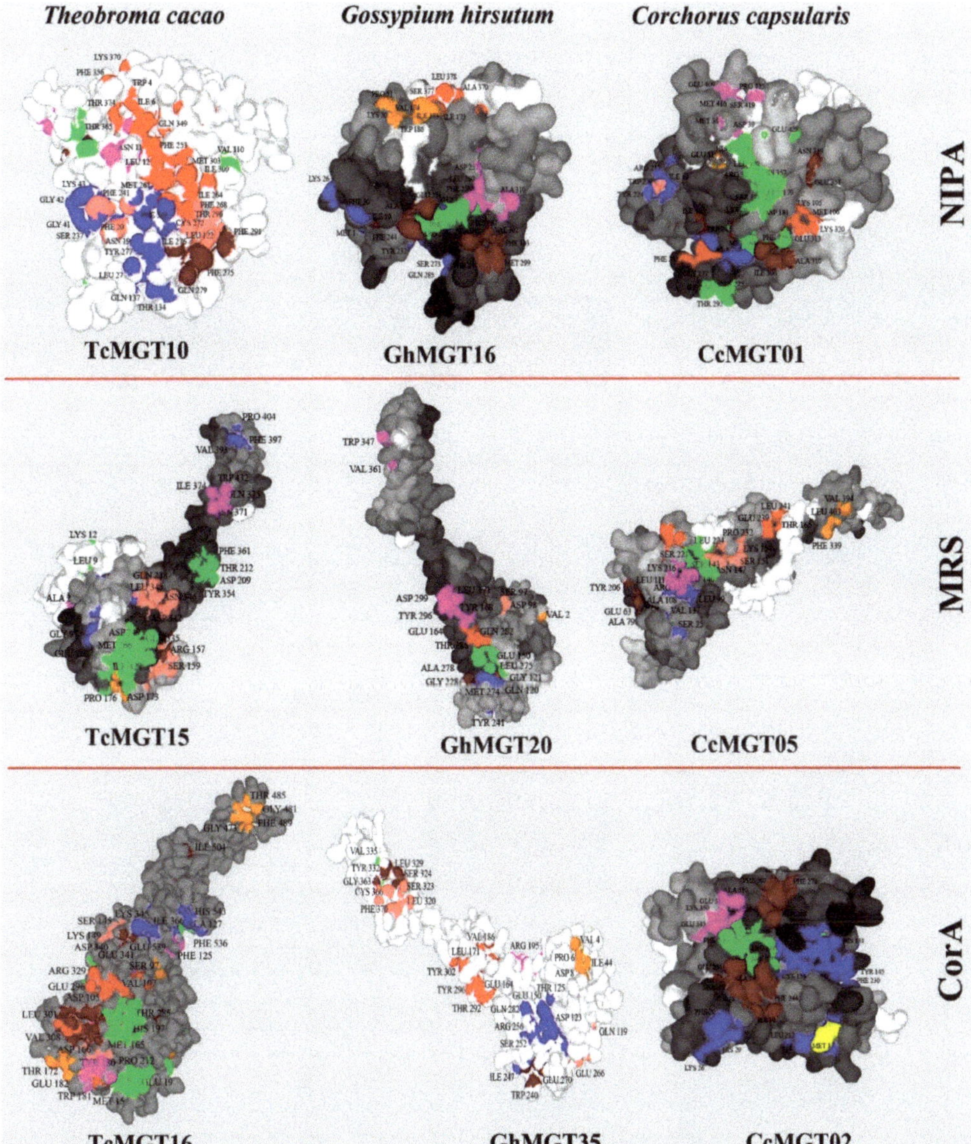

Figure 6. Docking analysis of candidates of sub-groups of MGTs, namely NIPA, MSR2, and CorA, in *Theobroma cacao*, *Gossypium hirsutum*, and *Corchorus capsularis*.

3.5. Distribution of Cis-Regulatory Elements in Promoter Region of MGT Genes

The promoter regions of *MGTs* in three plant species, comprising *T. cacao*, *G. hirsutum*, and *C. capsularis*, were analyzed and compared based on type and frequency of cis-regulatory elements. All recognized elements were classified into five groups: hormone-responsive elements, stress-responsive elements, light-responsive elements, growth-responsive elements, and binding sites of transcription factors. Our results revealed that *MGT'* promoters contain cis-regulatory elements related to stress response (Figure 7 and Table S4). In addition, all stress-responsive elements were grouped in six classes related to drought, wounding, anaerobic, low temperature, biotic stress, and general stresses. However, cis-regulatory elements responsive to low temperature, anaerobic stresses, and biotic stresses were observed most often in the promoter region of *TcMGT* genes (Figure 7b). In addition, regulatory elements related to response to abscisic acid (ABA), salicylic acid (SA), auxin, gibberellin (GA), and methyl jasmonate (MeJA) were observed among hormone-responsive elements (Figure 7b,f,i). We found that *MGT* genes might be more frequently induced in response to ABA hormone.

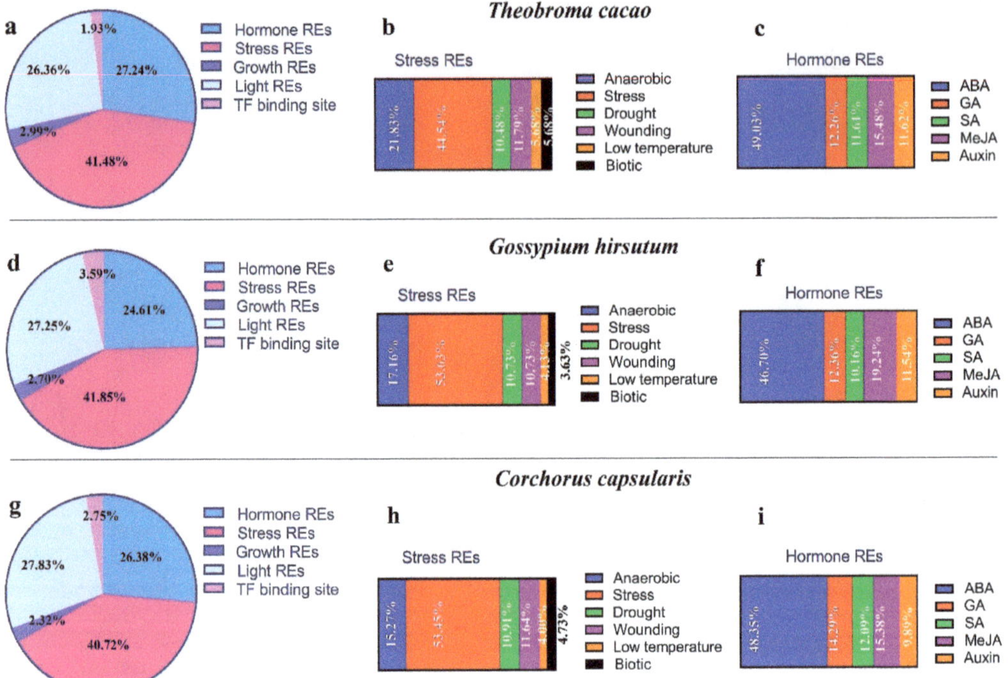

Figure 7. Proportion of cis-regulatory elements in promoter regions of *MGT* genes. Identified cis-regulatory elements were classified as hormone-responsive elements (REs), stress REs, growth REs, light REs, and transcription factor (TF) binding site in *Theobroma cacao* (**a**), *Gossypium hirsutum* (**d**), and *Corchorus capsularis* (**g**). Proportion of cis-regulatory elements related to stress responsiveness in *Theobroma cacao* (**b**), *Gossypium hirsutum* (**e**), and *Corchorus capsularis* (**h**). Proportion of different groups of hormone-related cis-regulatory elements in *Theobroma cacao* (**c**), *Gossypium hirsutum* (**f**), and *Corchorus capsularis* (**i**).

3.6. Expression Profile of TcMGT Genes

The expression levels of *TcMGTs* were also provided in response to *P. megakarya* after 0 h, 6 h, 24 h, and 72 h after infection using available RNA-seq data of two contrasting genotypes: *T. cacao* Nanay (fungal-susceptible cultivar) and Scavina (fungal-resistant cultivar) (Figure 8a,b). According to expression heatmaps, most *TcMGT* genes were less induced by a fungal infection, *P. megakarya*. A *NIPA* gene, *TcMGT02*, showed an upregulation after 48 h of fungal infection in both cultivars (Figure 8). In addition, two *MRS2* genes, *TcMGT18* and *TcMGT12*, were more expressed after 72 h in the fungal-resistant cultivar (Figure 8b).

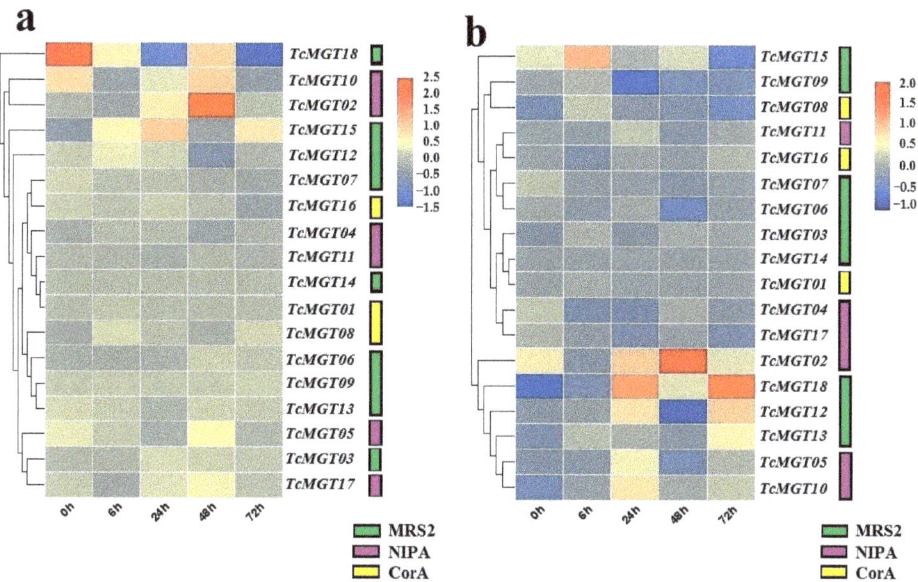

Figure 8. Expression profile of *TcMGT* genes in response *to P. megakarya* inoculation after 0 h, 6 h, 24 h, and 72 h. Nanay (NA-32) is the susceptible cultivar (**a**) and Scavina (SCA6) is the tolerant cultivar (**b**). The type of TcMGT protein sub-group is highlighted using different colors.

3.7. Expression Profile of GhMGT Genes

The expression profile of *GhMGT* genes was investigated in different tissues of *G. hirsutum* and under abiotic stresses, including cold, heat, drought, and salinity (Figure 9). The results illustrated that *GhMGTs* are involved in early and late responses to abiotic stresses (Figure 9a). For instance, a *MRS2* gene, *GhMGT32*, showed an upregulation in response to the temperature stresses of cold and heat after one hour (Figure 9a). Furthermore, *GhMGT24*, as a *NIPA* gene, was more upregulated in response to all studied abiotic stresses after 6 and 12 h. In addition, the expression profile of *GhMGT* genes showed that six *MRS2* genes, i.e., *GhMGT12*, *GhMGT33*, *GhMGT06*, *GhMGT41*, *GhMGT13*, and *GhMGT20*, are more expressed after 24 h of drought and salt stress. In the first hours of heat stress, *NIPA* genes are more expressed than *MRS2* genes in cotton. In addition, *CorA*- like genes of *GhMGT35* and *GhMGT14* showed more upregulation in response to drought and salt stress. Expression levels of *GhMGTs* were also evaluated in different tissues and organs of *G. hirsutum* (Figure 9b). The results show that *GhMGTs* were expressed in different organs for the proper distribution of magnesium throughout the cotton plant. In root tissues, *NIPA* genes are more expressed, while in leaf and torus tissues, *MRS2* genes are more expressed (Figure 9b). Furthermore, two *CorA*-like genes showed high expression in filament tissues. In addition, most *GhMGTs* are expressed in ovule tissues in 10 days post-anthesis (DPA) (Figure 9b).

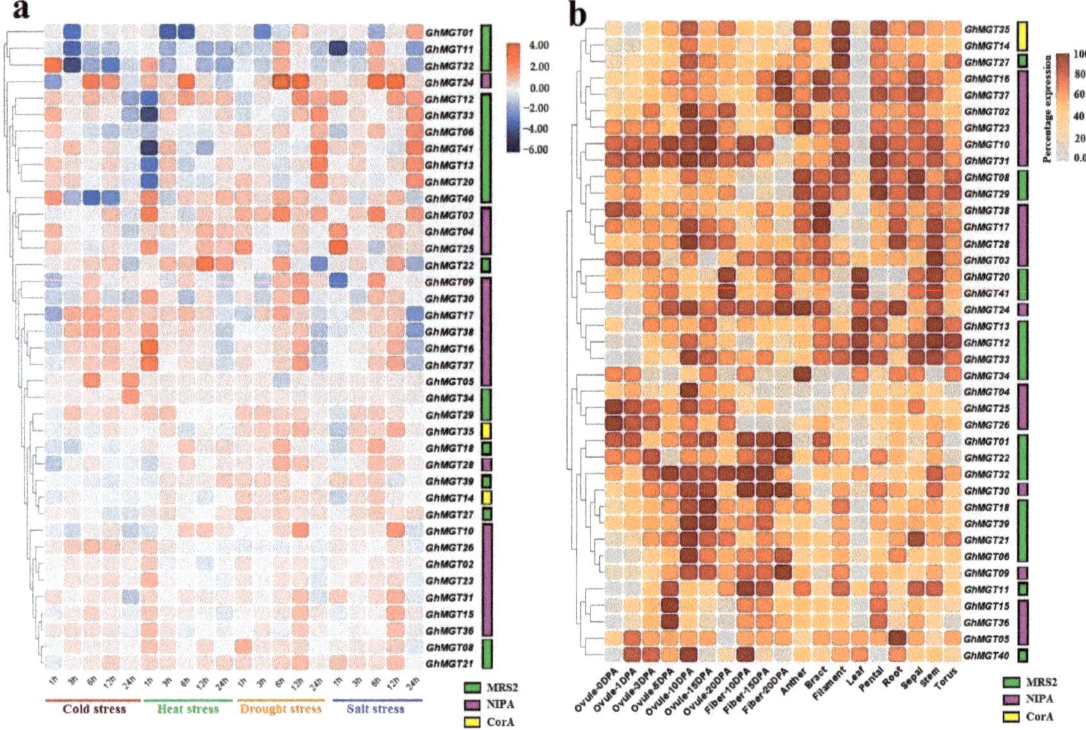

Figure 9. Expression profile of *GhMGT* genes in response to abiotic stresses, including cold, heat, drought, and salt stress (**a**), and in different tissues of cotton (**b**). The type of GhMGT protein sub-group is highlighted using different colors.

3.8. Co-Expression Network of MGT Genes

A co-expression network of *TcMGTs* was formed using their orthologs in the diploid model plant, *Arabidopsis* (Figure 10). All nodes in the co-expression network of *MGTs* were classified into four groups (I to IV) (Figure 10). In group I, three auxin-responsive genes, namely *SAUR4*, *7*, and *12*, two genes affecting lipid metabolism, *PLC* genes, two genes involving in disease resistance-responsive, namely *AT4G38700* and *AT2G21110*, and *CYP21-1* affecting protein folding showed high co-expression scores with *MGT* genes. In group II, a protein phosphatase 2C, *AT2G20050*, was found with high co-expression with three *MGT* genes. In group III, *MGT* genes showed high co-expression with *AT5G42070*, a hypothetical protein, *RPL15* (ribosomal protein), *AT2G23390* (acyl-CoA protein), and *CH1*. In group IV, high co-expression connections were observed between *MGT* genes and genes involving in cell wall organization/modification processes such as *S2LB*, *AGM1*, *AGM2*, *EXPA13*, *GXM2*, *GXM1*, and *QRT1*. In addition, *LrgB* involved in response to water deprivation and *MHX* encoding a mg/proton exchanger showed string co-expression with *MGTs*. Gene ontology (GO) analysis illustrated that the biological processes, including mg ion transporter, photoprotection, and cell wall pectin biosynthetic process, xylan biosynthetic process, and chloroplast organization, were significantly enriched based on all nodes of co-expression networks of *MGTs*.

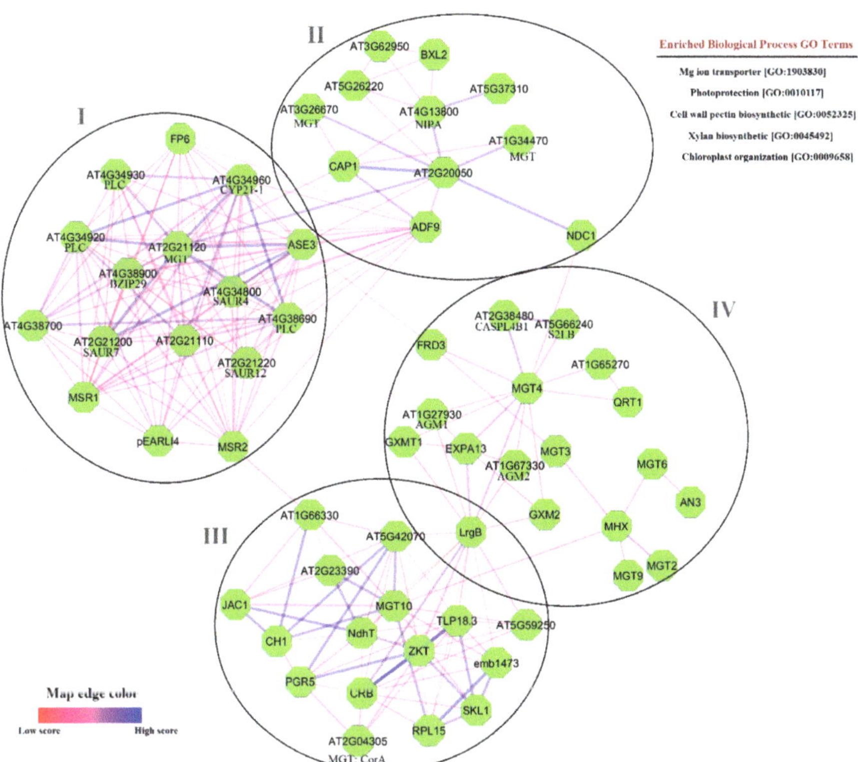

Figure 10. Co-expression networks of *MGT* genes based on the model plant *Arabidopsis thaliana*. Thickness and color intensity of each edge indicate the value of co-expression between two genes.

4. Discussion

Cocoa (*Theobroma cacao* L.) is an economical plant due to its wide use in producing chocolate, a popular commodity worldwide. Scientific research is underway to better understand the genome and metabolomes of this plant. In addition, several gene families, including *GASA* [40], *sucrose transporter* [65], *WRKY* [66], *NAC* [67], *desaturase* [68], and *sucrose synthase* [69], have been investigated and characterized in cacao. Due to the important role of magnesium ions in regulating plant growth and development, *magnesium transporters* (MGTs) have been investigated in cacao, upland cotton (*Gossypium hirsutum*), and white jute (*Corchorus capsularis*). We characterized 18 putative non-redundant *MGT* genes in cacao (*TcMGTs*) along with 41 *MGTs* in *G. hirsutum* (*GhMGTs*), and 16 *MGTs* from the genome of *C. capsularis* (*CcMGTs*). In previous studies, 62 *MGTs* in *Camelina sativa* [12], 41 *MGTs* in *Triticum turgidum* [12], 12 *MGTs* in *Zea mays* [70], 16 *MGTs* in *Pyrus bretschneideri* [23], 36 *MGTs* in *Brassica napus* [25], 12 *MGTs* in *Fagaria vesca* [27], and 8 *MGTs* in *Poncirus trifoliata* [24] were characterized. The number of *MGTs* is probably correlated with polyploidy events and genome size [12,71]. The prediction of the pI value of MGT proteins illustrated that CcMGTs are more acidophilic proteins than TcMGTs and GhMGTs, indicating that CcMGTs are mostly active under acidic conditions (pI < 6.50). This can be related to the optimal growing environment of white jute.

According to the phylogenetic analysis, MGT family members from *T. cacao*, *G. hirsutum*, and *C. capsularis* can be classified into three groups, and CcMGTs showed close relationships to TcMGTs. However, the MGT family proteins of *Arabidopsis* and rice were divided into five clusters based on phylogenetic analysis [11,72]. In the current study, MRS2 sub-group proteins showed more diversity than the other two sub-groups, indicating that *NIPA* and *CorA* genes may be derived from *MRS2* genes during the process of evolution. In addition, more *MGT* genes were identified in cotton, which is due to polyploidy in *G. hirsutum*. The analysis of gene structure showed that *TcMGTs* and *GhMGTs* contained 4 to 15 exons, while *CcMGTs* contained 4 to 21 exons. This suggests that under evolution events more insertions and deletions of introns occurred in *MGT* genes, especially in *CcMGTs*. In addition, the first duplication event was estimated to occur approximately 103 MYA between *CcMGT05*, as a *MRS2* gene, and *CcMGT04*, as a *NIPA* gene.

Several segmental/tandem duplication events were estimated to occur between *TcMGTs*. Notably, a gene cluster of three MRS2/*TcMGT* genes, namely *TcMGT013*, *TcMGT14*, and *TcMGT15*, was observed in chromosome 6, providing an avenue for further molecular investigations in cacao. In addition, the K_a/K_s ratios in most duplicated genes were less than one, suggesting that *MGTs* have evolved slowly under a purifying selection [54,71]. The comparisons between structures of MGT proteins revealed that NIPA proteins include the conserved structures with more transmembrane regions. More transmembrane regions could indicate a more important role for the NIPA protein group in transport of magnesium within the plant cell [12]. However, the transmembrane structure of MSR2 and CorA proteins in *T. cacao* and *C. capsularis* is highly conserved, while these proteins possess a different transmembrane structure in *G. hirsutum*, which may affect the ability of the magnesium transmembrane transport process [73,74]. Moreover, sub-groups of MGTs in the three studied plants showed diversity in the predicted pocket sites in the 3D structure. Overall, our findings suggest that PHE, GLU, LEU, GLY, and ILE, as the key binding sites, are associated with the function and interaction of MGTs in response to environmental stimuli and changes in ion/Mg concentration [54,75,76].

Magnesium transporters as well as other ion transporters are not only involved in response to Mg concentrations, but also their activity can be affected by changes in environmental conditions [10,12]. Analysis of gene promoter regions is one of the strategies to predict the response of target genes to various environmental factors [77,78]. *MGT* genes have a high potential to respond to stresses, both biotic and abiotic, as well as ABA based on the distribution of corresponding cis-regulatory elements in the promoter region. However, investigations that are more molecular are needed to confirm their functions. The black rod disease caused by the genus *Phytophthora* is a production limiting factor, reducing cacao production by around 20–25% [79]. On the other hand, expression studies of genes in beans of bulk cultivars, like disease resistance and fine flavor cocoa (disease susceptibility), may be of interest [80,81]. In the current study, the expression level of *TcMGT* genes was investigated using RNA sequencing data responses to *P. megakarya* in two contrasting cultivars of *T. cacao*, susceptible and fungal-resistant cultivars. Two TcMGTs, i.e., *TcMGT12* and *TcMGT18*, were identified as *P. megakarya* responsiveness genes by their upregulation specifically in the cacao-tolerant cultivar. These may be a good target for further molecular studies related to introducing the new cocao-resistant cultivars with high-quality, delicious chocolate. Furthermore, the expression profile of *GhMGTs* suggested that *MGT* genes are involved in response to abiotic stresses such as temperature stresses (cold and heat stress), drought, and salinity stress. Nevertheless, *GhMGTs* can be expressed in different plant tissues to regulate Mg homeostasis. Previous studies have indicated that *MGT* genes are associated with maintaining ion homeostasis in plant tissues during adverse conditions [20,21].

Mg transporter genes via affecting Na^+ transporters and K^+ transporters (HKTs) can improve salinity tolerance in plant species [82]. In addition, MGTs have been speculated to regulate the downstream pathways related to response to abiotic stresses by interacting with Ca^{2+} sensors [83]. Our findings revealed that the *MGT* duplicated gene pair could have diverse expression patterns, suggesting that these genes probably under some modifications or insertion/deletion in their sequence, CDS, or promoter regions have received novel functions [71,84]. Modifications, such as gains and losses of cis elements in promoters between duplicated gene pairs, e.g., parent and daughter genes, could occur after duplication events, affecting the expression levels [85,86]. By constructing a network of co-expression genes, it is possible to identify other molecular pathways in which target genes are involved to gain a better understanding of the function of genes [54,87]. In this study, *MGT* genes showed diverse co-expressions with genes involved in auxin-responsive processes, lipid metabolism, cell wall organization, photoprotection, and chloroplast organization. Magnesium is a critical element of chlorophyll, affecting photosynthesis rate and biomass production [88,89]. Overall, *MGTs* appear to be involved in various pathways to control plant growth and development and response to adverse conditions.

5. Conclusions

A genome-wide analysis of *MGT* family genes was performed in the genomes of *Theobroma cacao, Corchorus capsularis, and Gossypium hirsutum*. Our findings provide insight into certain aspects of the sequence structure, evolutionary events, regulatory systems, and function of *MGT* genes in three species of Malvaceae. Furthermore, our results show that MGTs are involved in diverse cellular pathways, and they can interact with proteins associated with growth and development as well as with response to environmental stimuli. Further functional-molecular analyses are required to improve our understanding of the role of MGTs in cacao resistance to stress and to increase the quality of chocolate.

Supplementary Materials: The following are available online at https://www.mdpi.com/article/10.3390/agronomy11081651/s1, Table S1. Details of each sequence of MGTs in *Theobroma cacao, Corchorus capsularis*, and *Gossypium hirsutum* analyzed in our study, including protein sequences, coding sequences, genomic sequences, and promoter regions. Table S2. A list and properties of the studied *MGT* genes in *Theobroma cacao, Corchorus capsularis*, and *Gossypium hirsutum*. Table S3. Predicted duplicated gene pairs in the MGT protein family in *Theobroma cacao, Corchorus capsularis*, and *Gossypium hirsutum*. Table S4. Cis-regulatory elements in promoter regions of the *MGT* gene family in *Theobroma cacao, Corchorus capsularis*, and *Gossypium hirsutum*. Figure S1. 3D structures of MGT proteins from various clades, namely NIPA, MRS2, and CorA, in *Theobroma cacao, Corchorus capsularis*, and *Gossypium hirsutum*.

Author Contributions: Conceptualization, P.H. and A.; methodology, P.H., A. and S.F.; formal analysis, P.H., A. and S.F.; investigation, P.H., A. and P.P.; writing—original draft preparation, P.H. and A.; writing—review and editing, P.H., A. and P.P.; funding acquisition, P.P. All authors have read and agreed to the published version of the manuscript.

Funding: This research received no external funding.

Institutional Review Board Statement: Not applicable.

Informed Consent Statement: Not applicable.

Data Availability Statement: A public data set is analyzed in the manuscript. All analyzed data are available in the main manuscript or as Supplementary Material.

Acknowledgments: Open access funding provided by University of Helsinki.

Conflicts of Interest: The authors have no conflict of interest to declare.

References

1. Ceylan, Y.; Kutman, U.B.; Mengutay, M.; Cakmak, I. Magnesium applications to growth medium and foliage affect the starch distribution, increase the grain size and improve the seed germination in wheat. *Plant Soil* **2016**, *406*, 145–156. [CrossRef]
2. Chen, Z.C.; Peng, W.T.; Li, J.; Liao, H. Functional dissection and transport mechanism of magnesium in plants. In *Proceedings of the Seminars in Cell & Developmental Biology*; Elsevier: Amsterdam, The Netherlands, 2018; Volume 74, pp. 142–152.
3. Hermans, C.; Conn, S.J.; Chen, J.; Xiao, Q.; Verbruggen, N. An update on magnesium homeostasis mechanisms in plants. *Metallomics* **2013**, *5*, 1170–1183. [CrossRef]
4. Guo, W.; Nazim, H.; Liang, Z.; Yang, D. Magnesium deficiency in plants: An urgent problem. *Crop J.* **2016**, *4*, 83–91. [CrossRef]
5. Marschner, H. *Marschner's Mineral Nutrition of Higher Plants*; Academic Press: New York, NY, USA, 2011; ISBN 0123849063.
6. Hermans, C.; Bourgis, F.; Faucher, M.; Strasser, R.J.; Delrot, S.; Verbruggen, N. Magnesium deficiency in sugar beets alters sugar partitioning and phloem loading in young mature leaves. *Planta* **2005**, *220*, 541–549. [CrossRef]

7. Tang, N.; Li, Y.; Chen, L. Magnesium deficiency–induced impairment of photosynthesis in leaves of fruiting Citrus reticulata trees accompanied by up-regulation of antioxidant metabolism to avoid photo-oxidative damage. *J. Plant Nutr. Soil Sci.* **2012**, *175*, 784–793. [CrossRef]

8. Peng, H.-Y.; Qi, Y.-P.; Lee, J.; Yang, L.-T.; Guo, P.; Jiang, H.-X.; Chen, L.-S. Proteomic analysis of Citrus sinensis roots and leaves in response to long-term magnesium-deficiency. *BMC Genom.* **2015**, *16*, 253. [CrossRef]

9. Li, H.; Du, H.; Huang, K.; Chen, X.; Liu, T.; Gao, S.; Liu, H.; Tang, Q.; Rong, T.; Zhang, S. Identification, and functional and expression analyses of the CorA/MRS2/MGT-type *magnesium transporter* family in maize. *Plant Cell Physiol.* **2016**, *57*, 1153–1168. [CrossRef]

10. Li, H.; Wang, N.; Ding, J.; Liu, C.; Du, H.; Huang, K.; Cao, M.; Lu, Y.; Gao, S.; Zhang, S. The maize CorA/MRS2/MGT-type mg transporter, ZmMGT10, responses to magnesium deficiency and confers low magnesium tolerance in transgenic Arabidopsis. *Plant Mol. Biol.* **2017**, *95*, 269–278. [CrossRef]

11. Gebert, M.; Meschenmoser, K.; Svidová, S.; Weghuber, J.; Schweyen, R.; Eifler, K.; Lenz, H.; Weyand, K.; Knoop, V. A root-expressed *magnesium transporter* of the MRS2/MGT gene family in Arabidopsis thaliana allows for growth in low-Mg2+ environments. *Plant Cell* **2009**, *21*, 4018–4030. [CrossRef] [PubMed]

12. Faraji, S.; Ahmadizadeh, M.; Heidari, P. Genome-wide comparative analysis of Mg transporter gene family between Triticum turgidum and Camelina sativa. *BioMetals* **2021**, *4*, 639–660. [CrossRef]

13. Mao, D.; Chen, J.; Tian, L.; Liu, Z.; Yang, L.; Tang, R.; Li, J.; Lu, C.; Yang, Y.; Shi, J. Arabidopsis transporter MGT6 mediates magnesium uptake and is required for growth under magnesium limitation. *Plant Cell* **2014**, *26*, 2234–2248. [CrossRef]

14. Chen, Z.C.; Yamaji, N.; Motoyama, R.; Nagamura, Y.; Ma, J.F. Up-regulation of a *magnesium transporter* gene OsMGT1 is required for conferring aluminum tolerance in rice. *Plant Physiol.* **2012**, *159*, 1624–1633. [CrossRef]

15. Drummond, R.S.M.; Tutone, A.; Li, Y.-C.; Gardner, R.C. A putative *magnesium transporter* AtMRS2-11 is localized to the plant chloroplast envelope membrane system. *Plant Sci.* **2006**, *170*, 78–89. [CrossRef]

16. Lenz, H.; Dombinov, V.; Dreistein, J.; Reinhard, M.R.; Gebert, M.; Knoop, V. Magnesium deficiency phenotypes upon multiple knockout of Arabidopsis thaliana MRS2 clade B genes can be ameliorated by concomitantly reduced calcium supply. *Plant Cell Physiol.* **2013**, *54*, 1118–1131. [CrossRef]

17. Chen, J.; Li, L.; Liu, Z.; Yuan, Y.; Guo, L.; Mao, D.; Tian, L.; Chen, L.; Luan, S.; Li, D. *Magnesium transporter* AtMGT9 is essential for pollen development in Arabidopsis. *Cell Res.* **2009**, *19*, 887–898. [CrossRef] [PubMed]

18. Li, J.; Huang, Y.; Tan, H.; Yang, X.; Tian, L.; Luan, S.; Chen, L.; Li, D. An endoplasmic reticulum *magnesium transporter* is essential for pollen development in Arabidopsis. *Plant Sci.* **2015**, *231*, 212–220. [CrossRef]

19. Bose, J.; Babourina, O.; Shabala, S.; Rengel, Z. Low-pH and aluminum resistance in Arabidopsis correlates with high cytosolic magnesium content and increased magnesium uptake by plant roots. *Plant Cell Physiol.* **2013**, *54*, 1093–1104. [CrossRef]

20. Deng, W.; Luo, K.; Li, D.; Zheng, X.; Wei, X.; Smith, W.; Thammina, C.; Lu, L.; Li, Y.; Pei, Y. Overexpression of an Arabidopsis magnesium transport gene, AtMGT1, in Nicotiana benthamiana confers Al tolerance. *J. Exp. Bot.* **2006**, *57*, 4235–4243. [CrossRef]

21. Chen, Z.C.; Yamaji, N.; Horie, T.; Che, J.; Li, J.; An, G.; Ma, J.F. A *magnesium transporter* OsMGT1 plays a critical role in salt tolerance in rice. *Plant Physiol.* **2017**, *174*, 1837–1849. [CrossRef]

22. Schock, I.; Gregan, J.; Steinhauser, S.; Schweyen, R.; Brennicke, A.; Knoop, V. A member of a novel Arabidopsis thaliana gene family of candidate Mg^{2+} ion transporters complements a yeast mitochondrial group II intron-splicing mutant. *Plant J.* **2000**, *24*, 489–501. [CrossRef]

23. Zhao, Z.; Wang, P.; Jiao, H.; Tang, C.; Liu, X.; Jing, Y.; Zhang, S.; Wu, J. Phylogenetic and expression analysis of the *magnesium transporter* family in pear, and functional verification of PbrMGT7 in pear pollen. *J. Hortic. Sci. Biotechnol.* **2018**, *93*, 51–63. [CrossRef]

24. Liu, X.; Guo, L.-X.; Luo, L.-J.; Liu, Y.-Z.; Peng, S.-A. Identification of the magnesium transport (MGT) family in Poncirus trifoliata and functional characterization of PtrMGT5 in magnesium deficiency stress. *Plant Mol. Biol.* **2019**, *101*, 551–560. [CrossRef] [PubMed]

25. Zhang, L.; Wen, A.; Wu, X.; Pan, X.; Wu, N.; Chen, X.; Chen, Y.; Mao, D.; Chen, L.; Luan, S. Molecular identification of the magnesium transport gene family in Brassica napus. *Plant Physiol. Biochem.* **2019**, *136*, 204–214. [CrossRef]

26. Wang, Y.; Hua, X.; Xu, J.; Chen, Z.; Fan, T.; Zeng, Z.; Wang, H.; Hour, A.-L.; Yu, Q.; Ming, R. Comparative genomics revealed the gene evolution and functional divergence of *magnesium transporter* families in Saccharum. *BMC Genom.* **2019**, *20*, 83. [CrossRef]

27. Regon, P.; Chowra, U.; Awasthi, J.P.; Borgohain, P.; Panda, S.K. Genome-wide analysis of *magnesium transporter* genes in Solanum lycopersicum. *Comput. Biol. Chem.* **2019**, *80*, 498–511. [CrossRef]

28. Christenhusz, M.J.M.; Byng, J.W. The number of known plants species in the world and its annual increase. *Phytotaxa* **2016**, *261*, 201–217. [CrossRef]

29. Xu, Z.; Deng, M. Malvaceae. In *Identification and Control of Common Weeds: Volume 1*; Springer: Dordrecht, The Netherlands, 2017; pp. 717–735. ISBN 9789402411577.

30. Bayer, C.; Kubitzki, K. Malvaceae. In *Flowering Plants Dicotyledons*; Springer: Berlin/Heidelberg, Germany, 2003; pp. 225–311.

31. Cinquanta, L.; Cesare, C.D.; Manoni, R.; Piano, A.; Salvatori, G.; Cinquanta, L.; Cesare, C.D.; Manoni, R.; Piano, A. Mineral essential elements for nutrition in different chocolate products. *Int. J. Food Sci. Nutr.* **2016**, *7486*, 773–778. [CrossRef]

32. Ahmad, S.; Hasanuzzaman, M. *Cotton Production and Uses: Agronomy, Crop Protection, and Postharvest Technologies*; Springer: Singapore, 2020; ISBN 9789811514722.

33. Gheesling, R.H.; Perkins, H.F. Critical Levels of Manganese and Magnesium in Cotton at Different Stages of Growth1. *Agron. J.* **1970**, *62*, 29–32. [CrossRef]
34. Uzilday, R.Ö.; Uzilday, B.; Yalçinkaya, T.; Türkan, İ. Mg deficiency changes the isoenzyme pattern of reactive oxygen species-relatedenzymes and regulates NADPH-oxidase-mediated ROS signaling in cotton. *Turk. J. Biol.* **2017**, *41*, 868–880. [CrossRef]
35. Rahman, S.; Kazal, M.M.H.; Begum, I.A.; Alam, M.J. Exploring the Future Potential of Jute in Bangladesh. *Agriculture* **2017**, *7*, 96. [CrossRef]
36. Argout, X.; Martin, G.; Droc, G.; Fouet, O.; Labadie, K.; Rivals, E.; Aury, J.M.; Lanaud, C. The cacao Criollo genome v2.0: An improved version of the genome for genetic and functional genomic studies. *BMC Genom.* **2017**, *18*, 1–9. [CrossRef] [PubMed]
37. Islam, M.S.; Saito, J.A.; Emdad, E.M.; Ahmed, B.; Islam, M.M.; Halim, A.; Hossen, Q.M.M.; Hossain, M.Z.; Ahmed, R.; Hossain, M.S.; et al. Comparative genomics of two jute species and insight into fibre biogenesis. *Nat. Plants* **2017**, *3*, 16223. [CrossRef] [PubMed]
38. Bolser, D.M.; Staines, D.M.; Perry, E.; Kersey, P.J. Ensembl plants: Integrating tools for visualizing, mining, and analyzing plant genomic data. In *Plant Genomics Databases*; Springer: Berlin/Heidelberg, Germany, 2017; pp. 1–31.
39. Yu, J.; Jung, S.; Cheng, C.-H.; Ficklin, S.P.; Lee, T.; Zheng, P.; Jones, D.; Percy, R.G.; Main, D. CottonGen: A genomics, genetics and breeding database for cotton research. *Nucleic Acids Res.* **2014**, *42*, D1229–D1236. [CrossRef]
40. Abdullah; Faraji, S.; Mehmood, F.; Malik, H.M.T.; Ahmed, I.; Heidari, P.; Poczai, P. The GASA Gene Family in Cacao (Theobroma cacao, Malvaceae): Genome Wide Identification and Expression Analysis. *Agronomy* **2021**, *11*, 1425. [CrossRef]
41. Gasteiger, E.; Hoogland, C.; Gattiker, A.; Duvaud, S.; Wilkins, M.R.; Appel, R.D.; Bairoch, A. Protein Identification and Analysis Tools on the ExPASy Server. In *The Proteomics Protocols Handbook*; Humana Press: Totowa, NJ, USA, 2005; pp. 571–607.
42. Krogh, A.; Larsson, B.; Von Heijne, G.; Sonnhammer, E.L.L. Predicting transmembrane protein topology with a hidden Markov model: Application to complete genomes. *J. Mol. Biol.* **2001**, *305*, 567–580. [CrossRef]
43. Yu, C.; Chen, Y.; Lu, C.; Hwang, J. Prediction of protein subcellular localization. *Proteins Struct. Funct. Bioinforma* **2006**, *64*, 643–651. [CrossRef]
44. Lescot, M.; Déhais, P.; Thijs, G.; Marchal, K.; Moreau, Y.; Van De Peer, Y.; Rouzé, P.; Rombauts, S. PlantCARE, a database of plant cis-acting regulatory elements and a portal to tools for in silico analysis of promoter sequences. *Nucleic Acids Res.* **2002**, *30*, 325–327. [CrossRef]
45. Kearse, M.; Moir, R.; Wilson, A.; Stones-Havas, S.; Cheung, M.; Sturrock, S.; Buxton, S.; Cooper, A.; Markowitz, S.; Duran, C.; et al. Geneious Basic: An integrated and extendable desktop software platform for the organization and analysis of sequence data. *Bioinformatics* **2012**, *28*, 1647–1649. [CrossRef]
46. Larkin, M.A.; Blackshields, G.; Brown, N.P.; Chenna, R.; McGettigan, P.A.; McWilliam, H.; Valentin, F.; Wallace, I.M.; Wilm, A.; Lopez, R. Clustal W and Clustal X version 2.0. *Bioinformatics* **2007**, *23*, 2947–2948. [CrossRef] [PubMed]
47. Nguyen, L.-T.; Schmidt, H.A.; von Haeseler, A.; Minh, B.Q. IQ-TREE: A fast and effective stochastic algorithm for estimating Maximum-likelihood phylogenies. *Mol. Biol. Evol.* **2015**, *32*, 268–274. [CrossRef]
48. Hoang, D.T.; Chernomor, O.; von Haeseler, A.; Minh, B.Q.; Vinh, L.S. UFBoot2: Improving the ultrafast bootstrap approximation. *Mol. Biol. Evol.* **2018**, *35*, 518–522. [CrossRef] [PubMed]
49. Kalyaanamoorthy, S.; Minh, B.Q.; Wong, T.K.F.; von Haeseler, A.; Jermiin, L.S. ModelFinder: Fast model selection for accurate phylogenetic estimates. *Nat. Methods* **2017**, *14*, 587–589. [CrossRef] [PubMed]
50. Darriba, D.; Taboada, G.L.; Doallo, R.; Posada, D. jModelTest 2: More models, new heuristics and parallel computing. *Nat. Methods* **2012**, *9*, 772. [CrossRef]
51. Letunic, I.; Bork, P. Interactive Tree of Life (iTOL) v4: Recent updates and new developments. *Nucleic Acids Res.* **2019**, *47*, W256–W259. [CrossRef] [PubMed]
52. Bailey, T.L.; Boden, M.; Buske, F.A.; Frith, M.; Grant, C.E.; Clementi, L.; Ren, J.; Li, W.W.; Noble, W.S. MEME Suite: Tools for motif discovery and searching. *Nucleic Acids Res.* **2009**, *37*, W202–W208. [CrossRef] [PubMed]
53. Hu, B.; Jin, J.; Guo, A.-Y.; Zhang, H.; Luo, J.; Gao, G. GSDS 2.0: An upgraded gene feature visualization server. *Bioinformatics* **2015**, *31*, 1296–1297. [CrossRef]
54. Musavizadeh, Z.; Najafi-Zarrini, H.; Kazemitabar, S.K.; Hashemi, S.H.; Faraji, S.; Barcaccia, G.; Heidari, P. Genome-Wide Analysis of Potassium Channel Genes in Rice: Expression of the OsAKT and OsKAT Genes under Salt Stress. *Genes* **2021**, *12*, 784. [CrossRef]
55. Chen, C.; Chen, H.; Zhang, Y.; Thomas, H.R.; Frank, M.H.; He, Y.; Xia, R. TBtools: An Integrative Toolkit Developed for Interactive Analyses of Big Biological Data. *Mol. Plant* **2020**, *13*, 1194–1202. [CrossRef]
56. Rozas, J.; Ferrer-Mata, A.; Sánchez-DelBarrio, J.C.; Guirao-Rico, S.; Librado, P.; Ramos-Onsins, S.E.; Sánchez-Gracia, A. DnaSP 6: DNA sequence polymorphism analysis of large data sets. *Mol. Biol. Evol.* **2017**, *34*, 3299–3302. [CrossRef] [PubMed]
57. Yang, S.; Zhang, X.; Yue, J.-X.; Tian, D.; Chen, J.-Q. Recent duplications dominate NBS-encoding gene expansion in two woody species. *Mol. Genet. Genom.* **2008**, *280*, 187–198. [CrossRef]
58. Krzywinski, M.; Schein, J.; Birol, I.; Connors, J.; Gascoyne, R.; Horsman, D.; Jones, S.J.; Marra, M.A. Circos: An information aesthetic for comparative genomics. *Genome Res.* **2009**, *19*, 1639–1645. [CrossRef]
59. Kelley, L.A.; Mezulis, S.; Yates, C.M.; Wass, M.N.; Sternberg, M.J.E. The Phyre2 web portal for protein modeling, prediction and analysis. *Nat. Protoc.* **2015**, *10*, 845–858. [CrossRef]
60. Jiménez, J.; Doerr, S.; Martínez-Rosell, G.; Rose, A.S.; De Fabritiis, G. DeepSite: Protein-binding site predictor using 3D-convolutional neural networks. *Bioinformatics* **2017**, *33*, 3036–3042. [CrossRef]

61. Tian, W.; Chen, C.; Lei, X.; Zhao, J.; Liang, J. CASTp 3.0: Computed atlas of surface topography of proteins. *Nucleic Acids Res.* **2018**, *46*, W363–W367. [CrossRef]

62. DeLano, W.L. Pymol: An open-source molecular graphics tool. *CCP4 Newsl. Protein Crystallogr.* **2002**, *40*, 82–92.

63. Pokou, D.N.; Fister, A.S.; Winters, N.; Tahi, M.; Klotioloma, C.; Sebastian, A.; Marden, J.H.; Maximova, S.N.; Guiltinan, M.J. Resistant and susceptible cacao genotypes exhibit defense gene polymorphism and unique early responses to Phytophthora megakarya inoculation. *Plant Mol. Biol.* **2019**, *99*, 499–516. [CrossRef] [PubMed]

64. Zhu, T.; Liang, C.; Meng, Z.; Sun, G.; Meng, Z.; Guo, S.; Zhang, R. CottonFGD: An integrated functional genomics database for cotton. *BMC Plant Biol.* **2017**, *17*, 1–9. [CrossRef] [PubMed]

65. Li, F.; Wu, B.; Qin, X.; Yan, L.; Hao, C.; Tan, L.; Lai, J. Molecular cloning and expression analysis of the sucrose transporter gene family from Theobroma cacao L. *Gene* **2014**, *546*, 336–341. [CrossRef]

66. Dayanne, S.M.D.A.; Oliveira Jordão Do Amaral, D.; Del-Bem, L.E.; Bronze Dos Santos, E.; Santana Silva Raner, J.; Peres Gramacho, K.; Vincentz, M.; Micheli, F. Genome-wide identification and characterization of cacao WRKY transcription factors and analysis of their expression in response to witches' broom disease. *PLoS ONE* **2017**, *12*, e0187346.

67. Shen, S.; Zhang, Q.; Shi, Y.; Sun, Z.; Zhang, Q.; Hou, S.; Wu, R.; Jiang, L.; Zhao, X.; Guo, Y. Genome-wide analysis of the NAC domain transcription factor gene family in Theobroma cacao. *Genes* **2020**, *11*, 35. [CrossRef] [PubMed]

68. Zhang, Y.; Maximova, S.N.; Guiltinan, M.J. Characterization of a stearoyl-acyl carrier protein desaturase gene family from chocolate tree, *Theobroma cacao* L. *Front. Plant Sci.* **2015**, *6*, 1–12. [CrossRef]

69. Li, F.; Hao, C.; Yan, L.; Wu, B.; Qin, X.; Lai, J.; Song, Y. Gene structure, phylogeny and expression profile of the sucrose synthase gene family in cacao (*Theobroma cacao* L.). *J. Genet.* **2015**, *94*, 461–472. [CrossRef] [PubMed]

70. Li, H.; Liu, C.; Zhou, L.; Zhao, Z.; Li, Y.; Qu, M.; Huang, K.; Zhang, L.; Lu, Y.; Cao, M. Molecular and functional characterization of the *magnesium transporter* gene ZmMGT12 in maize. *Gene* **2018**, *665*, 167–173. [CrossRef] [PubMed]

71. Heidari, P.; Faraji, S.; Ahmadizadeh, M.; Ahmar, S.; Mora-Poblete, F. New insights into structure and function of TIFY genes in Zea mays and Solanum lycopersicum: A genome-wide comprehensive analysis. *Front. Genet.* **2021**, *12*, 534. [CrossRef]

72. Saito, T.; Kobayashi, N.I.; Tanoi, K.; Iwata, N.; Suzuki, H.; Iwata, R.; Nakanishi, T.M. Expression and functional analysis of the CorA-MRS2-ALR-type *magnesium transporter* family in rice. *Plant Cell Physiol.* **2013**, *54*, 1673–1683. [CrossRef] [PubMed]

73. Fukao, Y. Protein-protein interactions in plants. *Plant Cell Physiol.* **2012**, *53*, 617–625. [CrossRef] [PubMed]

74. Braun, P.; Aubourg, S.; Van Leene, J.; De Jaeger, G.; Lurin, C. Plant protein interactomes. *Annu. Rev. Plant Biol.* **2013**, *64*, 161–187. [CrossRef]

75. Galili, G.; Höfgen, R. Metabolic engineering of amino acids and storage proteins in plants. *Metab. Eng.* **2002**, *4*, 3–11. [CrossRef]

76. Beauregard, M.; Hefford, M.A. Enhancement of essential amino acid contents in crops by genetic engineering and protein design. *Plant Biotechnol. J.* **2006**, *4*, 561–574. [CrossRef]

77. Ahmadizadeh, M.; Chen, J.-T.; Hasanzadeh, S.; Ahmar, S.; Heidari, P. Insights into the genes involved in the ethylene biosynthesis pathway in Arabidopsis thaliana and Oryza sativa. *J. Genet. Eng. Biotechnol.* **2020**, *18*, 1–20. [CrossRef] [PubMed]

78. Heidari, P.; Mazloomi, F.; Nussbaumer, T.; Barcaccia, G. Insights into the SAM synthetase gene family and its roles in tomato seedlings under abiotic stresses and hormone treatments. *Plants* **2020**, *9*, 586. [CrossRef]

79. Adeniyi, D. Diversity of Cacao Pathogens and Impact on Yield and Global Production. In *Theobroma Cacao—Deploying Science for Sustainability of Global Cocoa Economy*; IntechOpen: London, UK, 2019.

80. De Wever, J.; Tulkens, D.; Verwaeren, J.; Everaert, H.; Rottiers, H.; Dewettinck, K.; Lefever, S.; Messens, K. A combined RNA preservation and extraction protocol for gene expression studies in cacao beans. *Front. Plant Sci.* **2020**, *11*, 992. [CrossRef] [PubMed]

81. De Wever, J.; De Coninck, T.; Everaert, H.; Afoakwa, E.O.; Coppieters, F.; Rottiers, H.; Opoku, S.Y.; Lowor, S.; Dewettinck, K.; Vandesompele, J. Selection and validation of reference genes for accurate RT-qPCR gene expression normalization in cacao beans during fermentation. *Tree Genet. Genomes* **2021**, *17*, 1–14. [CrossRef]

82. Martínez-Atienza, J.; Jiang, X.; Garciadeblas, B.; Mendoza, I.; Zhu, J.-K.; Pardo, J.M.; Quintero, F.J. Conservation of the salt overly sensitive pathway in rice. *Plant Physiol.* **2007**, *143*, 1001–1012. [CrossRef]

83. Manishankar, P.; Wang, N.; Köster, P.; Alatar, A.A.; Kudla, J. Calcium signaling during salt stress and in the regulation of ion homeostasis. *J. Exp. Bot.* **2018**, *69*, 4215–4226. [CrossRef]

84. Faraji, S.; Filiz, E.; Kazemitabar, S.K.; Vannozzi, A.; Palumbo, F.; Barcaccia, G.; Heidari, P. The AP2/ERF Gene Family in Triticum durum: Genome-Wide Identification and Expression Analysis under Drought and Salinity Stresses. *Genes* **2020**, *11*, 1464. [CrossRef]

85. Zhang, J. Evolution by gene duplication: An update. *Trends Ecol. Evol.* **2003**, *18*, 292–298. [CrossRef]

86. Hahn, M.W. Distinguishing among evolutionary models for the maintenance of gene duplicates. *J. Hered.* **2009**, *100*, 605–617. [CrossRef]

87. Ahmadizadeh, M.; Heidari, P. Bioinformatics study of transcription factors involved in cold stress. *Biharean Biol.* **2014**, *8*, 83–86.

88. Cakmak, I.; Kirkby, E.A. Role of magnesium in carbon partitioning and alleviating photooxidative damage. *Physiol. Plant.* **2008**, *133*, 692–704. [CrossRef]

89. Tränkner, M.; Jákli, B.; Tavakol, E.; Geilfus, C.-M.; Cakmak, I.; Dittert, K.; Senbayram, M. Magnesium deficiency decreases biomass water-use efficiency and increases leaf water-use efficiency and oxidative stress in barley plants. *Plant Soil* **2016**, *406*, 409–423. [CrossRef]

Article

An Integration of Transcriptomic Data and Modular Gene Co-Expression Network Analysis Uncovers Drought Stress-Related Hub Genes in Transgenic Rice Overexpressing *OsAbp57*

Muhammad-Redha Abdullah-Zawawi [1], Lay-Wen Tan [2], Zuraida Ab Rahman [3], Ismanizan Ismail [2,4] and Zamri Zainal [2,4,*]

[1] UKM Medical Molecular Biology Institute (UMBI), Universiti Kebangsaan Malaysia, Kuala Lumpur 56000, Malaysia
[2] Institute of Systems Biology (INBIOSIS), Universiti Kebangsaan Malaysia, Bangi 43600, Malaysia
[3] Malaysian Agricultural Research and Development Institute (MARDI), Serdang 43400, Malaysia
[4] Faculty of Science and Technology, Universiti Kebangsaan Malaysia, Bangi 43600, Malaysia
* Correspondence: zz@ukm.edu.my; Tel.: +60-389-213-936

Citation: Abdullah-Zawawi, M.-R.; Tan, L.-W.; Ab Rahman, Z.; Ismail, I.; Zainal, Z. An Integration of Transcriptomic Data and Modular Gene Co-Expression Network Analysis Uncovers Drought Stress-Related Hub Genes in Transgenic Rice Overexpressing *OsAbp57*. *Agronomy* **2022**, *12*, 1959. https://doi.org/10.3390/agronomy12081959

Academic Editors: Roxana Yockteng, Andrés J. Cortés and María Ángeles Castillejo

Received: 12 July 2022
Accepted: 17 August 2022
Published: 19 August 2022

Abstract: Auxin receptor plays a significant role in the plant auxin signalling pathway in response to abiotic stress. Recently, we found that transgenic rice overexpressing *ABP57* had higher drought tolerance than the wild-type cultivar, MR219, due to the fact of its enhanced leaf photosynthetic rate and yields under drought stress. We performed a microarray study on this line to investigate the underlying mechanisms contributing to the observed phenotype. After microarray data filtering, 3596 genes were subjected to modular gene co-expression network (mGCN) development using CEMiTool, an R package. We identified highly related genes in 12 modules that could act to specific responses towards drought or any of the abiotic stress types. Gene set enrichment and overrepresentation analyses for modules extracted two highly upregulated modules that are involved in drought-related biological processes such as transmembrane transport of metal ions and response to oxidative stress. Finally, 123 hub genes were identified in all modules after integrating co-expression information with physical interaction data. In addition, the interplay of significant pathways between the metabolism of chlorophyll and flavonoid and the signalling pathways of MAPK, IAA, and SA inferred the concurrent involvement of stress tolerance response. Collectively, our findings seek new future directions for breeding strategies in rice tolerant improvements.

Keywords: transcriptomics; co-expression network; modular analysis; drought stress; hub gene; *Abp57*; rice improvement; bioinformatics

1. Introduction

Rice (*Oryza sativa* L.) is one of the primary staple food sources that contributes to over 20% of the daily calorie intake of more than 3.5 billion people. Hence, considerable effort has been made to ensure a sufficient rice yield supply to meet the rising population's demand. However, a recent study suggested that rice's global average yield improvement is still far behind the required rate to achieve the projected global demands by 2050 [1]. The slower pace of yield improvement is partly due to the diminishing returns to further intensification of irrigated rice systems [2]. In addition, the growing competition for water resources from other crops and industrialisation have also limited the expansion of irrigated rice areas.

In contrast to irrigated rice systems, which have almost achieved the full genetic potential of the high-yielding rice cultivars, the yield of rainfed rice remains low and inconsistent. This indicates an enormous upside potential for the yield of rice grown under a rainfed ecosystem. Numerous studies have pointed out that frequent occurrences of drought and the lack of drought-tolerant high-yielding rice cultivars are two of the major factors contributing to the suboptimal rice production in rainfed ecosystems [3,4]. Therefore, the development of drought-tolerant high-yielding rice is of paramount importance to ramping up the production of rainfed rice.

Auxin is a phytohormone that functions as a major coordinator in virtually all of the biological processes in plant growth and development. These biological processes include a plant's tropic response towards gravity and light, organ patterning, cell differentiation, seed dormancy, and vascular development [5]. Recent mounting evidence shows that auxin may also play a role in plant response to drought stress. Much of the evidence comes from the realisation that the gene expression of many auxins-related genes was significantly altered under drought conditions. In rice, an expression analysis on 31 rice indole-3-acetic acid (*OsIAA*) genes showed that more than 15 were upregulated under drought stress [6]. In addition, a genome-level microarray analysis in rice also demonstrated that more than 204 auxin-responsive genes exhibited altered expression under desiccation conditions [7]. Given the response of auxin-related genes towards drought stress, numerous studies have been conducted to overexpress these genes in various plant species to enhance their drought tolerance. In potatoes, the overexpression of *YUCCA6*, a flavin monooxygenase of tryptophan-dependent auxin biosynthetic pathways, has resulted in elevated endogenous auxin levels and enhanced drought resistance [8]. In addition, the overexpression of auxin-related genes, such as *OsPIN3t*, *TLD1/GH3.13*, and *OsIAA6*, have also been found to confer higher tolerance of rice towards drought conditions [9–11]. Taken together, auxin-related genes show great potential to be utilised in the attempt to develop rice cultivar with drought tolerance.

Auxin-binding protein (ABP) is a family of low-abundance proteins that bind reversibly to auxin with high specificity and affinity. Owing to their auxin-binding properties, ABP has long been suggested to function as an auxin receptor, mediating diverse cellular responses in response to different endogenous auxin levels. ABP57 is a 57 kDa ABP that was first isolated from the soluble protein fraction of the shoot of rice (*Oryza sativa* L.) seedlings by Kim et al. (1998) [12]. An in vitro experiment has proven that *ABP57* functions by activating plasma membrane H^+-ATPase through direct interaction in response to the concentration of auxin [13]. In contrast to the classical *ABP1*, *ABP57* appears insensitive to naphthaleneacetic acid (NAA) despite having a very high binding affinity with IAA [14]. Numerous overexpression studies have shown that a higher expression level of *OsAbp57* can lead to increased seed size, faster seed germination, and seedling growth [15]. Recently, we found that transgenic rice overexpressing *ABP57* had higher drought tolerance than wild-type cultivar, MR219, due to the fact of an enhanced leaf photosynthetic rate and yields under drought stress [16]. We performed a microarray study on this line to investigate the underlying mechanisms contributing to the observed phenotype.

Co-expression is a well-known biological network technique in predicting the gene function using a myriad of transcriptomics data from RNA sequencing or microarrays technology by connecting the genes based on their similar expression profiles [17,18]. Through the guilt-by-association principle, genes with similar mRNA expression profiles across the tissue, treatment, or developmental stage are predicted to share a similar function and are regulated via similar transcription factors [17,19]. The co-expression relationship can be represented as a network graph that connects nodes (i.e., co-expressed genes) by edges, indicating correlation-based evidence between genes. A group of nodes that are highly interactive with one another is known as a module. Within a module, genes with high connectivity are defined as hub genes. Modules are usually enriched to discover a set of co-expressed genes that are overrepresented in a similar biological process (i.e., gene function such as plant defence or signal transduction), molecular function (i.e., gene activity such as protein kinase or catalytic activity), and cellular component (i.e., gene location such as the cell wall or chloroplast). This information would be essential for understanding the function of a gene and on how genes are produced and operate during response to drought in rice.

In the present study, an assessment of the morpho-physiological traits exhibited by transgenic rice overexpressing *OsAbp57* grown under two watering regimes was performed to identify the possible traits that may contribute to the drought tolerance of transgenic rice. In addition, a microarray analysis was also conducted to identify differentially expressed genes in the transgenic plant compared to its parental line, MR219, under normal conditions. Here, we performed a modular gene co-expression network (mGCN) analysis of this in-house microarray dataset to examine the co-expressed gene modules, evaluate the module activity between samples of MR219 and transgenic rice overexpressing *OsAbp57* (*Abp57*-OE), and enrich a set of co-expressed gene modules to associate them with a particular function, activity, and cellular location. In this study, we represented the phenotypic class of MR219 and transgenic rice as wild-type (WT) and overexpressing *OsAbp57* (*Abp57*-OE), respectively. To grasp an understanding of the transcriptional activity of overexpressed *OsAbp57* under drought conditions, we integrated protein–protein interaction (PPI) information into the constructed modules to identify the most connected genes (i.e., hubs) within the network that may play a function as a critical regulator in response to drought stress. Finally, we conducted an expression analysis on a few selected genes to investigate their role in rice drought tolerance.

2. Materials and Methods

2.1. Sample Preparation and RNA Extraction

The T3 seeds of *Abp57*-overexpressing rice (OE) were obtained from our previous *Agrobacterium*-mediated transformation study [20] and were cultivated on MS media for ten days along with MR219 (WT) seed as a control. Each WT and *Abp57*-OE group consisted of three biological replicates. Total RNA was isolated from seedling tissues of *Abp57*-OE and WT using TRIzol reagent (Life Technology, Waltham, MA, USA), followed by DNase treatment via Ambion TURBO™ to eliminate DNA contamination. The quality of the extracted RNA was measured using NanoDrop ND-1000 spectrophotometer (Thermo Fisher Scientific, Waltham, MA, USA) in which samples with a concentration of 1.8 to 2.1 (A260/A280) were considered. We determined the integrity of the extracted RNA using a Bioanalyzer (Agilent Technologies, Santa Clara, CA, USA). All samples with an RNA Integrity Number (RIN) of 8.0 to 10.0 were subjected to microarray analysis.

2.2. Gene Expression Profiling, Differentially Expressed Genes (DEGs), and Enrichment Analyses

Gene expression profiling was conducted on the three replicates of seedling tissues of WT and *Abp57*-OE using Affymetrix Rice (Chinese Build) Gene 1.0 ST Array that consists of 41,770 transcripts (Affymetrix Inc., Santa Clara, CA, USA). The CEL file was imported into Affymetrix Expression Console™ Software to check for quality control of raw data and data normalisation following the manufacturer's protocol. The robust multiarray analysis (RMA) method was applied to ensure that the relative log expression signal (RLE) was comparable between samples. The microarray experiment was then previously deposited into the NCBI by Tan et al. (2017) [21] and is accessible through the NCBI GEO DataSets (DOI: https://www.ncbi.nlm.nih.gov/geo/query/acc.cgi?acc=GSE99055, accessed on 1 June 2017) [22].

The CHP files resulting from normalisation for DEGs analysis were then imported into the Transcriptome Analysis Console (TAC) software version 3.1 (Applied Biosystems, Foster City, CA, USA), following the manufacturer's specifications. Using TAC, a statistical method of one-way analysis of variance (ANOVA) was utilised to obtain the DEGs of WT and *Abp57*-OE with default parameters: a fold change > |2| and an ANOVA *p*-value < 0.05. One-way ANOVA was calculated based on F = Mean sum of square between group (MSB)/Mean sum of square within group (MSW), where F follows an F distribution with degrees of freedom between group (df) = K – 1 and within group (DFw) = N – K. K represents the number of groups, and N is the total number of observations across all groups.

For the MSB formula, MSB = $\sum(Xi \sum Xt)^2/K \sum 1$; where Xi is the mean of group i, and Xt = mean of all observations (all observations from all groups are combined to form a single group, and the mean is subsequently calculated). For the MSW formula,

$SSW = \sum(Xij - Xj)^2/N - K$; where Xij is an observation within group j, and Xj = mean of group j. Thus, overall, $F = MSW/MSB = (\sum(Xi - Xt)^2/K - 1)/(\sum(Xij - Xj)^2/N - K)$.

For functional annotation of DEGs, we searched against several databases, including the National Center for Biotechnology Information version 236 (NCBI; https://www.ncbi.nlm.nih.gov, accessed on 15 March 2020) [23], UniProt version 2020_04 (https://www.uniprot.org, accessed on 15 March 2020) [24], Rice Genome Annotation Project version 7.0 (RGAP; http://rice.plantbiology.msu.edu/, accessed on 15 March 2020) [25], The Rice Annotation Project Database version 1.0 (RAP-DB; https://rapdb.dna.affrc.go.jp, accessed on 15 March 2020) [26], and the Beijing Genomics Institute Rise Information System version 1.0 (BGI-RIS; http://rise.genomics.org.cn/, accessed on 15 March 2020) [27]. The Gene Ontology (GO) terms were analysed with the Singular Enrichment Analysis (SEA) program available on AgriGO version 2.0 (http://bioinfo.cau.edu.cn/agriGO/, accessed on 15 March 2020) [28] with the Affymetrix Genome Array (GPL2025) as background. Genes with 2-fold expression levels and p-values < 0.05 were used for SEA analysis.

2.3. Expression Analyses of Randomly Selected Expressed Genes by qRT-PCR

Six genes were randomly selected for further investigation to validate the reliability of the microarray data as follows: *ATPF1G* (Os07g0513000), *OsCAO1* (Os10g0567400), *OsEno5* (Os06g0136600), *OsCPS4* (Os04g0178300), *NAS1* (Os03g0307300), and *CHI* (Os12g0115700). Total RNA was extracted from one-week-old seedlings of MR219 using TRIzol reagent (Life Technology, Waltham, MA, USA) and then subjected to DNase treatment using Ambion TURBO^{TM} DNase (Thermo Fischer Scientific, Waltham, MA, USA). The total RNA was reverse-transcribed into first-strand cDNA with a Maxima First Strand cDNA synthesis kit (Thermo Fisher Scientific, Waltham, MA, USA). An equal amount of cDNA was used as the template for PCR amplification using SYBR Green Master Mix (Thermo Fisher Scientific, Waltham, MA, USA). Genes of interest were amplified using the specific primers listed in Supplementary Materials Table S1. The relative expression level of an individual gene was determined using the $\Delta\Delta$ Ct calculation [29]. The housekeeping genes *U6* and *UBQ5* were used as an internal control. One-way ANOVA was then performed to assess DEGs with a p-value < 0.05. The expression level of DEGs was later compared with gene profiling data to determine the validity of the microarray data.

2.4. Construction of Gene Co-Expression Network Modules

To further support the reliability of our *Abp57*-OE microarray data in drought, the co-expression module analysis was carried out using the CEMiTool R package version 1.20 [30]. The published microarray dataset, under the accession number GSE99055, was employed as a discovery set [21]. In this study, the correlation method of Pearson was selected to transform the gene expression matrix ($m \times n$) into an adjacency matrix, where m denotes as genes and n represents the samples. A soft-threshold power (β) selection algorithm was executed based on a scale-free topology criterion to construct an adjacency matrix, later associated with the relationship between co-expression modules. The adjacency matrices were transformed using the best threshold determined based on the "scale-freeness" of the constructed network generated by the default function of CEMiTool. Unsigned network analysis was conducted to infer whether modules represented general processes of drought stress or any stress-related responses in transgenic rice overexpressing *OsAbp57*. The dissimilarity threshold of 0.8 was employed as a cut-off for hierarchical clustering by the agglomerative method, where genes with common expression levels were grouped into clusters. The following criteria were used to build the network module: coefficient of determination for linear regression fit, R^2 ($R^2 > 0.8$); the number of gene modules (min_ngen) > 20; threshold similarity of eigengene (diss_thresh > 0.8); the number of high-connectivity genes in each module, hub genes (n = 5).

2.5. Gene Set Enrichment and Overrepresentation Analysis of Modules

The gene set analysis was performed for each module using the mean rank method implemented in the Fast Gene Set Enrichment Analysis (FSGEA) R package to identify modules of interest from constructed gene co-expression network [31]. The activity of the modules was evaluated by determining which modules were upregulated or downregulated in the samples of the wild-type MR219 and the *OsAbp57* overexpressing lines. The size of the gene set was generated by default in a range between 15 and 1000. To enrich the associated function for each module, the hypergeometric distribution test (overrepresentation analysis) was performed using the clusterProfiler R package [32]. The Gene Ontology (GO) annotation of gene modules was retrieved and downloaded from Phytozome version 12.0 (https://phytozome.jgi.doe.gov/, accessed on 10 April 2021) [33]. The significance cut-off of p-value < 0.05 was used to assign overrepresented biological process (BP), molecular function (MF), and cellular component (CC) of genes within modules.

2.6. Identification and Validation of Hub Genes

The hub genes were determined by identifying genes with higher connectivity within the modules. The hub genes were ranked based on the top 10 genes within the network modules. For instance, if the module size was 200, then 20 genes with the largest number of connectivity could be considered hub genes [34]. Integration between co-expression information and protein–protein interactions (PPIs) may help discover important hubs in the module. The rice PPI dataset from the STRING database version 11.5 was retrieved by the Cytoscape plug-in, StringApp version 1.7.0 [35,36], using the list of gene modules as search queries with the default confidence cut-off of 0.4 and above. A single protein with no interaction was discarded, and only interactions with evidence of known interaction (i.e., curated databases and experimentally determined) were kept for further analysis. Module graphs for gene-encoded protein networks from the rice PPI dataset on STRING were constructed by merging PPI information to visualise gene interaction in co-expression modules and identify potential hubs.

2.7. KEGG Pathway Mapping Analysis

Each module was used to conduct pathway mapping analysis using the Kyoto Encyclopedia of Genes and Genomes (KEGG) Mapper tool version 4.3 (https://www.genome.jp/kegg/mapper.html, accessed on 1 June 2021) [37]. The KEGG Orthology identifier (KO ID) for genes was retrieved from the Phytozome v12 database using the BioMart tool [33,38]. All genes in each module were then categorised into the following KEGG pathways: (i) metabolism, (ii) genetic, (iii) environmental information processing, (iv) cellular processes, and (v) organismal systems. The interactions between hub genes were then manually connected as red arrows on the pathway maps.

3. Results
3.1. Global Gene Expression Profiling of the Abp57-OE Line

To gain insight into the transcriptional changes triggered by the *OsAbp57* overexpression, a global gene expression analysis on WT and transgenic line (*Abp57*-OE) grown under normal conditions was performed through DNA microarrays. The microarray data analysis revealed that a total of 131 genes (90 upregulated and 41 downregulated) were differentially expressed (fold change > 2, p-value < 0.05) in transgenic lines as compared to MR219. The reliability of the microarray data was then assessed by quantitative real-time PCR on six randomly selected DEGs, and the differences were validated (Figure 1). The expression level of DEGs was largely consistent with the microarray result, implying that the data were highly reliable.

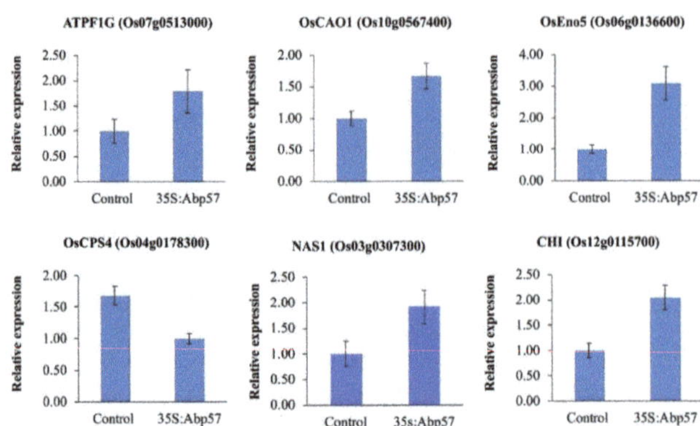

Figure 1. qPCR validations on six randomly selected DEGs. The bar plot represents the mean ± standard error for two biological replicates and three technical replicates. The difference in gene expression between the control (WT) and *Abp57*-OE was determined by a one-way ANOVA method with a cut-off *p*-value < 0.05. The error bars indicate the standard error.

Gene Ontology (GO) enrichment analysis revealed that response to stimulus, homeostasis process, transport, oxidation-reduction, lipid metabolic process, and biosynthetic process were among the significantly altered biological processes in transgenic rice. Meanwhile, for molecular function, DEGs were enriched considerably in metal ion binding, ATP binding, ion transmembrane transporter activity, antiporter activity, peroxidase activity, electron carrier activity, and catalytic activity. For cell components, we identified that the DEGs were significantly regulated in the membrane, chloroplast, and vacuole. The result of the GO enrichment analysis is summarised in Table 1.

Table 1. Functional classification of the differentially regulated genes (DEGs) in *OsAbp57*-overexpressing transgenic rice. GO term enrichment analysis was performed using the AgriGO analysis tool—Single Enrichment Analysis (http://bioinfo.cau.edu.cn/agriGO/analysis.php, accessed on 15 March 2020). The GO enrichment of biological process (B), molecular function (F), and cellular component (C) was generated based on a false discovery rate (FDR) and adjusted *p*-value cut-off of 1.0×10^{-5}.

GO Term	Ontology	Description	Number of DEGs	FDR
GO:0050896	P	Response to stimulus	32	2.20×10^{-24}
GO:0051234	P	Establishment of localisation	30	6.00×10^{-21}
GO:0051179	P	Localisation	30	6.00×10^{-21}
GO:0006810	P	Transport	29	4.60×10^{-20}
GO:0009987	P	Cellular process	51	8.40×10^{-18}
GO:0009607	P	Response to biotic stimulus	14	8.50×10^{-18}
GO:0006811	P	Ion transport	17	3.20×10^{-16}
GO:0008152	P	Metabolic process	47	1.70×10^{-15}
GO:0006950	P	Response to stress	17	5.80×10^{-12}
GO:0042592	P	Homeostatic process	10	1.00×10^{-11}
GO:0051704	P	Multi-organism process	10	2.30×10^{-10}
GO:0003824	F	Catalytic activity	50	6.30×10^{-14}
GO:0022857	F	Transmembrane transporter activity	17	7.60×10^{-14}
GO:0005488	F	Binding	48	2.60×10^{-12}
GO:0015075	F	Ion transmembrane transporter activity	13	2.60×10^{-12}
GO:0005215	F	Transporter activity	17	5.00×10^{-11}
GO:0022891	F	Substrate-specific transmembrane transporter activity	13	5.00×10^{-11}
GO:0046872	F	Metal ion binding	25	6.60×10^{-11}
GO:0043167	F	Ion binding	25	1.20×10^{-10}
GO:0043169	F	Cation binding	25	1.20×10^{-10}
GO:0022892	F	Substrate-specific transporter activity	13	2.80×10^{-10}

3.2. Modular Gene Co-Expression Analyses of the OsAbp57-OE Line

To further support the reliability of our *Abp57*-OE microarray data in drought stress, we performed the mGCN analysis. In total, 41,770 genes across six samples were used to construct an mGCN to determine key modules of drought-tolerant transgenic rice, *OsAbp57*-OE. In Figure 2A, the dendrogram demonstrates the clustering of six samples, whereby the samples were clustered based on wild-type samples (turquoise) and *OsAbp57*-OE samples (red). Using the default parameters, the soft-threshold value (β) of 14 was selected by a scale-free topology fit (R^2) of 0.824, indicating the best threshold for a scale-free network model (Figure 2B). A total of 13 co-expressed modules containing 3596 genes (i.e., M1 to M12, including one noncorrelated module, M13) were identified using the dissimilarity threshold of 0.8 as a hierarchical clustering cut-off. Module M1 holds the highest number of co-expressed genes (2004), followed by other modules in the following order, and the smallest number of genes (41) was discovered in M11 and M12. Of these, only 11 co-expressed modules (M2 to M12) with a *p*-value less than 0.05 were subjected to significant module activity. All module enrichment plots discovered in co-expression analysis were shown to be upregulated in the class of *OsAbp57*-OE samples (OE) compared with the wild-type samples (WT), except for module M6 of *Abp57*-OE, which featured a general downregulation when compared with the WT sample (Figure 2C). The upregulation in *Abp57*-OE class and downregulation in WT class has suggested the sensitivity of transgenic rice, *OsAbp57*, towards drought stress.

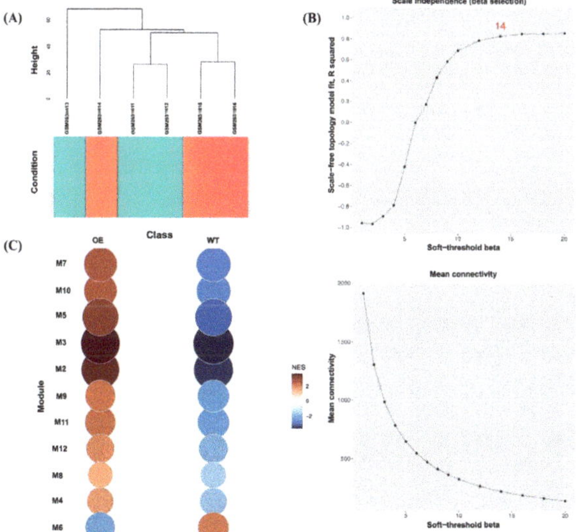

Figure 2. CEMiTool outputs for the *OsAbp57*-OE line microarray dataset. (**A**) Clustering dendrogram of genes based on the expression profiles. The turquoise colour indicates MR219, and the red colour represents the *OsAbp57*-OE line. (**B**) Scale-free topology (R^2) and mean connectivity to identify the soft-threshold power (β) between 1 and 20. The scale-freeness of the network was determined at a soft threshold of 14, above the R^2 threshold of 0.8. (**C**) Gene set enrichment analysis for module activity of the *OsAbp57* OE line (OE) and MR219 (WT). The size and colour of the modules represent the normalised enrichment score (NES). All modules were upregulated in OE except M6, with downregulation in OE and upregulation in WT.

In OE, modules M2 (OE; NES = 3.39; WT; NES = −3.46) and M3 (OE; NES = 3.58; WT; NES = −3.63) featured strong upregulation and downregulation in module activity (Figure 2C). The enrichment analysis of each module demonstrated that co-expressed genes in modules M2 and M3 were involved in GO terms transmembrane transport including metal ion transport and response to oxidative stress (Figure 3A,B). Module M2 was enriched in defence response (adjusted $p = 0.30099$; 11/1117 genes), cell redox homeostasis (adjusted $p = 0.37956$; 10/1117 genes), and proteolysis (adjusted $p = 0.3183$; 51/1117 genes). However, genes in module M3 were involved in the fatty acid biosynthetic process (adjusted $p = 0.88612$; 18/1117 genes) and signal transduction (adjusted $p = 0.88612$; 13/1117 genes). Overall, other modules were also strongly enriched in response to oxidative stress, such as modules M4, M5, M7, M8, M10, and M11 (Supplementary Materials Table S2).

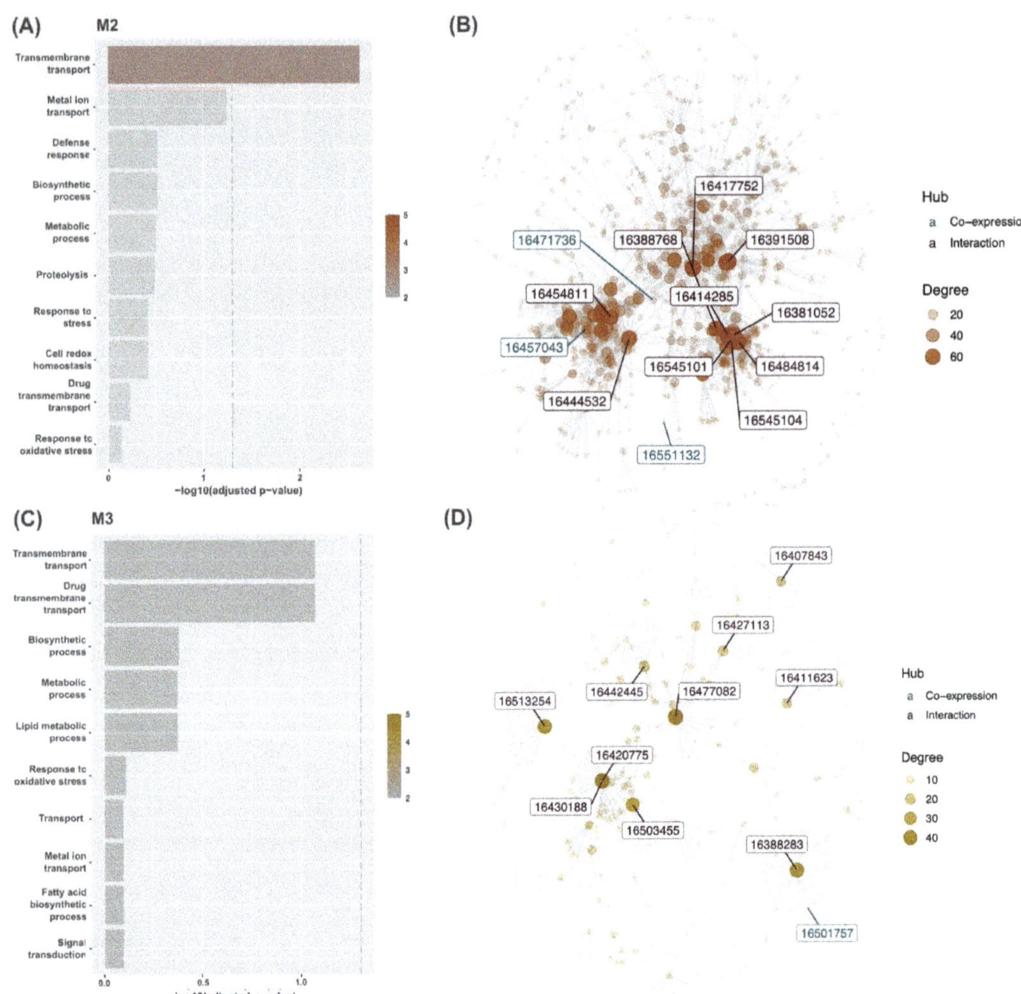

Figure 3. Bar graph for the top ten GO terms enriched between genes in modules (**A**) M2 and (**C**) M3, respectively, and gene sets from the Phytozome database. The dashed line represents the -log10 adjusted *p*-value of 0.01. Interaction network of modules (**B**) M2 and (**D**) M3. The top ten hubs' colours are shown based on the originality of hubs present in the CEMiTool co-expression module (blue) or rice PPI dataset (red) on the STRING database.

3.3. Hub Genes' Identification in OsAbp57-OE Associated with Drought Stress

A hub gene is a high-degree gene in the module and is presumed to be a vital candidate that controls other genes in the network. The top 10 most connected genes in each module were determined using the abovementioned methods. We identified 123 hub genes in 12 modules (i.e., M1 to M12). Detailed information on the hub genes of each module is listed in Supplementary Materials File S1. The hub gene analysis discovered 15 origin hubs, which were initially found in the CEMiTool module, including *UGDH* and *ONI3* in module M1; *KINUA, TPR*, and *ABC-1* in M2; *HRZ2* in M3; *SUPM1* and *BAK1* in M5; *AAO2* and *4CL1* in M7; *PEPC-2* and *NPR5* in M10; *PPR* and *OAT* in M11; *OsMutY* in M12. Moreover, we found that origin hubs associated with drought stress mechanisms were upregulated in the network, particularly *HRZ2* in response to Fe^{3+} starvation and JA-mediated signalling pathway, *AAO2* in response to water distress, and *OAT* in various responses to abiotic stress tolerance such as to water distress, osmotic and salt stress, hormonal stimuli of ABA, JA, BR, and IAA.

Figure 3B,D demonstrate the gene network for interesting modules M2 and M3, respectively. We discovered several upregulated genes in enriched *OsAbp57*-OE modules M2 and M3 such as *HSP70, KINUA, TPR, ABC-1, PK, Os01g0655500, PRR1, MDH, HRZ2, TIG, PTR, PRF1*, and *KAR* (Table 2). Interestingly, the Gene Ontology annotation shows that some hub genes are significantly involved in stress-related biological processes, for instance, in response to the metal ion, Cd^{2+} (i.e., *HSP70*), Fe^{3+} starvation and JA-mediated signalling pathway (i.e., *HRZ2*), and reactive oxygen species (ROS) (i.e., *MDH*), including *PRR1*, are reported to play a role in the phosphorelay signal transduction system and the regulation of circadian rhythm, *PK* in glycolysis, *KAR* in elongation of fatty acid, and *TIG* and *PTR* in transport function of protein and oligopeptide, respectively. However, although several defence-related biological processes have been reported through the GO annotation, the potential mechanisms underlying the regulation of drought tolerance genes by *OsAbp57* are not well understood. A further experiment must be conducted to determine how these hub genes may navigate the function of defence during drought conditions in rice.

Table 2. Hub genes from interesting modules M2 and M3.

Probe ID	Gene ID	Gene Name	Description	Gene Ontology	Trait Ontology	Log2FC	Up or Down
16381052	Os01g0100700	*RPS5-1*	40S ribosomal protein S5-1	Defence response to fungus, response to H_2O_2, ABA, salt stress, water deprivation	Oxidative stress, sheath blight disease resistance, drought and salt tolerance, ABA sensitivity	0.04197709	Down
16388768	Os01g0686800	*RACK1A*	WD repeat-containing protein	Shoot and root development, seed germination, circadian rhythm, detection of redox state, defence response to fungus, ET- and SA-activated signalling pathway, SA-mediated signalling pathway, photoperiodism, response to SA, ET, salt stress, H_2O_2, and Cd^{2+}	ABA content, blast disease, disease resistance, ET sensitivity, photoperiod sensitivity, salt tolerance	0.0606787	Down
16391508	Os01g0840100	*HSP70*	Heat shock protein Hsp70	Response to Cd^{2+} and stress	–	0.09918908	Up
16454811	Os02g0797400	*MCM5*	MCM family protein	Cell cycle, DNA replication	–	0.0735361	Down
16444532	Os02g0805200	*PCNA*	Proliferating cell nuclear antigen	Regulation of DNA replication, double-stranded break repair, response to gamma ray, DNA damage stimulus, UV, radiation, and H_2O_2	UV light sensitivity	0.0467894	Down
16457043	Os03g0152900	*KINUA*	Kinesin-like protein KIN-UA	Microtubule-based movement	–	0.13444718	Up
16471736	Os03g0308800	*TPR*	Tetratricopeptide-like helical domain containing protein	RNA processing, regulation of translation	–	0.1440686	Up
16484814	Os04g0598200	*RPL12*	60S ribosomal protein L12	Translation, ribosomal large subunit assembly, defence response to bacterium, response to heat, cold, SA, JA, H_2O_2, and water deprivation	Heat, cold & drought tolerance, JA sensitivity, bacterial blight disease resistance, oxidative stress	0.1768968	Down
16545101	Os08g0117200	*RPS13*	40S ribosomal protein S13	Translation, defence response to fungus, response to water deprivation, salt stress, ABA, H_2O_2	Drought & salt tolerance, oxidative stress, sheath blight disease resistance, ABA sensitivity.	0.93240096	Down

Table 2. *Cont.*

Probe ID	Gene ID	Gene Name	Description	Gene Ontology	Trait Ontology	Log2FC	Up or Down
16545104	Os08g0117300	RPS13	40S ribosomal protein S13	Translation	–	0.0503529	Down
16551132	Os09g0250700	ABC-1	ABC-1 domain containing protein.	–	–	0.1204065	Up
16414285	Os10g0466700	RPL17	60S ribosomal protein L17	Translation	–	0.0262824	Down
16417752	Os11g0216000	PK	Pyruvate kinase family protein	Glycolysis	–	0.09241819	Up
16388283	Os01g0655500	Os01g0655500	Protein kinase, core domain containing protein.	Protein phosphorylation	–	0.27208653	Up
16442445	Os02g0618200	PRR1	Two-component response regulator-like PRR1 (Pseudo-response regulator 1)	Phosphorelay signal transduction system, regulation of circadian rhythm	–	0.10612373	Up
16477082	Os03g0773800	MDH	Malate dehydrogenase	Response to ROS, metabolic processes of malate and carbohydrate, cell redox homeostasis, TCA cycle	–	0.11776116	Up
16503455	Os05g0143800	GH3.6	GH3 auxin-responsive promoter domain containing protein	Response to IAA, light stimulus, salt stress	Salt tolerance	0.2140365	Down
16501757	Os05g0551000	HRZ2	Hemerythrin motif-containing RING- and Zn-finger protein2	Response to Fe^{3+} starvation, JA-mediated signalling pathway	Fe^{3+} sensitivity	0.24291249	Up
16513254	Os06g0308000	TIG	Trigger factor-like protein	Protein folding and transport	–	0.09959792	Up
16411623	Os10g0111700	PTR	POT family protein, peptide transporter	Oligopeptide transport	–	0.20793006	Up
16407843	Os10g0323600	PRF1	Profilin A	Cytoskeleton organisation, sequestering of actin monomers	–	0.16372392	Up
16420775	Os11g0106700	FER1	FERRITIN 1	Response to Fe^{3+}, Zn^{2+}, H_2O_2, ABA, bacterdium and cold; leaf and flower development; Fe^{3+} transport and homeostasis; photosynthesis	–	0.1615892	Down
16430188	Os12g0106000	FER2	FERRITIN 2	Response to Fe^{3+}, Zn^{2+}, Cu^{2+}, H_2O_2, cold, bacterium, ABA; leaf and flower development; Fe^{3+} transport and homeostasis; photosynthesis	Temperature response trait, Cu^{2+} sensitivity	0.1529454	Down
16427113	Os12g0242700	KAR	Beta-ketoacyl reductase	Fatty acid elongation	–	0.48300481	Up

To validate the hub gene, Figure 4 demonstrates the interaction network for all hub gene modules: M1 to M12. The network consisted of 125 nodes and 584 edges of which 14 were identified as single nodes. Figure 4B showed β-ketoacyl reductase (*KAR*), a fatty acid biosynthetic gene, which was highly upregulated in *OsAbp57*-OE. We found that stress-related hub genes, including *HSP70*, *HRZ2*, and *MDH*, significantly interacted with other upregulated hub genes (Figure 4B). For instance, *HSP70* with *EF2*; *PK*, *PPR*, and *MDH*, *HRZ2* with *Os01g0655500*; and *MDH* with *CFR*, *ACP*, *ACP-1*, *ACP3*, *ppc4*, *PK*, *cytME2*, *HSP70*, and *PPR1*. Several downregulated genes were also identified to interact with *HSP70* and *MDH*, suggesting its potential negative regulation of drought responses in *OsAbp57*-OE.

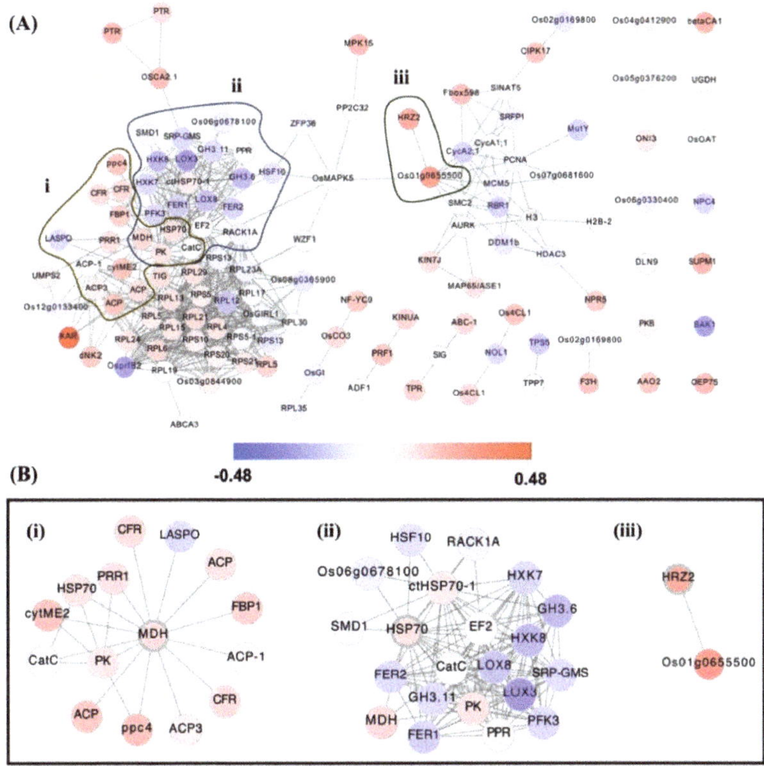

Figure 4. Interaction network for all hub genes in *OsAbp57*-OE: (**A**) modules M1–M12; (**B**) stress-related hub genes: (**i**) *MDH*, (**ii**) *HSP70*, and (**iii**) *HRZ2*.

3.4. Pathway Mapping Analysis of OsAbp57-OE

The number of genes from network modules M1 to M12 was assigned according to their presence in the KEGG pathway maps. The pathway maps are categorised into metabolism, organismal systems, cellular processes, and environmental and genetic information processing. The modules M1 to M11 demonstrated active involvement in metabolism (Supplementary File S2). Genes with a putative function in cellular processes were absent in modules M5 to M7, M9, M10, and M12, including the absence of environmental information processing in M4, M5, M7 to M9, and M12, organismal systems in M7 to M12, and genetic and environmental processing in M3, M4, M8, and M10. The hub genes showed high participation in the genetic information process (30 genes) and metabolism (29 genes) including participation in the environmental information process (4 genes), cellular processes (3 genes), and organismal systems (2 genes) (Figure 5A). In addition, integration with protein–protein interaction data validated the correlation of hubs between the pathway maps. The hubs under the phenylpropanoid biosynthesis were correlated to the porphyrin and chlorophyll metabolism map in module M7 (Figure 5B).

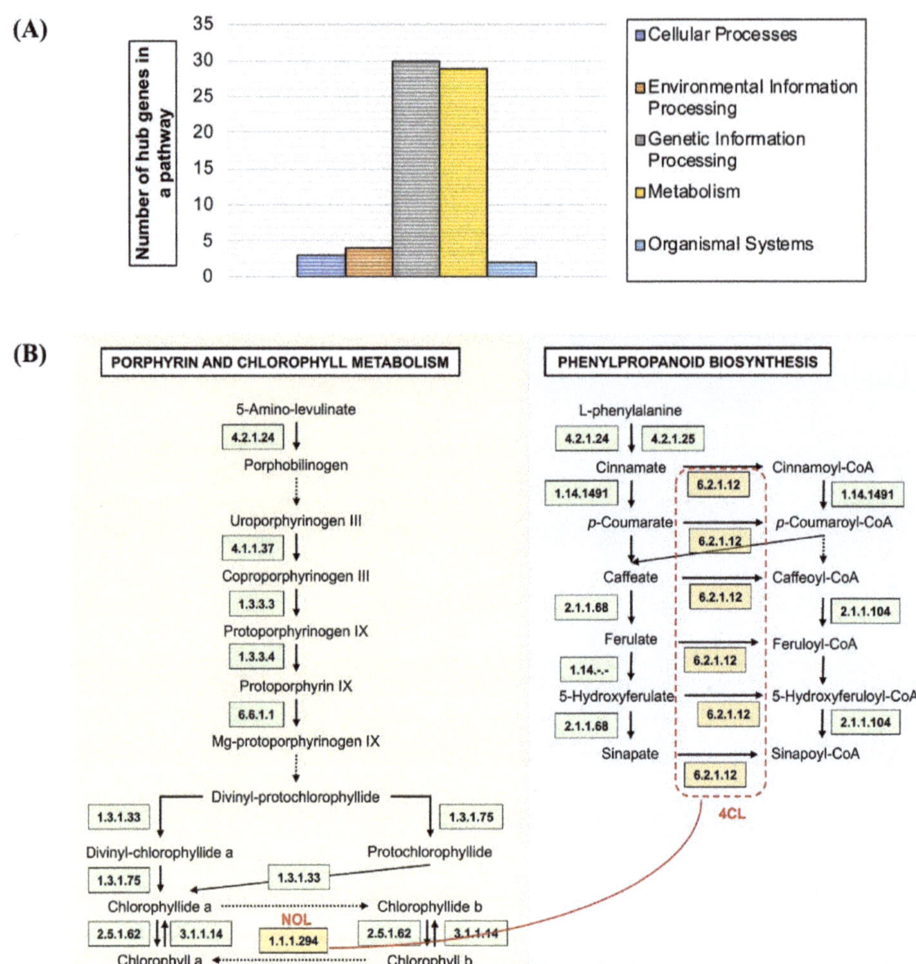

Figure 5. Pathway mapping analysis of *OsAbp57*-OE. (**A**) Classification of hub genes in the modular network into five main pathway maps: cellular processes, environmental information processing, genetic information processing, and organismal systems; (**B**) interaction between M7 hubs in the porphyrin and chlorophyll metabolism pathway, and phenylpropanoid biosynthesis maps are represented with red lines.

No correlation of hubs within other modules was identified between the pathway maps. Pathway analysis of several hubs under plant hormone signal transduction showed interaction with the MAPK signalling pathway map. Two hubs of *GH3* in M3 and M6, *GH3.6* and *GH3.11* under plant hormone signal transduction for plant growth, showed interplay to disease resistance *NPR5* and stress-tolerant response *CatC* under the MAPK signalling pathway map (Figure 6).

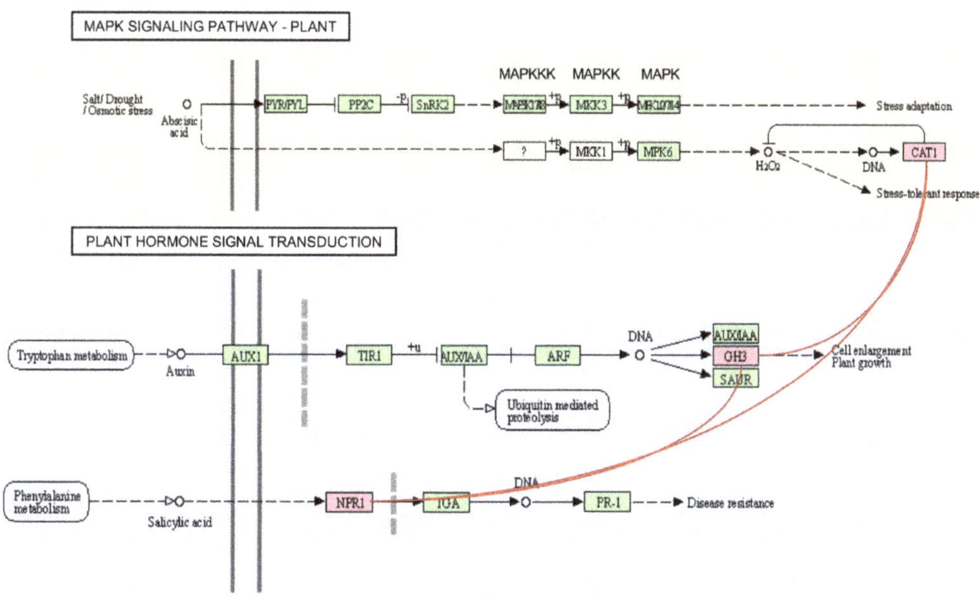

Figure 6. Interaction between hub genes in MAPK signalling pathway and plant hormone signal transduction maps. Interaction is represented in red lines.

4. Discussion

In the present study, we demonstrated that the overexpression of an auxin-binding protein, *OsAbp57*, can enhance the drought tolerance of a popular Malaysian rice cultivar, MR219. A comparative analysis of drought tolerance among two *OsAbp57* overexpressing lines and MR219 revealed a much delayed drought-induced leaf senescence in *OsAbp57* overexpressing lines compared to MR219. Numerous previous studies have reported that delayed drought-induced leaf senescence is one of the important traits that can enhance the tolerance of a plant toward drought stress [8,39,40]. This has been further supported by Liu et al. (2016) [41], who reported that transgenic rice with accelerated drought-induced senescence exhibited higher sensitivity to drought stress than its nontransgenic parental line.

Further examination of the physio-morphological traits also revealed that the performance of both *OsAbp57* overexpressing lines was superior to MR219 in most of the traits evaluated under both watering regimes. We noticed that the root dry weight of *OsAbp57* overexpressing exhibited the greatest responsiveness towards drought conditions. Compared to its counterparts grown under the normal condition, the root dry weight of *OsAbp57* overexpressing lines grown under drought conditions and the root weight of both *OsAbp57* overexpressing lines grown under drought conditions exhibited more than 47% of increments (OE1: 47%; OE2: 52%). Being the primary organ involved in plant water uptake and the first organ to perceive drought stress, the root traits have long been known as one of the determining factors in plant productivity under drought stress [42]. This is particularly true for the plant species employing drought avoidance mechanisms, such as rice. Previous studies on rice have also reported that the root mass and length are good predictors of rice yield under drought due to improved contact between the root and the shrinking soil water [43]. Therefore, higher responsiveness of the root weight of *OsAbp57* overexpressed in response to drought stress may be one of the factors contributing to the drought resistance of transgenic rice, which allows the transgenic plant to avoid a drastic reduction of plant water potential despite a shortage of soil moisture.

Although *OsAbp57* overexpressing lines exhibited a higher drought tolerance than MR219, we also observed that the overexpression of *OsAbp57* led to approximately a 23–48% reduction in the leaf photosynthetic rate compared to MR219. This implies that the *OsAbp57* overexpression may have a negative effect on the photosynthetic activities of rice. Suboptimal performance of transgenic rice under well-irrigated conditions has also been reported in many previous studies. One of the examples was the transgenic plant overexpressing *OsPYL/RCAR5* [44], whereby the overexpression had resulted in dwarf phenotype and yield reduction. In addition, the transgenic plant overexpressing *OsNAC6* produced shoot growth retardation under control conditions [45]. Therefore, employing drought-inducible promoters may avoid the adverse effects imposed by the overexpression of *OsAbp57* under well-irrigated conditions.

Our transcriptome data showed that *OsAbp57* regulated redox metabolism in rice. The overexpressing of *OsAbp57* led to changes in redox activity and increased iron ion uptake in transgenic rice. Iron is essential for plant growth, but excess Fe^{2+} can generate ROS via the Fenton reaction. These results signified the involvement of ROS metabolism in plants. *OsAbp57* has been proven to bind to IAA and directly activate plasma membrane H^+-ATPase. These findings agree with previous studies, whereby plant growth and development activities governed by auxin are closely associated with ROS. The application of auxin to cells rapidly induced ROS generation. Although ROS caused considerable cellular damage, these molecules are also important signalling molecules for stress tolerance. The possibility of utilising a gene involved in Fe^{2+} homeostasis in establishing drought-tolerant rice has been demonstrated by a recent journal. According to the author, the expression of the nicotianamine biosynthetic genes (*OsNAS1* and *OsNAS2*) was upregulated significantly in several drought-tolerant transgenic rice. Previous ChIP-seq analysis showed that *OsNAS1* and *OsNAS2* are the direct targets of a stress-responsive transcription factor called *OsNAC6*, which is necessary for drought response. The overexpression of *OsNAS1* and *OsNAS2* has also demonstrated the importance of the accumulation of NA for drought tolerance in rice [46].

The ROS in plants must be kept at a safe level to avoid cellular damage and death. Previous studies showed that drought stress would lead to a burst of different ROS in different cellular compartments, such as mitochondria, chloroplast, and peroxisome [47]. Increasing evidence indicates that enhanced production of ROS may play a role in plant stress signalling, which can facilitate plants to perceive stress levels in different organs. However, the unregulated production of ROS is detrimental to the cell. Therefore, an increase in ROS production must be accompanied by upregulation of the ROS scavenging system. Our microarray data shows that biosynthesis of flavonoids was upregulated in transgenic rice overexpressing *OsAbp57*. Flavonoid has solid antioxidative properties by reducing the production and quenching ROS, especially those derived from photosynthetic apparatus. This is done through several mechanisms, such as suppression of singlet oxygen, inhibiting enzymes that generate ROS, and chelate ions of transition metals that may catalyse ROS production.

Decreased plant height and increased branching are essential characteristics for rice crop improvement and are also part of a plant's acclimatisation strategy to diminish stress exposure (stress-induced morphogenic responses, SIMRs). The interactive network of auxin, ROS, and antioxidants has been proposed to form a redox signalling module that links plant development and environmental cues. According to Xia et al. (2020) [48], *OsWUS* plays a significant function in tiller development and weak apical dominance, and the loss of *OsWUS* function influenced rice plant morphology. Furthermore, auxin response is significantly enhanced due to the inhibition of auxin-associated gene *ASP1*, a physical interactor of *OsWUS* that suppressed the formation of tiller buds in *OsWUS* loss-of-function mutant, decreased culm number 1 (dc1).

Our results showed that overexpression of *OsAbp57* enhanced drought tolerance in transgenic rice. This might be due to the altered auxin homeostasis in the plant. Previous studies indicated that endogenous and exogenous auxin positively regulated ROS metabolism and antioxidant activity. Likewise, transgenic potato overexpressing *AtYUC6* (member of the *YUCCA* family of flavin-containing monooxygenase) showed high auxin and enhanced drought-tolerant phenotype through a regulated ROS homeostasis [8]. Enhanced ROS production or ROS signalling is also associated with various abiotic stresses, such as drought, salt stress, oxidative stress, UV-radiation stress, and heavy metal stress. The disturbed ROS homeostasis during stress conditions could act as a signal to activate a stress response pathway, such as mitogen-activated protein kinase (MAPK) cascades. Moreover, our microarray analysis also showed downregulation of gibberellin biosynthesis activity. According to Zawaski and Busov (2014) [49], gibberellin (GA) catabolism and repressive signalling mediate shoot growth inhibition and physiological adaptation in response to drought. The transgenic plants with GA-deficiency or GA-insensitive displayed more excellent resistance to drought. In addition, GA-deficiency in response to Fe homeostasis could also lead to dwarfism of the shoot, where Fe-deficient causes foliar chlorosis and decreased leaf biomass. The concentrations of Fe and CHL, an indicator of Fe status, were higher in the leaves of dwarfed transgenic rice [50].

The modular co-expression network (mGCN) constructed based on our microarray data, showed strong upregulation of module activity in *OsAbp57*. Modules M2 and M3 refer to transmembrane transport, which significantly adjusts water scarcity by allocating various molecules through the root, stomata, and cuticle [51]. Although the transport of several metal ions, including K^+, Na^+, and Cl^-, has been reported to be critical in counteracting drought stress [51], the role of Fe^{3+} transport in rice is still not well understood. In fact, the presence of Fe causes the activation of reactive oxygen species (ROS) scavenging enzymes, such as catalase and peroxidase, which are known to regulate the expression of the stress-responsive gene to confer tolerance to environmental stress [52,53]. Also, GSEA revealed promising insights into the tolerance mechanisms of *OsAbp57* against drought. For instance, the activity of most of the modules, which are associated with oxidative stress, suggests an intense regulation of ROS and ROS-mediated signalling pathways during drought. Our mGCN study revealed the dynamics of genes involved in drought tolerance due to the possible increase of reactive oxygen species (ROS) levels when oxidative stress is induced during water depletion. This leads to redox homeostasis, thus activating redox-dependent signalling that could initiate the adaptive plant response. To escape such water scarcity, membrane transport will take place to transfer molecules, metal ions, and water in the context of root response to drought. However, further experiments are much needed to assess the importance of Fe^{3+} transport during a water shortage.

Integration of mGCN with protein–protein interaction (PPI) data revealed the putative involvement of *OsAbp57* hubs in drought stress tolerance. Interestingly, the origin hub that is unlikely present in the PPI data has shown to be upregulated under drought stress. Under drought conditions, Fe micronutrients play a key role in enhancing stress tolerance as it produces assimilates [52]. *HRZ2* functions to negatively regulate the response of Fe deficiency and activate JA signalling at the early stages of Fe sufficiency [54,55]. The *OsAbp57* tolerance of Fe-sufficient conditions may have incurred as manifest in module M3 by origin hub *HRZ2*, which was reported previously to increase the expression of genes involved in Fe uptake and translocation in the *HRZ*-knockdown shoots and roots [56]. The other predicted origin hub, *AAO2*, is specifically expressed in shoot and shows drastic changes in transcript accumulation under drought and salinity stress [57].

Meanwhile, *OAT*, controlled by a stress-responsive transcription factor, *SNAC2*, confers good tolerance to drought and osmotic stress through activation of ROS-scavenging enzymes and ABA-mediated pathways [58]. The regulation of origin hubs identifies new potential genes that may have functioned to increase drought tolerance in *OsAbp57*. The presence of *MDH* and *PK* further explains the possible occurrence of carbohydrate synthesis and glycolysis, respectively [59,60]. In contrast, *TIG*, a distant FKBP family, contains a targeting region that binds to ribosomes and helps to determine the subcellular localisation of mature protein during water deficit [61]. The presence of *PTR* infers the oligopeptide transport events [62] and explains possibilities for nutrient uptake and transport in rice, such as nitrogen and arsenic [63,64]. In rice, *PRR1*, a component of the circadian clock, plays a vital role in regulating the photoperiod of flowering response. However, a recent study by Wei et al. (2021) [65] reported that *PRR73* positively regulated salt tolerance by co-joint with *HDAC10* to repress the transcription of Na$^+$ transporter *HKT2;1* in transgenic rice.

We noticed several important hubs physically interacting with each other based on the PPI data. From the interesting modules, we discussed the results based on the upregulated hub associated with drought stress and hub interaction between the pathway maps.

In module M2, *HSP70* functions in protein folding and preventing DNA degradation or fragmentation under stressed and unstressed conditions [66,67]. We found *HSP70* to be upregulated in *OsAbp57*, considering this gene confers drought stress tolerance by maintaining the protein structure or DNA of the plant cell. The interaction of *HSP70* with *MDH* and *PK*, which are involved in carbohydrate synthesis and glycolysis, may be regarded as a crucial trait for plant-life-sustaining activities. Drought causes changes in sucrose and amino acid content, which was revealed by an increased phosphoenolpyruvate carboxylase (PEPC) expression in starch and sucrose metabolism [68]. In poplar, the *PPR* gene may be involved in environmental adaptation as it was confirmed to respond to cold, salinity, and JA conditions [69] and under drought in Arabidopsis [70]. In M3, the interacted genes, *HRZ2* and *Os01g0655500*, are related in function. As *HRZ2* plays a dominant role in Fe uptake and translocation, *Os01g0655500*, a protein kinase-containing domain gene, is regulated in response to Fe deficiency and excess [71]. Among the *MDH* neighbours, *ACP* may suggest an essential role in type II fatty acid synthesis and mitochondrial protection against drought stress [72]. In mitochondrial protection, *ACP* has been identified as a transmitter for nutrient status concerning mitochondria biogenesis [73]. We found ACPs, including *ACP1* and *ACP3*, were upregulated in *OsAbp57*, which has previously been reported to improve salt tolerance through alterations of fatty acid composition and control the concentration of Na$^+$/K$^+$ [74]. In addition, by scavenging ROS at higher concentrations of Na$^+$, another interesting hub, *CFR*, may enable the protection of chloroplasts from chlorophyll degradation and photodamage of photosystem II [75].

Of all interesting hubs, the flavonoid biosynthesis pathway seems to be functionally correlated to porphyrin and chlorophyll metabolism through the correlation between *4CL1* and *NOL* genes. The relationship between these two pathways is likely to co-exist during adverse environmental conditions that cause leaf colour to shift due to changes in pigment ratios. In a study by Shen et al. (2018) [76], the synergistic effect of flavonoid reduction and porphyrin and chlorophyll enhancement resulted in a change in the leaf colour of tea plants. Hub genes encoding proteins of the MAPK signalling pathway were also associated with plant hormone signal transduction maps. Upon water scarcity or drought conditions, the accumulation of ABA and H$_2$O$_2$ activates the MAPK signalling pathway, thereby promoting the catalase activity, which is responsible for maintaining the optimal level of H2O2 in plant cells [77]. Our findings found that *GH3* genes (i.e., *GH3.11* and *GH3.6*) were involved in plant growth and development via the IAA signal transduction pathway. However, a *GH3* family gene, *GH3-2*, modulated ET-IAA crosstalk to confer drought and cold tolerance in rice [78]. The interplay of the SA signal transduction pathway with the MAPK signalling pathway and IAA signal transduction pathway inferred the possible event of combined biotic and abiotic stresses. Abiotic stress weakens plant immunity and enhances plant susceptibility to pathogenic organisms [79].

5. Conclusions

This work provided new insight into the emergence of drought tolerance in *OsAbp57*. We presented a solid integration study that can analyse transcriptomics data to discover potential candidates for the drought-responsive gene via modular gene correlations. Integration of our present results and the literature search portrayed the hub gene function of mGCN involved in important biological processes associated with drought stress tolerance. We demonstrated possible hub genes underpinning potential drought tolerance by increased reactive oxygen species (ROS) level at the state of water depletion, therefore causing the occurrence of metal ion transport, redox homeostasis, and activation of redox-dependent signalling that may trigger the adaptive response in rice. The hub genes also link the chlorophyll and flavonoid metabolism pathways and feature the interplay between the MAPK signalling pathway with IAA and SA signal transduction pathways. These concurrent events might have occurred in *OsAbp57* due to the effect of colour changes in plant cells, stimulated by *4CL* and *NOL* genes, and the possible occurrence of combined stresses activated by *GH3*, *CATC*, and *NPR* genes through the pathogenic susceptibility via weakened immunity during drought stress. The hub genes discovered in our mGCN are crucial in rice breeding strategies to enhance yield and produce drought-tolerant rice varieties. Some of the hub genes and pathways indicated in this study are established candidates associated with drought stress, confirming the functional validity of these findings. However, other poorly investigated genes could necessitate novel mechanisms in *OsAbp57*, worthwhile of further investigations in future research.

Supplementary Materials: The following supporting information can be downloaded at: https://www.mdpi.com/article/10.3390/agronomy12081959/s1, Table S1: Primers used in this study for qPCR validation; Table S2: List of overrepresented biological processes in modules (M1–M12) determined by hypergeometric test; File S1: Detailed information on the hub genes of each module; File S2: Detailed information on modular genes mapped on KEGG pathways.

Author Contributions: Conceptualisation, M.-R.A.-Z. and Z.Z.; methodology, M.-R.A.-Z. and Z.Z.; formal analysis, M.-R.A.-Z. and L.-W.T.; investigation, M.-R.A.-Z. and L.-W.T.; writing—original draft preparation, M.-R.A.-Z.; writing—review and editing, M.-R.A.-Z., Z.A.R., I.I. and Z.Z.; supervision, Z.Z.; funding acquisition, Z.Z. All authors have read and agreed to the published version of the manuscript.

Funding: This research was funded by Geran Universiti Penyelidikan (GUP), grant number GUP-2021 044 awarded to Z.Z. by Universiti Kebangsaan Malaysia (UKM). The PhD scholarship to L.W.T. was funded by MyBrain15 from the Ministry of Higher Education Malaysia (MoHE).

Institutional Review Board Statement: Not applicable.

Informed Consent Statement: Not applicable.

Data Availability Statement: Not applicable.

Conflicts of Interest: The authors declare no conflict of interest.

References

1. Ray, D.K.; Mueller, N.D.; West, P.C.; Foley, J.A. Yield trends are insufficient to double global crop production by 2050. *PLoS ONE* **2013**, *8*, e66428. [CrossRef] [PubMed]
2. Li, Y.; Zhang, W.; Ma, L.; Wu, L.; Shen, J.; Davies, W.J.; Oenema, O.; Zhang, F.; Dou, Z. An analysis of China's grain production: Looking back and looking forward. *Food Energy Secur.* **2014**, *3*, 19–32. [CrossRef]
3. Pandey, S.; Bhandari, H. Drought: Economic costs and research implications. In *Drought Frontiers in Rice: Crop Improvement for Increased Rainfed Production*; Serraj, R., Bennett, J., Hardy, B., Eds.; World Scientific: Singapore, 2007; pp. 3–17.
4. Swain, P.; Raman, A.; Singh, S.P.; Kumar, A. Breeding drought tolerant rice for shallow rainfed ecosystem of eastern India. *Field Crops Res.* **2017**, *209*, 168–178. [CrossRef] [PubMed]
5. Woodward, A.W.; Bartel, B. Auxin: Regulation, action, and interaction. *Ann. Bot.* **2005**, *95*, 707–735. [CrossRef] [PubMed]
6. Song, Y.; Wang, L.; Xiong, L. Comprehensive expression profiling analysis of OsIAA gene family in developmental processes and in response to phytohormone and stress treatments. *Planta* **2009**, *229*, 577–591. [CrossRef] [PubMed]
7. Jain, M.; Khurana, J.P. Transcript profiling reveals diverse roles of auxin-responsive genes during reproductive development and abiotic stress in rice. *FEBS J.* **2009**, *276*, 3148–3162. [CrossRef]
8. Kim, J.I.; Baek, D.; Park, H.C.; Chun, H.J.; Oh, D.H.; Lee, M.K.; Cha, J.Y.; Kim, W.Y.; Kim, M.C.; Chung, W.S.; et al. Overexpression of Arabidopsis YUCCA6 in potato results in high-auxin developmental phenotypes and enhanced resistance to water deficit. *Mol. Plant.* **2013**, *6*, 337–349. [CrossRef]
9. Jung, H.; Lee, D.K.; Choi, Y.D.; Kim, J.K. OsIAA6, a member of the rice Aux/IAA gene family, is involved in drought tolerance and tiller outgrowth. *Plant Sci.* **2015**, *236*, 304–312. [CrossRef]
10. Zhang, S.W.; Li, C.H.; Cao, J.; Zhang, Y.C.; Zhang, S.Q.; Xia, Y.F.; Sun, D.Y.; Sun, Y. Altered architecture and enhanced drought tolerance in rice via the down-regulation of indole-3-acetic acid by TLD1/OsGH3.13 activation. *Plant Physiol.* **2009**, *151*, 1889–1901. [CrossRef]

11. Zhang, Q.; Li, J.; Zhang, W.; Yan, S.; Wang, R.; Zhao, J.; Li, Y.; Qi, Z.; Sun, Z.; Zhu, Z. The putative auxin efflux carrier OsPIN3t is involved in the drought stress response and drought tolerance. *Plant J.* **2012**, *72*, 805–816. [CrossRef]

12. Kim, Y.S.; Kim, D.; Jung, J. Isolation of a novel auxin receptor from soluble fractions of rice (*Oryza sativa* L.) shoots. *FEBS Lett.* **1998**, *438*, 241–244. [CrossRef]

13. Kim, Y.S.; Min, J.K.; Kim, D.; Jung, J. A soluble auxin-binding protein, ABP57. Purification with anti-bovine serum albumin antibody and characterization of its mechanistic role in the auxin effect on plant plasma membrane H$^+$-ATPase. *J. Biol. Chem.* **2001**, *276*, 10730–10736. [CrossRef]

14. Kim, Y.S.; Kim, D.; Jung, J. Two isoforms of soluble auxin receptor in rice (*Oryza sativa* L.) plants: Binding property for auxin and interaction with plasma membrane H$^+$-ATPase. *Plant Growth Regul.* **2000**, *32*, 143–150. [CrossRef]

15. Kim, D.H.; Lee, K.P.; Kim, M.I.; Kwon, Y.J.; Kim, Y.S. Gene encoding auxin receptor protein derived from rice and use thereof. US Patent 12/935,483, 24 May 2012.

16. Kamarudin, Z.S.; Shamsudin, N.A.A.; Othman, M.H.C.; Shakri, T.; Tan, L.W.; Sukiran, N.L.; Isa, N.M.; Rahman, Z.A.; Zainal, Z. Morpho-physiology and antioxidant enzyme activities of transgenic rice plant overexpressing ABP57 under reproductive stage drought condition. *Agronomy* **2020**, *10*, 1530. [CrossRef]

17. Hansen, B.O.; Vaid, N.; Musialak-Lange, M.; Janowski, M.; Mutwil, M. Elucidating gene function and function evolution through comparison of co-expression networks of plants. *Front. Plant Sci.* **2014**, *19*, 394. [CrossRef]

18. Usadel, B.; Obayashi, T.; Mutwil, M.; Giorgi, F.M.; Bassel, G.W.; Tanimoto, M.; Chow, A.; Steinhauser, D.; Persson, S.; Provart, N.J. Co-expression tools for plant biology: Opportunities for hypothesis generation and caveats. *Plant Cell Environ.* **2009**, *32*, 1633–1651. [CrossRef]

19. Allocco, D.J.; Kohane, I.S.; Butte, A.J. Quantifying the relationship between co-expression, co-regulation and gene function. *BMC Bioinform.* **2004**, *5*, 18. [CrossRef]

20. Tan, L.W.; Rahman, Z.A.; Goh, H.H.; Hwang, D.J.; Ismail, I.; Zainal, Z. Production of transgenic rice (*indica* cv. MR219) overexpressing *Abp57* gene through *Agrobacterium*-mediated transformation. *Sains Malays* **2017**, *46*, 703–711. [CrossRef]

21. Tan, L.W.; Tan, C.S.; Rahman, Z.A.; Goh, H.H.; Ismail, I.; Zainal, Z. Microarray dataset of transgenic rice overexpressing *Abp57*. *Data Brief* **2017**, *14*, 267–271. [CrossRef]

22. Barrett, T.; Edgar, R. Gene expression omnibus: Microarray data storage, submission, retrieval, and analysis. *Methods Enzymol.* **2006**, *411*, 352–369.

23. Sayers, E.W.; Beck, J.; Bolton, E.E.; Bourexis, D.; Brister, J.R.; Canese, K.; Comeau, D.C.; Funk, K.; Kim, S.; Klimke, W.; et al. Database resources of the National Center for Biotechnology Information. *Nucleic Acids Res.* **2021**, *49*, D10–D17. [CrossRef]

24. UniProt Consortium. UniProt: The universal protein knowledgebase in 2021. *Nucleic Acids Res.* **2021**, *49*, D480–D489. [CrossRef]

25. Kawahara, Y.; de la Bastide, M.; Hamilton, J.P.; Kanamori, H.; McCombie, W.R.; Ouyang, S.; Schwartz, D.C.; Tanaka, T.; Wu, J.; Zhou, S.; et al. Improvement of the *Oryza sativa* Nipponbare reference genome using next generation sequence and optical map data. *Rice* **2013**, *6*, 4. [CrossRef]

26. Ohyanagi, H.; Tanaka, T.; Sakai, H.; Shigemoto, Y.; Yamaguchi, K.; Habara, T.; Fujii, Y.; Antonio, B.A.; Nagamura, Y.; Imanishi, T.; et al. The Rice Annotation Project Database (RAP-DB): Hub for *Oryza sativa* ssp. japonica genome information. *Nucleic Acids Res.* **2006**, *34*, D741–D744. [CrossRef]

27. Zhao, W.; Wang, J.; He, X.; Huang, X.; Jiao, Y.; Dai, M.; Wei, S.; Fu, J.; Chen, Y.; Ren, X.; et al. BGI-RIS: An integrated information resource and comparative analysis workbench for rice genomics. *Nucleic Acids Res.* **2004**, *32*, D377–D382. [CrossRef]

28. Tian, T.; Liu, Y.; Yan, H.; You, Q.; Yi, X.; Du, Z.; Xu, W.; Su, Z. agriGO v2.0: A GO analysis toolkit for the agricultural community, 2017 update. *Nucleic Acids Res.* **2017**, *45*, W122–W129. [CrossRef]

29. Schmittgen, T.D.; Livak, K.J. Analyzing real-time PCR data by the comparative C(T) method. *Nat. Protoc.* **2008**, *3*, 1101–1108. [CrossRef]

30. Russo, P.S.T.; Ferreira, G.R.; Cardozo, L.E.; Bürger, M.C.; Arias-Carrasco, R.; Maruyama, S.R.; Hirata, T.D.C.; Lima, D.S.; Passos, F.M.; Fukutani, K.F.; et al. CEMiTool: A Bioconductor package for performing comprehensive modular co-expression analyses. *BMC Bioinform.* **2018**, *19*, 56. [CrossRef]

31. Korotkevich, G.; Sukhov, V.; Budin, N.; Shpak, B.; Artyomov, M.N.; Sergushichev, A. Fast gene set enrichment analysis. *bioRxiv* **2021**. [CrossRef]

32. Yu, G.; Wang, L.G.; Han, Y.; He, Q.Y. clusterProfiler: An R package for comparing biological themes among gene clusters. *Omics* **2012**, *16*, 284–287. [CrossRef]

33. Goodstein, D.M.; Shu, S.; Howson, R.; Neupane, R.; Hayes, R.D.; Fazo, J.; Mitros, T.; Dirks, W.; Hellsten, U.; Putnam, N.; et al. Phytozome: A comparative platform for green plant genomics. *Nucleic Acids Res.* **2012**, *40*, D1178–D1186. [CrossRef]

34. Lin, C.T.; Xu, T.; Xing, S.L.; Zhao, L.; Sun, R.Z.; Liu, Y.; Moore, J.P.; Deng, X. Weighted Gene Co-expression Network Analysis (WGCNA) Reveals the Hub Role of Protein Ubiquitination in the Acquisition of Desiccation Tolerance in *Boea hygrometrica*. *Plant Cell Physiol.* **2019**, *60*, 2707–2719. [CrossRef] [PubMed]

35. Doncheva, N.T.; Morris, J.H.; Gorodkin, J.; Jensen, L.J. Cytoscape StringApp: Network Analysis and Visualization of Proteomics Data. *J. Proteome Res.* **2019**, *18*, 623–632. [CrossRef] [PubMed]

36. Szklarczyk, D.; Gable, A.L.; Lyon, D.; Junge, A.; Wyder, S.; Huerta-Cepas, J.; Simonovic, M.; Doncheva, N.T.; Morris, J.H.; Bork, P.; et al. STRING v11: Protein-protein association networks with increased coverage, supporting functional discovery in genome-wide experimental datasets. *Nucleic Acids Res.* **2019**, *47*, D607–D613. [CrossRef] [PubMed]

37. Kanehisa, M.; Sato, Y.; Kawashima, M. KEGG mapping tools for uncovering hidden features in biological data. *Protein Sci.* **2022**, *31*, 47–53. [CrossRef] [PubMed]

38. Guberman, J.M.; Ai, J.; Arnaiz, O.; Baran, J.; Blake, A.; Baldock, R.; Chelala, C.; Croft, D.; Cros, A.; Cutts, R.J.; et al. BioMart Central Portal: An open database network for the biological community. *Database* **2011**, *2011*, bar041. [CrossRef]

39. Jan, A.; Maruyama, K.; Todaka, D.; Kidokoro, S.; Abo, M.; Yoshimura, E.; Shinozaki, K.; Nakashima, K.; Yamaguchi-Shinozaki, K. OsTZF1, a CCCH-tandem zinc finger protein, confers delayed senescence and stress tolerance in rice by regulating stress-related genes. *Plant Physiol.* **2013**, *161*, 1202–1216. [CrossRef] [PubMed]

40. Sakuraba, Y.; Kim, Y.S.; Han, S.H.; Lee, B.D.; Paek, N.C. The Arabidopsis Transcription Factor NAC016 Promotes Drought Stress Responses by Repressing AREB1 Transcription through a Trifurcate Feed-Forward Regulatory Loop Involving NAP. *Plant Cell* **2015**, *27*, 1771–1787. [CrossRef]

41. Liu, J.; Shen, J.; Xu, Y.; Li, X.; Xiao, J.; Xiong, L. Ghd2, a CONSTANS-like gene, confers drought sensitivity through regulation of senescence in rice. *J. Exp. Bot.* **2016**, *67*, 5785–5798. [CrossRef]

42. Comas, L.H.; Becker, S.R.; Cruz, V.M.; Byrne, P.F.; Dierig, D.A. Root traits contributing to plant productivity under drought. *Front. Plant Sci.* **2013**, *4*, 442. [CrossRef]

43. Fageria, N.K.; Moreira, A. The role of mineral nutrition on root growth of crop plants. In *Advances in Agronomy*; Sparks, D.L., Ed.; Elsevier Inc.: Amsterdam, The Netherlands, 2011; Volume 110, pp. 251–331.

44. Kim, H.; Lee, K.; Hwang, H.; Bhatnagar, N.; Kim, D.Y.; Yoon, I.S.; Byun, M.O.; Kim, S.T.; Jung, K.H.; Kim, B.G. Overexpression of PYL5 in rice enhances drought tolerance, inhibits growth, and modulates gene expression. *J. Exp. Bot.* **2014**, *65*, 453–464. [CrossRef]

45. Nakashima, K.; Tran, L.S.; Van Nguyen, D.; Fujita, M.; Maruyama, K.; Todaka, D.; Ito, Y.; Hayashi, N.; Shinozaki, K.; Yamaguchi-Shinozaki, K. Functional analysis of a NAC-type transcription factor OsNAC6 involved in abiotic and biotic stress-responsive gene expression in rice. *Plant J.* **2007**, *51*, 617–630. [CrossRef]

46. Lee, D.K.; Chung, P.J.; Jeong, J.S.; Jang, G.; Bang, S.W.; Jung, H.; Kim, Y.S.; Ha, S.H.; Choi, Y.D.; Kim, J.K. The rice OsNAC6 transcription factor orchestrates multiple molecular mechanisms involving root structural adaptions and nicotianamine biosynthesis for drought tolerance. *Plant Biotechnol. J.* **2017**, *15*, 754–764. [CrossRef]

47. Cruz de Carvalho, M.H. Drought stress and reactive oxygen species: Production, scavenging and signaling. *Plant Signal. Behav.* **2008**, *3*, 156–165. [CrossRef]

48. Xia, T.; Chen, H.; Dong, S.; Ma, Z.; Ren, H.; Zhu, X.; Fang, X.; Chen, F. OsWUS promotes tiller bud growth by establishing weak apical dominance in rice. *Plant J.* **2020**, *104*, 1635–1647. [CrossRef]

49. Zawaski, C.; Busov, V.B. Roles of gibberellin catabolism and signaling in growth and physiological response to drought and short-day photoperiods in Populus trees. *PLoS ONE* **2014**, *9*, e86217. [CrossRef]

50. Wang, B.; Wei, H.; Xue, Z.; Zhang, W.H. Gibberellins regulate iron deficiency-response by influencing iron transport and translocation in rice seedlings (*Oryza sativa*). *Ann. Bot.* **2017**, *119*, 945–956. [CrossRef]

51. Nieves-Cordones, M.; García-Sánchez, F.; Pérez-Pérez, J.G.; Colmenero-Flores, J.M.; Rubio, F.; Rosales, M.A. Coping with Water Shortage: An Update on the Role of K$^+$, Cl$^-$, and Water Membrane Transport Mechanisms on Drought Resistance. *Front. Plant Sci.* **2019**, *10*, 1619. [CrossRef]

52. Tripathi, D.K.; Singh, S.; Gaur, S.; Singh, S.; Yadav, V.; Liu, S.; Singh, V.P.; Sharma, S.; Srivastava, P.; Prasad, S.M.; et al. Acquisition and homeostasis of iron in higher plants and their probable role in abiotic stress tolerance. *Front. Environ. Sci.* **2018**, *5*, 86. [CrossRef]

53. You, J.; Chan, Z. ROS Regulation during Abiotic Stress Responses in Crop Plants. *Front. Plant Sci.* **2015**, *6*, 1092. [CrossRef]

54. Kobayashi, T.; Nagasaka, S.; Senoura, T.; Itai, R.N.; Nakanishi, H.; Nishizawa, N.K. Iron-binding haemerythrin RING ubiquitin ligases regulate plant iron responses and accumulation. *Nat. Commun.* **2013**, *4*, 2792. [CrossRef]

55. Kobayashi, T.; Itai, R.N.; Senoura, T.; Oikawa, T.; Ishimaru, Y.; Ueda, M.; Nakanishi, H.; Nishizawa, N.K. Jasmonate signaling is activated in the very early stages of iron deficiency responses in rice roots. *Plant Mol. Biol.* **2016**, *91*, 533–547. [CrossRef] [PubMed]

56. Aung, M.S.; Kobayashi, T.; Masuda, H.; Nishizawa, N.K. Rice HRZ ubiquitin ligases are crucial for response to excess iron. *Physiol. Plant* **2018**, *163*, 282–296. [CrossRef] [PubMed]

57. Batth, R.; Singh, K.; Kumari, S.; Mustafiz, A. Transcript Profiling Reveals the Presence of Abiotic Stress and Developmental Stage Specific Ascorbate Oxidase Genes in Plants. *Front. Plant Sci.* **2017**, *17*, 198. [CrossRef] [PubMed]

58. You, J.; Hu, H.; Xiong, L. An ornithine δ-aminotransferase gene OsOAT confers drought and oxidative stress tolerance in rice. *Plant Sci.* **2012**, *197*, 59–69. [CrossRef] [PubMed]

59. Kim, S.G.; Lee, J.S.; Bae, H.H.; Kim, J.T.; Son, B.Y.; Kim, S.L.; Baek, S.-B.; Shin, S.; Jeon, W.-T. Physiological and proteomic analyses of Korean F1 maize (*Zea mays* L.) hybrids under water-deficit stress during flowering. *Appl. Biol. Chem.* **2019**, *62*, 32. [CrossRef]

60. Rodrigues, F.A.; de Laia, M.L.; Zingaretti, S.M. Analysis of gene expression profiles under water stress in tolerant and sensitive sugarcane plants. *Plant Sci.* **2009**, *176*, 286–302. [CrossRef]

61. Ahn, J.C.; Kim, D.W.; You, Y.N.; Seok, M.S.; Park, J.M.; Hwang, H.; Kim, B.G.; Luan, S.; Park, H.S.; Cho, H.S. Classification of rice (*Oryza sativa* L. Japonica nipponbare) immunophilins (FKBPs, CYPs) and expression patterns under water stress. *BMC Plant Biol.* **2010**, *10*, 253. [CrossRef]

62. Newstead, S. Molecular insights into proton coupled peptide transport in the PTR family of oligopeptide transporters. *Biochim. Biophys. Acta* **2015**, *1850*, 488–499. [CrossRef]

63. Tang, Z.; Chen, Y.; Chen, F.; Ji, Y.; Zhao, F.J. OsPTR7 (OsNPF8.1), a Putative Peptide Transporter in Rice, is Involved in Dimethylarsenate Accumulation in Rice Grain. *Plant Cell Physiol.* **2017**, *58*, 904–913. [CrossRef]

64. Yang, X.; Xia, X.; Zeng, Y.; Nong, B.; Zhang, Z.; Wu, Y.; Tian, Q.; Zeng, W.; Gao, J.; Zhou, W.; et al. Genome-wide identification of the peptide transporter family in rice and analysis of the PTR expression modulation in two near-isogenic lines with different nitrogen use efficiency. *BMC Plant Biol.* **2020**, *20*, 193. [CrossRef] [PubMed]

65. Wei, H.; Wang, X.; He, Y.; Xu, H.; Wang, L. Clock component OsPRR73 positively regulates rice salt tolerance by modulating OsHKT2;1-mediated sodium homeostasis. *EMBO J.* **2021**, *40*, e105086. [CrossRef] [PubMed]

66. Cho, E.K.; Choi, Y.J. A nuclear-localized HSP70 confers thermoprotective activity and drought-stress tolerance on plants. *Biotechnol. Lett.* **2009**, *31*, 597–606. [CrossRef] [PubMed]

67. Sarkar, N.K.; Kundnani, P.; Grover, A. Functional analysis of Hsp70 superfamily proteins of rice (*Oryza sativa*). *Cell Stress Chaperones* **2013**, *18*, 427–437. [CrossRef]

68. Pan, L.; Zhang, X.; Wang, J.; Ma, X.; Zhou, M.; Huang, L.; Nie, G.; Wang, P.; Yang, Z.; Li, J. Transcriptional Profiles of Drought-Related Genes in Modulating Metabolic Processes and Antioxidant Defenses in *Lolium multiflorum*. *Front. Plant Sci.* **2016**, *7*, 519. [CrossRef]

69. Xing, H.; Fu, X.; Yang, C.; Tang, X.; Guo, L.; Li, C.; Xu, C.; Luo, K. Genome-wide investigation of pentatricopeptide repeat gene family in poplar and their expression analysis in response to biotic and abiotic stresses. *Sci. Rep.* **2018**, *8*, 2817. [CrossRef]

70. Jiang, S.C.; Mei, C.; Liang, S.; Yu, Y.T.; Lu, K.; Wu, Z.; Wang, X.F.; Zhang, D.P. Crucial roles of the pentatricopeptide repeat protein SOAR1 in Arabidopsis response to drought, salt and cold stresses. *Plant Mol. Biol.* **2015**, *88*, 369–385. [CrossRef]

71. Bashir, K.; Hanada, K.; Shimizu, M.; Seki, M.; Nakanishi, H.; Nishizawa, N.K. Transcriptomic analysis of rice in response to iron deficiency and excess. *Rice* **2014**, *7*, 18. [CrossRef]

72. Yang, X.; Liu, X.; Zhou, Y.; Zhang, F.; Huang, L.; Wang, J.; Song, J.; Qiu, L. New insights on the function of plant acyl carrier proteins from comparative and evolutionary analysis. *Genomics* **2021**, *113*, 1155–1165. [CrossRef]

73. Masud, A.J.; Kastaniotis, A.J.; Rahman, M.T.; Autio, K.J.; Hiltunen, J.K. Mitochondrial acyl carrier protein (ACP) at the interface of metabolic state sensing and mitochondrial function. *Biochim. Biophys. Acta Mol. Cell Res.* **2019**, *1866*, 118540. [CrossRef]

74. Huang, J.; Xue, C.; Wang, H.; Wang, L.; Schmidt, W.; Shen, R.; Lan, P. Genes of ACYL CARRIER PROTEIN Family Show Different Expression Profiles and Overexpression of ACYL CARRIER PROTEIN 5 Modulates Fatty Acid Composition and Enhances Salt Stress Tolerance in Arabidopsis. *Front. Plant Sci.* **2017**, *8*, 987. [CrossRef]

75. Chatterjee, J.; Patra, B.; Mukherjee, R.; Basak, P.; Mukherjee, S.; Ray, S.; Bhattacharyya, S.; Maitra, S.; GhoshDastidar, K.; Ghosh, S.; et al. Cloning, characterization and expression of a chloroplastic fructose-1, 6-bisphosphatase from *Porteresia coarctata* conferring salt-tolerance in transgenic tobacco. *Plant Cell Tissue Organ Cult.* **2013**, *114*, 395–409. [CrossRef]

76. Shen, J.; Zou, Z.; Zhang, X.; Zhou, L.; Wang, Y.; Fang, W.; Zhu, X. Metabolic analyses reveal different mechanisms of leaf color change in two purple-leaf tea plant (*Camellia sinensis* L.) cultivars. *Hortic. Res.* **2018**, *5*, 7. [CrossRef]

77. Ye, N.; Zhu, G.; Liu, Y.; Li, Y.; Zhang, J. ABA controls H_2O_2 accumulation through the induction of OsCATB in rice leaves under water stress. *Plant Cell Physiol.* **2011**, *52*, 689–698. [CrossRef]

78. Greco, M.; Chiappetta, A.; Bruno, L.; Bitonti, M.B. In *Posidonia oceanica*, cadmium induces changes in DNA methylation and chromatin patterning. *J. Exp. Bot.* **2012**, *63*, 695–709. [CrossRef]

79. Pandey, P.; Irulappan, V.; Bagavathiannan, M.V.; Senthil-Kumar, M. Impact of Combined Abiotic and Biotic Stresses on Plant Growth and Avenues for Crop Improvement by Exploiting Physio-morphological Traits. *Front. Plant Sci.* **2017**, *8*, 537. [CrossRef]

Article

Photosystem II Repair Cycle in Faba Bean May Play a Role in Its Resistance to *Botrytis fabae* Infection

María Ángeles Castillejo [1,2,*], Ángel M. Villegas-Fernández [2], Tamara Hernández-Lao [1] and Diego Rubiales [2]

[1] Agroforestry and Plant Biochemistry, Proteomics and Systems Biology, Department of Biochemistry and Molecular Biology, University of Cordoba, UCO-CeiA3, 14014 Córdoba, Spain; b42helat@uco.es

[2] Institute for Sustainable Agriculture, Spanish National Research Council (CSIC), Avda. Menéndez Pidal, s/n, 14004 Córdoba, Spain; avillegas@ias.csic.es (Á.M.V.-F.); diego.rubiales@ias.csic.es (D.R.)

* Correspondence: bb2casam@uco.es

Abstract: Chocolate spot, which is caused by the necrotrophic fungus *Botrytis fabae*, is a major foliar disease occurring worldwide and dramatically reducing crop yields in faba bean (*Vicia faba*). Although chemical control of this disease is an option, it has serious economic and environmental drawbacks that make resistant cultivars a more sensible choice. The molecular mechanisms behind the defense against *B. fabae* are poorly understood. In this work, we studied the leave proteome in two faba bean genotypes that respond differently to *B. fabae* in order to expand the available knowledge on such mechanisms. For this purpose, we used two-dimensional gel electrophoresis (2DE) in combination with Matrix-Assisted Laser Desorption/Ionization (MALDI-TOF/TOF). Univariate statistical analysis of the gels revealed 194 differential protein spots, 102 of which were identified by mass spectrometry. Most of the spots belonged to proteins in the energy and primary metabolism, degradation, redox or response to stress functional groups. The MS results were validated with assays of protease activity in gels. Overall, they suggest that the two genotypes may respond to *B. fabae* with a different PSII protein repair cycle mechanism in the chloroplast. The differences in resistance to *B. fabae* may be the result of a metabolic imbalance in the susceptible genotype and of a more efficient chloroplast detoxification system in the resistant genotype at the early stages of infection.

Keywords: *Botrytis fabae*; faba bean; resistance; proteomic analysis; photosystem II repair cycle

Citation: Castillejo, M.Á.; Villegas-Fernández, Á.M.; Hernández-Lao, T.; Rubiales, D. Photosystem II Repair Cycle in Faba Bean May Play a Role in Its Resistance to *Botrytis fabae* Infection. *Agronomy* **2021**, *11*, 2247. https://doi.org/10.3390/agronomy11112247

Academic Editor: Giovanni Caruso

Received: 11 September 2021
Accepted: 3 November 2021
Published: 6 November 2021

Publisher's Note: MDPI stays neutral with regard to jurisdictional claims in published maps and institutional affiliations.

1. Introduction

By virtue of its high nutritional value, faba bean (*Vicia faba* L.) is an important food legume for human consumption and livestock feeding [1]. In fact, it is regarded as an excellent protein crop on the basis of its ability to provide nitrogen inputs into temperate agricultural systems, and also because of its increased yield potential and nitrogen-fixing capacity relative to other grain legumes [2,3]. Faba bean is the fourth most widely grown cool season grain legume (pulse) globally after pea (*Pisum sativum*), chickpea (*Cicer arietinum*) and lentil (*Lens culinaris*) (FAOSTAT 2019; https://www.fao.org, accessed on 15 September 2021). However, its yield is greatly affected by some environmental conditions, including biotic and abiotic stresses [3,4].

The necrotrophic fungus *Botrytis fabae* Sard. (teleomorph: *Botryotinia fabae* Lu & Wu) causes chocolate spot, which is one of the most destructive diseases for faba bean plants worldwide [5,6]. Infected plants exhibit chocolate-colored lesions on aboveground parts and, especially, on leaves. The disease, which starts in bean crops where inoculum is present in residues from previous years or in contaminated seeds [7], may be especially aggressive under high humidity and temperature conditions. Such conditions can lead to extensive necrosis of plant tissues and severe damage, and can also favor the spread of the pathogen to other plants [8]. Prolonged favorable conditions for *B. fabae* growth can result in considerable economic crop losses through reduced grain yields and quality [9]. The severity of chocolate spot epidemics can be mitigated with integrated disease management strategies, such as the use of clean seeds, crop rotations, lower planting densities, application of fungicides and selection of more resistant plant varieties [6,10]. Although some germplasm accessions have shown moderate to high levels of resistance [11], resistant cultivars adapted to different cultivation areas are scarcely available.

As a necrotrophic pathogen, *B. fabae* must kill and decompose host cells in order to feed on them. The fungus can infect plants via a variety of mechanisms mediated by lytic enzymes, toxins, stress-induced reactive oxygen species (ROS), necrosis-secreted proteins and a wide variety of secondary metabolites [12–16]. On the other hand, plants can stop the progression of the fungus by using constitutive or infection-induced mechanisms. Such mechanisms can be of the physical (cuticle and cell wall) [17,18] or chemical type (phytoanticipins and phytoalexins) [19], but can also involve induction of pathogenesis-related proteins or defensins, or accumulation of antimicrobial compounds [20,21].

Plants are known to accumulate ROS in response to necrotrophic fungi [22–27] and faba bean cultivars have been found to respond to *B. fabae* with differential ROS accumulation, lipid peroxidation and enzymatic ROS scavenging activity [28]. Recently, enhanced functionality in photosystem II (PSII), probably resulting from ROS accumulation in response to short-time exposure to *B. cinerea*, was reported in tomato plants [27]. However, little is known about the specific molecular mechanisms by which plants respond to *B. fabae*. A transcription factor (TF) analysis of the response of *M. truncatula* to *B. fabae* and *B. cinerea* [29] revealed some TFs to be involved in differential responses and others to be responsible for resistance to the two pathogens.

To our knowledge, few omics studies have focused on plant responses to *Botrytis*. As confirmed by using mutants at the transcriptomic level, *Arabidopsis thaliana* and tomato possess some genes whose expression is related to *B. cinerea* resistance. Such defense-related genes include some encoding PR protein 1 (PR1), β-1,3-glucanase and subtilisin-like protease, and other proteins involved in secondary metabolite synthesis (reviewed in [15]), but still others are involved in responses to abiotic stresses, such as signaling hormone pathways, which affect photosynthesis, and protein synthesis and transport [15,30,31]. Even fewer proteomics studies have addressed plant–*Botrytis* interactions [32,33] and most have focused on the pathogen *B. cinerea* (reviewed in [15]). Thus, Marra et al. used 2DE coupled to Matrix-Assisted Laser Desorption/Ionization (MALDI-TOF) analysis [32] to examine the interaction of beans with *B. cinerea* and *Trichoderma*. They found pathogenesis-related proteins and other disease-related factors, such as potential resistance genes, to be seemingly associated with interactions with both the pathogen and *Trichoderma*. In addition, a shotgun proteomic study of *B. cinerea*-infected tomato fruit at different ripening stages identified a substantial number of proteins responsible for pathogenicity (mainly PR and disease resistance proteins, proteases and peroxidases), as well as others protecting the fruit from the oxidative stress response by the host [33].

In the absence of a reference genome assembly for *Vicia faba* owing to its enormous size (13 Gbp) and complexity (e.g., abundance of transposable elements), high-throughput methods, such as transcriptome analysis, have proved efficient for enriching genomic resources (reviewed in [1]). However, only limited DNA sequence data from reported transcriptome datasets have been made available on public databases [34]. Using high-throughput omic technology can no doubt help expand existing knowledge of the plant–pathogen interaction and provide a basis for developing improved crop breeding programs. The main aim of this work was to go deeper into the knowledge of the molecular mechanisms underlying the defense against *B. fabae* in faba bean. For this purpose, we studied the leave proteome in two faba bean genotypes that respond differently to *B. fabae* by using two-dimensional gel electrophoresis (2DE) in combination with (MALDI-TOF/TOF) mass spectrometry (MS). Some results of the MS analysis were validated by assays of protease activity in gels. Overall, the results suggest that the key to stopping the spread of the pathogen onto leaves is mainly a regulatory ROS production mechanism occurring in the chloroplast.

2. Materials and Methods

2.1. Plant Material and Sample Collection

Two faba bean genotypes known to exhibit a contrasting response to *B. fabae* were used, namely: Baraca as susceptible genotype and BPL710 as resistant genotype [10,35]. Seeds of the two genotypes were grown in 1 L pots filled with a (1:1) sand–peat mixture under controlled conditions: $(20 \pm 2)\,^{\circ}$C, 12 h dark/12 h light photoperiod and a photon flux density of 150 µmol m^{-2} s^{-1}.

A total 27 plants per genotype were used, with leaves being sampled from 12 inoculated plants and 12 uninoculated (control) plants of each. Three other plants were inoculated and were used to score disease symptoms. Plants were inoculated according to Villegas-Fernandez et al. [35], with *B. fabae* local monosporic isolate (Bf-CO-05) being grown on Petri dishes containing V8 medium and spores suspended at a 4.5×10^5 spores/mL concentration in a glucose/water solution (1.2% *w/v*). Three-week-old plants were then sprayed with the suspension at a rate of 1.5 mL/plant and incubated in a growth chamber at a relative humidity above 95% in the dark. By contrast, control (uninoculated) plants were sprayed with a glucose solution containing no spores. Disease symptoms were evaluated after restoring the photoperiod 24 h later but keeping the relative humidity above 90%. Evaluations were done two and six days after inoculation (dai). Disease severity (DS) was calculated by visual estimation of the proportion of plant surface covered with chocolate spots, with estimates being corrected by increasing the weight of the aggressive lesions by 50% with the formula DS = (% nonaggressive lesions) + 1.5 × (% aggressive lesions). The results thus obtained were analyzed statistically with the software Statistix 8 (Analytical Software, Tallahassee, FL, USA). Data were subjected to the arcsin $\sqrt{x}$ transformation in order to offset evaluation bias and to increase normality in their distribution prior to analysis of variance (ANOVA).

Faba leaves for proteomic analyses were collected at two different times while both control and inoculated plants were still under incubation in the dark [35]. Sampling was done 6 h post-inoculation (hpi)—an early time at which no symptoms were apparent— and then 12 hpi—when the earliest symptoms of chocolate spots became macroscopically visible. All of the leaves from six individual plants (three biological replicates from two plants each) per condition (treatment and sampling time) were collected, frozen in liquid nitrogen and stored at $-80\,^{\circ}$C until protein extraction.

2.2. Protein Extraction and Gel Electrophoresis

Faba leaves samples (ca. 0.5 g fresh weight) from three independent replicates per treatment, sampling time and genotype were crushed with liquid nitrogen in a precooled mortar to obtain a fine powder. Proteins were extracted into TCA–phenol [36] and the resulting pellets were resuspended in a solubilization buffer containing 7 M urea (Merck, Kenilworth, NJ, USA), 2 M thiourea (Sigma–Aldrich, St. Louis, MI, USA), 2% (*w/v*) CHAPS (Sigma–Aldrich), 2% (*v/v*) Bio-Lyte 3–10 carrier ampholytes (BioRad, Hercules, CA, USA), 2% (*w/v*) DTT (Sigma–Aldrich) and Bromophenol Blue traces (Sigma–Aldrich). Protein concentrations were determined with the Bradford assay (BioRad) and proteins were then separated in 2D electrophoresis gels.

For 2DE analysis, 18-cm IPG DryStrips (Amersham Biosciences, Amersham, UK) were used with nonlinear pH gradients over the range 3–10. Strips were rehydrated passively for 6 h and then actively at 50 V for a further 6 h with 300 µL of sample buffer containing an amount of 400 µg of protein. Strips were loaded onto a PROTEAN IEF System (BioRad), focused at 20 °C with an increasing linear voltage and equilibrated according to Castillejo et al. [37]. They were then transferred onto vertical slabs of 10% SDS polyacrylamide gels. Electrophoresis runs were done at 30 V at 15 °C for 30 min, and then at 60 V for about 14 h until the dye front reached the bottom of the gel. The gels were loaded with broad-range molecular markers (Bio-Rad) and, after electrophoresis, stained with Coomassie Brilliant Blue G-250 [38].

2.3. Image Acquisition and Statistical Analysis

Gels were scanned with the Molecular imager FX ProPlus Multi-imager system (BioRad) and the images thus obtained (Supplementary Figure S1) were analyzed with the software PDQuest Advanced v. 8.0.1 (BioRad), using 10 times the background signal as the presence threshold for spots. The quantitative data gathered from the spots in each gel (viz., normalized spot volumes given as individual spot intensity/normalization factor) were used to designate differences when comparing gel images. A multivariate statistical analysis of the entire data set was performed by using the web-based software tool NIA array [39]. Those spots showing significant differences ($p \leq 0.05$) in intensity, exhibiting a minimum change of ± 2 and being consistently present among replicates were selected for further MS/MS analysis.

2.4. Protein Identification by Mass Spectrometry (MALDI-TOF/TOF)

Differential gel spots were excised for digestion with trypsin [40] and peptide fragments from digested proteins were analyzed by mass spectrometry. For that purpose, peptides were crystallized in an α-cyano-4-hydroxycinnamic acid matrix and subjected to MALDI-TOF/TOF analysis over the m/z range 800–4000 by using a 4800 Proteomics Analyzer (Applied Biosystems, Foster City, CA, USA) at an accelerating voltage of 20 kV. Spectra were internally calibrated against peptides from trypsin autolysis ($M + H^+ = 842.509$, $M + H^+ = 2211.104$) and the five most abundant peptide ions in each spectrum were used for fragmentation analysis to elucidate peptide sequences. A combined peptide mass fingerprinting (PMF)/tandem mass spectrometry (MS + MSMS) search was performed by using the software GPS Explorer™.5 (Applied Biosystems) over the nonredundant NCBInr database restricted to *Viridiplantae* taxonomy in combination with the MASCOT search engine (Matrix Science, London; http://www.matrixscience.com accessed on 15 September 2019). The following parameters were allowed: a minimum of two peptides matches and a single trypsin miscleavage, and peptide modifications by carbamidomethylcysteine and methionine oxidation. The maximum tolerance for peptide mass matching was limited to 20 ppm. The score level and a minimum of four peptides per protein were chosen as PMF confidence parameters. Proteins were characterized in functional terms against the NCBInr database (https://www.ncbi.nlm.nih.gov/guide/proteins/, accessed on 15 October 2021). In addition, BLAST analysis (tblastn) was performed for all the identified proteins using the reference transcriptome *Vicia faba RefTrans v2* (2017), with 37,378 sequences deposited in Pulse Crops Database (https://www.pulsedb.org/, accessed on 15 October 2021). Only matches with an expectation (E) value of $\leq 1 \times 10^{-6}$ were considered. Mass spectrometry analyses were conducted at the Proteomics Facility of the Central Research Support Service (SCAI) of the University of Córdoba (Spain).

2.5. Zymography

Proteins from leaves (200 mg of frozen powdered tissue) were extracted with a mixture of 200 mM TrisHCl at pH 7.4, 3% (w/v) insoluble polyvinylpolypyrrolidone (PVPP), 10% (v/v) glycerol, 5 mM diethiothreitol (DTT) and 0.25% (v/v) Triton X-100. Samples were allowed to stand on ice for at least 10 min and were then centrifuged at $16,000 \times g$ at 4 °C for 30 min, with the proteins present in the supernatant then being quantified with the Bradford assay [41].

SDS-PAGE slabs containing 0.1% gelatin and 9% acrylamide were analyzed according to Heussen and Dowdle [42]. Thus, samples containing 100 μg of protein were diluted with a nondenaturing Laemmli buffer [62.5 mM TrisHCl, 10% (v/v) glycerol, 0.001% (w/v) Bromophenol Blue] and loaded onto 1 mm thick gel slabs for electrophoresis at 50 V at 4 °C for 30 min, with the voltage being raised to 80 V until the front reached the end of the gel. The gels were loaded with Spectra Multicolor Broad Range Protein Ladder (Thermo Scientific). After electrophoresis, gels were incubated in 2.5% (v/v) Triton X-100 at room temperature under constant agitation for 30 min to remove SDS. They were then washed with distilled water three times to remove Triton X-100 and incubated in a proteolysis buffer (100 mM citrate buffer, Na_2HPO_4/citric acid pH 6.8, 4 mM DTT and 10 mM cysteine) under constant agitation at 35 °C overnight. Proteolysis was stopped by transferring the gels to a solution containing 0.1% (*w/v*) Coomassie Brilliant Blue R-250 [43]. Finally, the gels were destained in a solution containing 40% methanol and 10% acetic acid until clear bands formed over a dark blue background.

3. Results

3.1. Disease Assessment

As can be seen in Figure 1, twelve hpi symptoms of *B. fabae* infection were already visible in the susceptible genotype (Baraca), but not in the resistant genotype (BPL710). Analyses of variance of the DS results revealed that the Baraca genotype was strongly affected both 2 and 6 dpi (average DS 23.9 and 52.5, respectively). On the other hand, the BPL710 genotype was highly resistant (ANOVA $p \leq 0.05$) in both samplings (average DS 11.6 and 14.6, respectively) (Figure 1; Supplementary Figure S2).

Figure 1. Contrasting response to *B. fabae* infection of three-week-old faba bean plants of Baraca (susceptible genotype) and BPL710 (resistant genotype) 12 h after inoculation.

3.2. Two-Dimensional Gel Electrophoresis and MSMS Analysis

Image analysis with the software PDQuest allowed, on average, 224 individual protein spots to be detected (Figure 2a; Supplementary Table S1). In addition, a hierarchical clustering analysis clearly separated the genotypes into two clusters (Figure 2b), thus confirming the reproducibility of the experiment. Principal component analysis (PCA) allowed 194 differential protein spots from the entire dataset to be identified by comparing genotypes (susceptible and resistant) and treatments (uninoculated and inoculated) in both samplings (6 and 12 hpi) (Figure 2c). The first two principal components (PCs) jointly explained 70% of the total variability in the data, PC1 separating genotypes. The PCs for individual genotypes explained 74% and 86% of variability in the susceptible and resistant genotype, respectively (Figure 2d,e). In both cases, samples clustered by sampling times.

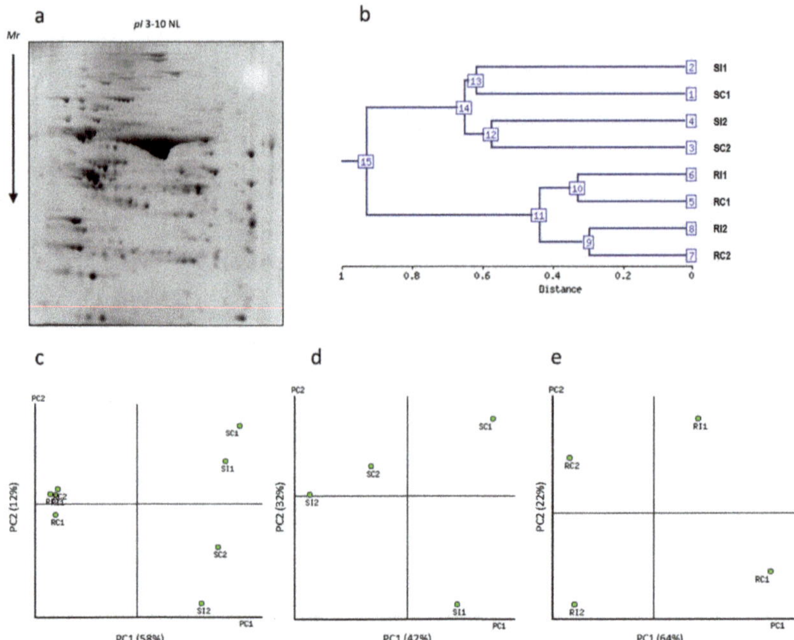

Figure 2. Typical Coomassie Brilliant Blue 2DE gel results for the susceptible genotype. (**a**) Dendrogram showing hierarchical clustering of experimental conditions. (**b**) Two-dimensional biplots showing associations between all experimental conditions in both genotypes (**c**), or independent genotypes (**d,e**), as generated by principal component analysis (PCA). Dendrogram and PCA data were obtained from average values under each set of experimental conditions: Susceptible (S) and resistant (R) genotypes; control (C) and *B. fabae* inoculated (I); 6 hpi (1) and 12 hpi (2).

Spots were classified as variable if they met the following criteria: (a) being consistently present or absent in the three replicates under each set of experimental conditions; (b) exhibiting at least a two-fold change in abundance ratio; and (c) exhibiting statistically significant differences ($p \leq 0.05$) between genotypes or treatments. A total of 129 protein spots were thus selected for MALDI-TOF/TOF analysis.

3.3. Protein Identification and Abundance Pattern Analysis

A protein search against the *Viridiplantae* index in the nonredundant NCBI database was performed and a total of 102 proteins were thus identified with high confidence (Table 1; Supplementary Figure S3), 70% of which matched legume species. Most of the proteins met the confidence identification criteria [viz., a score higher than 70 and at least four peptides per protein except for three spot proteins (viz., SSP 2802, 5002 and 11, which should be considered with caution)]. The proteins thus identified belonged to the main functional groups of energy and primary metabolism [photosynthesis (30) and other energy metabolism (6), carbohydrate (12) and amino acid metabolism (9)], followed by proteins of degradation (10), redox and response to stress groups (8) (Figure 3a).

Table 1. Proteins identified by MALDI-TOF MS analysis.

SSP [a]	Protein Name (% Identity to *V. faba* Transcriptome Entry) [b]	NCBI Accession	Score [c]	Species	PM [c] / Coverage %	*Mr/pI* Experimental (Theoretical) [d]	Functional Category	More/Less Abundance Change Ratio (FDR) [e]					
								S1	S2	R1	R2	$\frac{R}{S}$ 1	$\frac{R}{S}$ 2
3504	Enolase (93.3% *Vf_0033189*)	gi|42521309	261	*Glycine max*	10/28	57.2/5.7 (48.0/5.3)	Carbohydrate met.	3.0	6.3	1.0	1.0	0.0	0.0
6302	Isocitrate dehydrogenase [NADP], chloroplastic (85.1% *Vf_0028303*)	gi|2497259	153	*Medicago sativa*	17/40	44.9/6.2 (48.7/6.2)	Carbohydrate met.	0.4	9.0	0.8	0.6	1.2	0.6
2204	Fructose-1,6-bisphosphatase (93.2% *Vf_0032848*)	gi|5305145	109	*Pisum sativum*	4/14	40.0/5.5 (36.3/6.3)	Carbohydrate met.	16.4	6.7	1.7	0.2	0.4	0.9
6203	Fructose-bisphosphate aldolase, cytoplasmic isozyme 1 (77.9% *Vf_0020212*)	gi|1168408	352	*Pisum sativum*	15/54	39.4/6.5 (38.7/6.4)	Carbohydrate met.	1.1	3.8	1.5	1.5	0.3	0.1
4104	Fructose-bisphosphate aldolase 2, chloroplastic (80.6% *Vf_0020212*)	gi|461501	183	*Pisum sativum*	10/37	38.4/5.8 (38.0/5.5)	Carbohydrate met.	0.6	1.2	27.6	0.2	0.5	1.6
5103	Fructose-bisphosphate aldolase 2, chloroplastic (80.6% *Vf_0020212*)	gi|461501	474	*Pisum sativum*	17/51	36.8/6.3 (38.0/5.5)	Carbohydrate met.	3.8	1.9	27.7	0.0	0.3	1.5
105	N-glyceraldehyde-2-phosphotransferase-like (84.2% *Vf_0017868*)	gi|8885622	133	*Arabidopsis thaliana*	7/22	33.4/5.2 (32.0/5.1)	Carbohydrate met.	6.4	0.8	0.5	1.0	0.2	0.0
1007	Triose-phosphate isomerase (75.1% *Vf_0012679*)	gi|15226479	205	*Arabidopsis thaliana*	9/25	29.5/5.6 (33.6/7.7)	Carbohydrate met.	3.6	2.3	1.1	1.2	0.5	0.2
2002	Triose-phosphate isomerase (75.1% *Vf_0012679*)	gi|15226479	292	*Arabidopsis thaliana*	11/29	29.1/5.8 (33.6/7.7)	Carbohydrate met.	10.4	5.1	1.9	1.1	0.5	0.2
3004	Triose-phosphate isomerase (89.1% *Vf_0014793*)	gi|57283985	165	*Phaseolus vulgaris*	7/31	21.9/5.6 (27.4/5.9)	Carbohydrate met.	2.2	4.7	1.0	0.4	0.0	1.6
2705	Phosphoglucomutase, cytoplasmic (97.9% *Vf_0002214*)	gi|12585296	109	*Pisum sativum*	10/8	74.3/5.6 (63.5/5.5)	Carbohydrate met.	9.5	0.0	1.0	1.0	0.0	0.0
3702	Phosphoglucomutase, cytoplasmic (97.9% *Vf_0002214*)	gi|12585296	222	*Pisum sativum*	10/23	70.8/5.7 (63.5/5.5)	Carbohydrate met.	6.3	19.0	0.4	0.9	2.3	8.8
603	Beta-amylase (88.5% *Vf_0019893*)	gi|3913031	149	*Medicago sativa*	7/17	61.3/5.0 (56.5/5.3)	Major CHO met.	3.1	8.2	1.0	0.5	1.0	1.6
4808	Methionine synthase/Cobalamin-independent synthase family protein (94.5% *Vf_0005324*)	gi|219522337	104	*Cicer arietinum*	12/19	92.8/5.9 (84.6/6.0)	Amino acid met.	6.9	190.5	1.5	0.7	3.4	20.1
5805	Methionine synthase/Cobalamin-independent synthase family protein (94.5% *Vf_0005324*)	gi|219522337	312	*Cicer arietinum*	15/26	93.2/6.0 (84.6/6.0)	Amino acid met.	20.8	50.3	0.5	0.7	5.2	4.0
5808	Methionine synthase/Cobalamin-independent synthase family protein (94.5% *Vf_0005324*)	gi|219522337	344	*Cicer arietinum*	20/31	92.2/6.1 (84.6/6.0)	Amino acid met.	3.2	25.4	0.3	2.3	3.9	1.0
2402	Alanine aminotransferase (82.2% *Vf_0008300*)	gi|29569153	140	*Oryza sativa*	8/17	54.1/5.6 (54.0/8.0)	Amino acid met.	8.1	17.6	0.7	2.9	2.4	0.8
3402	Alanine aminotransferase 2 (82.2% *Vf_0008300*)	gi|29569153	104	*Oryza sativa*	4/9	54.1/5.7 (54.0/8.0)	Amino acid met.	63.4	0.0	1.3	1.4	51.7	11.9
1402	5-enolpyruvylshikimate 3-phosphate synthase (79.5% *Vf_0032499*)	gi|55740769	73	*Camptotheca acuminate*	8/14	52.1/5.2 (56.1/8.2)	Amino acid met.	0.0	1.0	1.0	1.0	0.0	1.0
3613	Ketol-acid reductoisomerase chloroplastic (97.3% *Vf_0004275*)	gi|6225542	560	*Pisum sativum*	23/43	61.5/5.7 (63.2/6.6)	Amino acid met.	12.0	0.5	0.7	0.4	2.2	0.4
7501	Serine hydroxymethyltransferase 2 (91.7% *Vf_0037308*)	gi|222142531	203	*Glycine max*	13/29	57.1/6.9 (55.0/8.2)	Amino acid met.	2.1	1.7	1.6	1.0	0.3	0.0

Table 1. *Cont.*

SSP [a]	Protein Name (% Identity to *V. faba* Transcriptome Entry) [b]	NCBI Accession	Score [c]	Species	PM [c]/ Coverage %	*Mr/pI* Experimental (Theoretical) [d]	Functional Category	More/Less Abundance Change Ratio (FDR) [e]					
								S1	S2	R1	R2	$\frac{R}{S}$ 1	$\frac{R}{S}$ 2
1104	Putative lactoylglutathione lyase (78.0% *Vf_0028431*)	gi|15810219	183	*Arabidopsis thaliana*	8/26	33.4/5.3 (32.0/5.1)	Amino acid met.	3.7	1.2	0.6	0.4	0.5	0.4
7602	Chain A, Dihydrolipoamide Dehydrogenase (99.6% *Vf_0005095*)	gi|9955321	337	*Pisum sativum*	22/62	57.4/6.4 (50.0/6.1)	Lipid metabolism	51.2	25.1	0.9	0.8	7.7	1.9
2102	Pyruvate dehydrogenase E1 component subunit beta, mitochondrial (96.1% *Vf_0005095*)	gi|1709454	171	*Pisum sativum*	6/18	39.3/5.5 (39.0/5.9)	Lipid metabolism	0.7	6.7	1.0	1.0	0.0	0.0
3705	Zeaxanthin epoxidase, chloroplastic (72.1% *Vf_0030326*)	gi|5902706	92	*Solanum lycopersicum*	5/9	74.3/5.8 (73.6/6.2)	Hormone met.	1.0	1.0	0.6	1.1	∞	∞
5601	Polyphenol oxidase A1, chloroplastic (98.5% *Vf_0006701*)	gi|1172586	254	*Vicia faba*	8/8	64.9/6.0 (68.9/7.0)	Pigment biosynt.	0.0	1.0	1.0	1.0	0.0	1.0
5603	Polyphenol oxidase A1, chloroplastic (98.5% *Vf_0006701*)	gi|1172586	253	*Vicia faba*	9/13	64.7/6.0 (68.9/7.0)	Pigment biosynt.	9.5	0.0	1.0	1.0	0.0	0.0
4607	Polyphenol oxidase A1, chloroplastic (98.5% *Vf_0006701*)	gi|1172586	274	*Vicia faba*	9/12	64.4/5.9 (68.9/7.0)	Pigment biosynt.	1.0	1.0	0.0	1.0	∞	1.0
4605	Polyphenol oxidase A1, chloroplastic (98.5% *Vf_0006701*)	gi|1172586	151	*Vicia faba*	11/13	61.8/5.8 (68.9/7.0)	Pigment biosynt.	8.7	0.6	1.0	1.0	0.0	0.0
2103	Coproporphyrinogen-III oxidase, chloroplastic (70.9% *Vf_0022525*)	gi|2493810	148	*Nicotiana tabacum*	4/11	38.6/5.6 (45.3/7.6)	Co-factor and vitamine met.	7.7	1.2	1.0	1.0	0.0	0.0
2601	ATP synthase CF1 alpha subunit (96.7% *Vf_0021629*)	gi|219673973	546	*Trifolium subterraneum*	19/36	59.9/5.4 (55.7/5.1)	Energy metabolism	2.1	2.7	28.2	0.0	1.5	4.6
1601	ATP synthase CF1 alpha subunit (95.3% *Vf_0021629*)	gi|139387459	126	*Phaseolus vulgaris*	9/20	57.0/5.3 (55.7/5.2)	Energy metabolism	0.6	1.2	1.5	1.3	22.4	2.3
604	ATP synthase CF1 beta subunit ATP synthase alpha/beta family protein (97.6% *Vf_0007913*)	gi|295136979	572	*Pisum sativum*	18/48	58.9/5.2 (53.2/5.1)	Energy metabolism	0.7	2.7	1.0	1.6	9.0	2.3
505	ATP synthase CF1 beta subunit ATP synthase alpha/beta family protein (97.6% *Vf_0007913*)	gi|295136979	882	*Pisum sativum*	26/65	57.9/5.2 (53.2/5.1)	Energy metabolism	2.0	17.9	1.2	1.9	5.8	9.5
1503	ATP synthase CF1 alpha subunit (95.4% *Vf_0021629*)	gi|295137014	573	*Pisum sativum*	21/38	56.5/5.3 (54.7/5.7)	Energy metabolism	4.5	3.2	1.2	0.9	3.9	2.7
1501	ATP synthase CF1 beta subunit ATP synthase alpha/beta family protein (97.6% *Vf_0007913*)	gi|295136979	755	*Pisum sativum*	21/51	56.2/5.2 (53.2/5.1)	Energy metabolism	2.4	1.5	1.3	1.2	1.5	1.2
9701	Sulfite reductase (96.4% *Vf_0029426*)	gi|119225844	129	*Pisum sativum*	17/25	75.6/8.8 (77.3/9.1)	S metabolism	∞	1.0	1.0	1.0	1.0	1.0
3704	Transketolase (95.5% *Vf_0028016*)	gi|4586600	214	*Cicer arietinum*	3/30	82.8/5.8 (17.1/5.8)	Photosynthesis	11.3	7.0	1.2	1.0	35.9	3.2
3706	Transketolase (95.5% *Vf_0028016*)	gi|4586600	91	*Cicer arietinum*	4/44	82.7/5.8 (17.1/5.8)	Photosynthesis	4.0	4.0	1.0	0.5	18.0	2.8
4704	Transketolase (95.5% *Vf_0028016*)	gi|4586600	86	*Cicer arietinum*	4/44	82.5/5.9 (17.1/5.8)	Photosynthesis	3.8	17.1	0.6	0.7	6.4	1.7
4702	Transketolase (95.5% *Vf_0028016*)	gi|4586600	189	*Cicer arietinum*	5/48	81.6/5.8 (17.1/5.8)	Photosynthesis	3.8	3.1	1.0	1.1	8.1	1.5
1604	RuBisCO large subunit-binding protein subunit beta chloroplastic/chaperonin subunit beta (96.3% *Vf_0035079*)	gi|2506277	423	*Pisum sativum*	16/39	67.8/5.5 (63.3/5.8)	Photosynthesis	19.1	30.5	0.9	0.4	11.5	9.4
2602	RuBisCO large subunit-binding protein subunit beta chloroplastic/chaperonin subunit beta (96.3% *Vf_0035079*)	gi|2506277	72	*Pisum sativum*	11/27	66.8/5.6 (63.3/5.8)	Photosynthesis	1.8	2.4	2.2	0.5	1.9	1.5

Table 1. *Cont.*

SSP [a]	Protein Name (% Identity to *V. faba* Transcriptome Entry) [b]	NCBI Accession	Score [c]	Species	PM [c]/ Coverage %	Mr/pI Experimental (Theoretical) [d]	Functional Category	More/Less Abundance Change Ratio (FDR) [e]					
								S1	S2	R1	R2	$\frac{R}{S}$ 1	$\frac{R}{S}$ 2
1606	RuBisCO large subunit-binding protein subunit beta chloroplastic/chaperonin subunit beta (96.3% *Vf*_0035079)	gi|2506277	257	*Pisum sativum*	10/26	65.7/5.5 (63.3/5.8)	Photosynthesis	0.9	3.7	1.5	0.5	5.9	4.1
601	RuBisCO large subunit-binding protein subunit alpha chloroplastic /chaperonin-60alpha (98.4% *Vf*_0081659)	gi|1710807	480	*Pisum sativum*	15/36	65.8/5.0 (62.0/5.2)	Photosynthesis	2.4	5.2	0.7	0.2	1.0	3.3
4501	Ribulose-1,5-bisphosphate carboxylase/oxygenase large subunit (95.9% *Vf*_0007913)	gi|825737	556	*Carya illinoinensis*	20/40	57.1/5.8 (51.6/6.1)	Photosynthesis	26.5	7.5	0.5	37.7	4.3	2.8
4505	Ribulose-1,5-bisphosphate carboxylase/oxygenase large subunit (96.5% *Vf*_0007913)	gi|33113311	584	*Carya ovate*	23/42	55.8/5.9 (51.4/6.1)	Photosynthesis	5.1	1.7	6.2	1.4	41.3	3.3
4406	Ribulose-1,5-bisphosphate carboxylase/oxygenase large subunit (95.6% *Vf*_0007913)	gi|21634071	485	*Cressa depressa*	23/44	55.5/5.8 (50.6/6.7)	Photosynthesis	3.3	4.3	0.1	17.6	38.4	17.3
5507	Ribulose-1,5-bisphosphate carboxylase/oxygenase large subunit (96.4% *Vf*_0007913)	gi|225544093	709	*Caragana camilli-schneideri*	25/45	54.4/6.0 (52.8/6.3)	Photosynthesis	1.9	1.4	0.8	2.4	41.8	2.0
5408	Ribulose-1,5-bisphosphate carboxylase/ oxygenase (95.7% *Vf*_0007913)	gi|74179244	619	*Aristolochia arborea*	23/44	52.5/6.1 (52.0/6.1)	Photosynthesis	14.6	1.3	0.9	0.7	21.7	7.2
5407	Ribulose-1,5-bisphosphate carboxylase/ oxygenase large subunit (95.3% *Vf*_0007913)	gi|62861204	759	*Paracroton zeylanicus*	25/50	50.8/6.2 (52.0/6.0)	Photosynthesis	14.1	0.9	0.6	0.9	11.2	3.7
2304	Phosphoglycerate kinase chloroplastic (86.1% *Vf*_0006851)	gi|129915	192	*Triticum aestivum*	9/19	49.6/5.6 (50.0/6.6)	Photosynthesis	96.2	2.6	1.0	1.0	0.0	0.0
2301	Phosphoglycerate kinase chloroplastic (86.1% *Vf*_0006851)	gi|129915	254	*Triticum aestivum*	7/16	46.1/5.6 (50.0/6.6)	Photosynthesis	12.8	3.9	6.9	0.8	6.3	4.6
2303	Phosphoglycerate kinase chloroplastic (85.6% *Vf*_0006851)	gi|2499497	571	*Nicotiana tabacum*	15/33	44.5/5.6 (50.3/8.5)	Photosynthesis	3.2	3.7	2.5	1.5	3.7	1.7
7503	Ribulose-1,5-bisphosphate carboxylase/ oxygenase large (95.9% *Vf*_0007913)	gi|825737	372	*Carya illinoinensis*	21/42	56.2/6.7 (51.6/6.1)	Photosynthesis	3.5	6.8	0.9	1.8	98.3	3.2
6418	Ribulose-1,5-bisphosphate carboxylase/ oxygenase (95.7% *Vf*_0007913)	gi|74179244	589	*Aristolochia arborea*	24/42	53.7/6.3 (52.0/6.1)	Photosynthesis	15.2	1.7	0.8	1.4	1.3	2.5
8308	Geranylgeranyl hydrogenase (91.3% *Vf*_0052793)	gi|19749359	267	*Glycine max*	18/38	48.6/8.7 (51.7/9.1)	Photosynthesis	1.7	5.0	0.9	0.4	2.1	5.9
203	Sedoheptulose-1,7-bisphosphatase (81.8% *Vf*_0007079)	gi|229597543	230	*Cucumis sativus*	9/21	42.8/5.2 (42.5/5.9)	Photosynthesis	2.4	3.2	1.0	0.7	1.6	1.4
3202	Sedoheptulose-1,7-bisphosphatase (81.8% *Vf*_0007079)	gi|229597543	215	*Cucumis sativus*	12/34	42.6/5.6 (42.5/5.9)	Photosynthesis	5.5	11.2	1.5	3.3	1.7	1.5
1204	Phosphoribulokinase (99.1% *Vf*_0003052)	gi|1885326	350	*Pisum sativum*	14/46	41.4/5.4 (39.2/5.4)	Photosynthesis	2.4	2.5	4.8	0.9	2.3	2.7

Table 1. *Cont.*

SSP [a]	Protein Name (% Identity to *V. faba* Transcriptome Entry) [b]	NCBI Accession	Score [c]	Species	PM [c]/ Coverage %	Mr/pI Experimental (Theoretical) [d]	Functional Category	More/Less Abundance Change Ratio (FDR) [e]					
								S1	S2	R1	R2	$\frac{R}{S}$ 1	$\frac{R}{S}$ 2
2202	Phosphoribulokinase (99.1% _0003052)	gi|1885326	160	*Pisum sativum*	7/27	41.1/5.5 (39.2/5.4)	Photosynthesis	1.2	1.5	1.2	0.8	3.4	2.0
1201	Photosystem II stability/assembly factor HCF136, chloroplast precursor (78.8% *Vf*_0033934)	gi|255559812	237	*Ricinus communis*	8/20	41.1/5.3 (43.4/7.1)	Photosynthesis	1.3	2.2	1.8	2.0	0.5	0.3
3107	Aldolase (80.8% *Vf*_0017749)	gi|169039	137	*Pisum sativum*	9/27	38.5/5.6 (38.0/5.5)	Photosynthesis	2.3	1.6	5.3	1.2	0.3	1.7
3105	Aldolase (80.8% *Vf*_0017749)	gi|169039	88	*Pisum sativum*	8/29	38.0/5.6 (38.0/5.5)	Photosynthesis	0.0	0.1	∞	0.8	0.0	0.1
1001	Chloroplast chlorophyll a/b binding protein (99.2% *Vf*_0037012)	gi|157786302	265	*Pisum sativum*	10/34	30.3/5.2 (28.4/5.5)	Photosynthesis	0.9	0.8	6.8	0.9	0.3	1.4
4306	Transaminase mtnE, putative (80.9% *Vf*_0021772)	gi|255562088	159	*Ricinus communis*	5/11	46.2/5.9 (50.9/6.9)	Photosynthesis	1.2	4.4	1.0	1.0	0.0	0.0
401	Chloroplast ribulose-1,5-bisphosphate carboxylase activase (81.0% *Vf*_0005564)	gi|115392208	122	*Triticum aestivum*	6/21	51.7/5.1 (40.3/6.5)	Photosynthesis	0.9	0.0	1.0	1.0	0.0	0.0
1502	UDP-glucose pyrophosphorylase (89.7% *Vf*_0034269)	gi|12585472	271	*Astragalus penduliflorus*	12/29	54.1/5.3 (51.6/5.9)	Protein synthesis	1.1	2.3	1.0	1.0	0.0	0.0
2801	ClpC protease (92.1% *Vf*_0007069)	gi|4105131	70	*Spinacia oleracea*	12/16	96.6/5.6 (99.6/8.8)	Protein degrad.	6.7	8.8	0.5	0.4	8.9	2.7
2804	ClpC protease (98.5% *Vf*_0007069)	gi|461753	145	*Pisum sativum*	19/26	96.1/5.6 (102.8/6.6)	Protein degrad.	5.3	1.6	0.3	1.4	20.0	1.2
1802	ATP-dependent Clp protease (98.5% *Vf*_0007069)	gi|461753	383	*Pisum sativum*	26/33	95.3/5.3 (102.8/6.6)	Protein degrad.	2.3	6.4	1.5	1.2	1.1	1.7
2703	Cell division protease ftsH homolog, chloroplastic (91.1% *Vf*_0034616)	gi|17865463	262	*Medicago sativa*	16/30	75.4/5.6 (75.8/5.6)	Protein degrad.	5.9	7.0	1.1	1.6	0.9	0.2
2707	Cell division protease ftsH homolog chloroplastic (91.1% *Vf*_0034616)	gi|17865463	140	*Medicago sativa*	10/19	74.8/5.7 (75.8/5.6)	Protein degrad.	14.1	60.9	0.6	0.8	4.0	3.5
2706	Cell division protease ftsH homolog, chloroplastic (91.1% *Vf*_0034616)	gi|17865463	315	*Medicago sativa*	21/38	70.9/5.7 (75.8/5.6)	Protein degrad.	2.9	6.0	0.6	1.2	2.7	0.9
1702	Putative zinc dependent protease/FTSH protease 8 (87.0% *Vf*_0002195)	gi|84468324	206	*Trifolium pretense*	11/22	74.5/5.3 (75.4/5.5)	Protein degrad.	∞	8.0	0.7	1.6	∞	1.2
2101	Serine-type endopeptidase (96.5% *Vf*_0035526)	gi|270342123	70	*Phaseolus vulgaris*	5/15	39.5/5.4 (45.2/7.7)	Protein degrad.	∞	1.0	1.0	1.0	1.0	1.0
2802	Ubiquitin-specific-processing protease 8 (62.7% *Vf*_0023012)	gi|257050978	61	*Arabidopsis thaliana*	13/27	99.1/5.2 (90.7/5.5)	Protein degrad.	1.0	10.4	1.0	1.0	1.0	0.0
3803	ATP-dependent Clp protease/CLPC homologue 1 (98.5% *Vf*_0007069)	gi|461753	308	*Pisum sativum*	19/26	83.8/5.7 (102.8/6.6)	Protein degrad.	7.3	13.4	1.7	1.6	8.2	3.0
702	Chaperone DnaK (stromal 70 kDa heat shock-related protein, chloroplastic) (93.7% *Vf*_0024557)	gi|92870233	955	*Medicago truncatula*	27/38	81.5/5.0 (75.8/5.2)	Stress response	3.5	4.5	1.6	0.6	4.2	3.0
1701	Heat shock protein 70 (92.1% *Vf*_0006213)	gi|56554972	562	*Medicago sativa*	23/40	81.8/5.3 (71.4/5.1)	Stress response	3.6	4.5	2.1	0.8	1.7	5.3
2701	Heat shock 70 kDa protein mitochondrial (93.1% *Vf*_0016658)	gi|585272	216	*Pisum sativum*	14/26	76.2/5.6 (72.4/5.8)	Stress response	13.1	11.1	1.0	1.0	0.0	0.0
4302	Monodehydroascorbate reductase I (92.0% *Vf*_0006284)	gi|51860738	167	*Pisum sativum*	10/24	47.9/5.8 (47.4/5.8)	Stress response	3.7	2.2	6.4	1.0	1.4	0.3
3001	L-ascorbate peroxidase, cytosolic (60.9% *Vf*_0035596)	gi|1351963	171	*Pisum sativum*	8/34	31.0/6.1 (27.2/5.5)	Stress response	1.9	0.5	1.1	0.6	0.5	2.1

Table 1. *Cont.*

SSP [a]	Protein Name (% Identity to *V. faba* Transcriptome Entry) [b]	NCBI Accession	Score [c]	Species	PM [c]/ Coverage %	*Mr/pI* Experimental (Theoretical) [d]	Functional Category	More/Less Abundance Change Ratio (FDR) [e]					
								S1	S2	R1	R2	$\frac{R}{S}$ 1	$\frac{R}{S}$ 2
3103	CDSP32 protein (Chloroplast Drought-induced Stress Protein of 32 kDa) (69.4% *Vf*_0033087)	gi l 2582822	133	*Solanum tuberosum*	4,'14	33.8/6.0 (33.8/8.1)	Stress response	0.9	0.8	0.3	0.2	0.4	0.7
1206	NADPH-dependent alkenal/one oxidoreductase, chloroplastic (85.3% *Vf*_0022282)	XP_003532009.1	190	*Glycine max*	6,'28	40.6/5.4 (31.2/9.2)	Redox	1.0	1.0	1.0	∞	1.0	1.0
11	Thioredoxin peroxidase (80.1% *Vf*_0034677)	gi l 21912927	108	*Nicotiana tabacum*	3,'16	21.6/5.2 (30.1/8.2)	Redox	2.4	0.3	16.3	0.6	0.0	0.7
5305	GDP-D-Mannose 3',5'-Epimerase (90.2% *Vf*_0021019)	gi l 15241945	108	*Arabidopsis thaliana*	8,'26	49.6/6.0 (43.1/5.8)	Cell wall	8.0	19.0	1.0	1.0	0.0	0.0
502	Hydroxyproline-rich glycoprotein family protein	gi l 18411523	147	*Arabidopsis thaliana*	9,'13	60.7/4.8 (49.4/5.2)	Cell wall	3.8	2.0	0.8	0.7	0.8	0.6
4404	Gdp-Mannose-3', 5'-Epimerase (89.9% *Vf*_0021019)	gi l 83754656	112	*Arabidopsis thaliana*	9,'22	50.6/5.9 (43.2/5.8)	Cell wall	0.5	3.7	1.0	1.0	0.0	0.0
101	PAP fibrillin (84.1% *Vf*_0025094)	gi l 87240799	268	*Medicago truncatula*	8,'18	34.4/4.9 (34.1/4.9)	Cell organization	1.4	1.7	1.8	0.5	0.2	1.5
1308	Actin (99.7% *Vf*_0005527)	gi l 34541966	560	*Trifolium pretense*	16/51	47.9/5.5 (41.9/5.3)	Cell organization	0.7	12.9	0.4	7.7	3.1	0.5
1404	Actin (99.1% *Vf*_0028038)	gi l 1498334	113	*Glycine max*	8,'39	49.3/5.3 (37.3/5.5)	Cell organization	2.6	11.5	1.0	1.0	0.0	0.0
1304	Actin (99.7% *Vf*_0005527)	gi l 34541966	431	*Trifolium pretense*	16/53	49.0/5.3 (41.9/5.3)	Cell organization	0.8	8.7	1.0	1.8	0.5	0.6
3205	Actin (94.3% *Vf*_0008358)	gi l 1498384	172	*Zea mays*	5,'19	39.7/5.6 (37.3/5.5)	Cell organization	0.0	6.4	1.0	1.0	0.0	0.0
3302	Elongation factor Tu (97.5% *Vf*_0005994)	gi l 6015084	284	*Pisum sativum*	17/35	48.1/5.6 (53.1/6.6)	Transcription/ Translation	∞	2.8	2.5	1.2	∞	0.8
3309	Elongation factor Tu (97.5% *Vf*_0005994)	gi l 6015084	480	*Pisum sativum*	22/43	47.9/5.7 (53.1/6.6)	Transcription/ Translation	0.3	13.0	11.7	2.2	1.7	1.4
3307	Elongation factor Tu (97.5% *Vf*_0005994)	gi l 6015084	584	*Pisum sativum*	22/46	47.7/5.7 (53.1/6.6)	Transcription/ Translation	7.6	7.5	1.0	0.9	11.6	8.2
9401	Elongation factor 1 alpha (96.9% *Vf*_0022300)	gi l 61741088	201	*Actinidia deliciosa*	16/37	54.0/9.0 (49.6/9.2)	Transcription/ Translation	4.4	4.8	1.3	2.4	7.0	4.9
3603	Tic62 protein (89.8% *Vf*_0007594)	gi l 21616072	99	*Pisum sativum*	7,24	61.8/5.7 (57.1/8.8)	Signaling	1.0	0.0	1.0	1.0	1.0	0.0
1708	V-type proton ATPase catalytic subunit A (95.9% *Vf*_0002956)	gi l 12585490	146	*Citrus unshiu*	17/42	74.4/5.5 (68.9/5.3)	Transport	8.9	54.4	0.7	1.2	3.8	5.3
5002	Carbonate dehydratase (64.2% *Vf*_0022150)	gi l 47606728	105	*Flaveria bidentis*	3,17	30.3/6.2 (35.9/5.8)	Miscellaneous	1.3	5.8	1.3	0.6	0.5	3.1
6003	Carbonic anhydrase (85.8% *Vf*_0022150)	gi l 270342124	153	*Phaseolus vulgaris*	7,34	29.0/6.5 (35.9/8.1)	Miscellaneous	0.9	0.5	3.3	∞	0.2	0.0

[a] Standard spot number assigned to each spot protein (SSP) by PDQuest software (BioRad). [b] Percentages of identity to *V. faba* transcriptome entries [*Vicia faba RefTrans V2* (2017) (https://www.pulsedb.org/, accessed on 15 October 2021)] obtained by Blast (tblastn) analysis are displayed in brackets. The coding "v.faba_CSFL_reftransV2_number" has been simplified by "*Vf*_number" for each transcript. [c] PM: number of peptides matched (from peptide mass fingerprinting) with the homologous protein from the database. Some of these peptides were automatically MSMS fragmented. [d] Experimental mass (Mr, kDa) and pI were calculated with PDQuest software and standard molecular mass markers. Theoretical values were retrieved from the protein database (NCBInr). [e] Values are given as normalized volume (calculated with PDQuest software) and represent change ratios in response to *B. fabae* inoculation of each genotype (S: Baraca; R: BPL710), and between uninoculated genotypes (R/S) at both sampling times (1: 6 hpi; 2: 12 hpi).

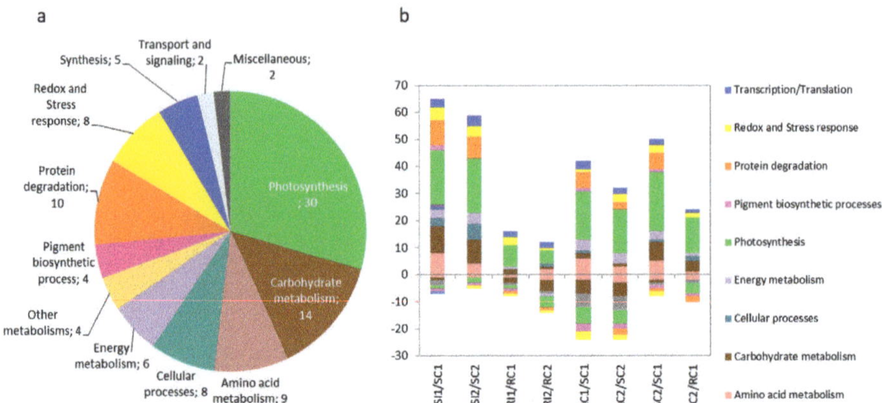

Figure 3. Functional groups of differential proteins. (**a**) Proteins up- or down-accumulating in response to *B. fabae* inoculation as grouped by functional category. (**b**) Susceptible (S) and resistant (R) genotype; control (C) and *B. fabae* inoculated (I); 6 hpi (1) and 12 hpi (2).

Figure 3b compares treatments, genotypes and sampling times. The susceptible genotype (S) showed the largest number of changes in proteins in response to inoculation, but mainly in those of the primary and energy metabolism groups (Figure 3b). In fact, a strikingly large number of degradation proteins showed changes in S genotype in response to inoculation that were not observed in the resistant genotype (R). The group of degradation proteins comprised ten proteases, namely: three cell division proteases ftsH chloroplastic (gi | 17865463), two ATP-dependent Clp proteases (gi | 461753), two Clp proteases (gi | 4105131, gi | 461753), one ubiquitin-specific protease 5 (gi | 257050978), one zinc dependent protease/FTSH protease 8 (gi | 84468324) and one serine-type endopeptidase (gi | 270342123). Interestingly, most of the identified proteases were significantly increased in response to inoculation in the susceptible genotype but, as revealed by the heat map (Figure 4a, Table 1), none was in the resistant genotype. A comparison of the uninoculated leaf proteome revealed a much greater number of proteins of the energy metabolism and protein degradation groups in R than in S, mainly in the first sampling (Figures 3b and 4a). In addition, these functional groups were increased in control S plants in the second sampling.

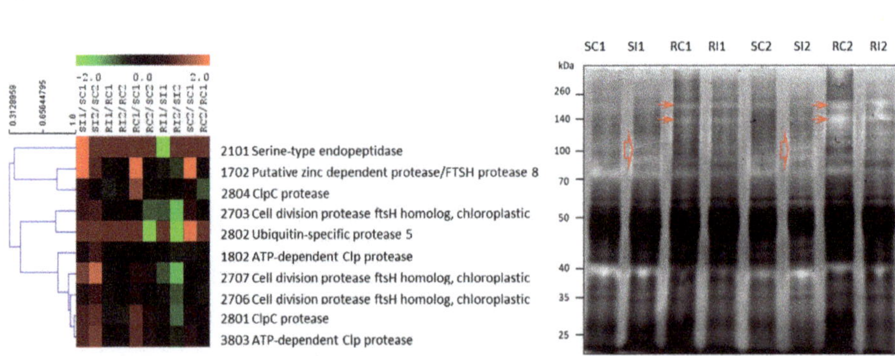

Figure 4. (**a**) Hierarchical clustering of proteases identified by MALDI-TOF analysis based on protein abundance as determined by 2DE. Susceptible genotype (S) and resistant genotype (R); control (C) and *B. fabae* inoculated (I); 6 hpi (1) and 12 hpi (2). (**b**) Zymogram of proteases in faba leaves separated by SDS-PAGE bearing gelatin under nondenaturing conditions. Arrowheads denote differential bands between genotypes or treatments.

3.4. Protease Gel Activity

The results of the protease gel activity assay confirmed those previously obtained by MS. Protease bands corresponding to high molecular weights (~90 to 120 kDa) were stronger in inoculated plants of the susceptible genotype in both samplings, coinciding with the molecular weight of some proteases identified by MALDI-TOF (viz., Clp proteases, SSPs 1802, 2801, 2804, 3803; and Ubiquitin-specific-processing protease 8, SSP 2802) (Figure 4b; Supplementary Figure S4). Gel activity assay also exposed a strong band at 40 kDa not changing substantially with the specific conditions and potentially corresponded to the serine-type endopeptidase identified by MS analysis (SSP 2101). In any case, the most striking result was the presence of two well-defined protease bands at high molecular weights (140–250 kDa) that were especially strong in the resistant genotype (and, particularly, in the second sampling). The area of the protease activity bands were estimated using ImageJ software (ImageJ.JS (imjoy.io)), and the data are presented in Supplemental Figure S5. Since activity gels were used under native conditions and 2DE-MS analyses conducted under denaturing conditions, these bands may well correspond to protein complexes not identified with the denaturing gels.

4. Discussion

Botrytis fabae is a necrotrophic plant pathogen causing chocolate spot, which is one of the most devastating diseases for faba bean production worldwide [5,6]. The mechanism by which plants counteract infection by this pathogen is of great agronomic interest. ROS production (especially H_2O_2 induction by *Botrytis*) is known to occur in a wide variety of plants [23,25–27] and to be one of the earliest plant responses to fungal infection [22,24]. *Botrytis fabae* reportedly increases lipid peroxidation, and the levels of ROS and antioxidant enzymes (superoxide dismutase, catalase and ascorbate peroxidase), substantially in faba bean [28]. In fact, ROS were found to accumulate rapidly in leaf tissue of a resistant cultivar at early stages of infection, but more markedly and over longer periods at later stages in its susceptible counterpart [28].

4.1. The Role of Chloroplasts as Redox Sensors Eliciting an Acclimatory Response to Stressing Conditions

ROS accumulation at an early stage of infection is triggered by plasma membrane-bound NADPH oxidases and typically occurs in the apoplast [44]. However, chloroplastic and peroxisomal ROS production have been reported to contribute to plant immunity as well [45,46]. Although ROS can also be produced by other organelles (notably peroxisomes and mitochondria), the chloroplast is possibly a major source. Some proteins in the chloroplast are involved in cross-talk with the cytosol and nucleus to govern the outcome of defense signaling [47]. Besides triggering ROS signals, chloroplasts can perceive, mediate or even amplify ROS signals originating in the apoplast [48]. In addition, there is evidence that the role of chloroplastic ROS production in coordinating cell death or modulating defense outputs is highly specific in targeting various types of invading pathogens. Thus, some chloroplastic components may be specific targets for microbial effector molecules, which suggest that chloroplasts communicate through these target molecules to elicit ROS production in the apoplast, presumably to contain spread of the lesion [49].

Chloroplast-derived ROS has been shown to play a role in plant resistance against *B. cinerea* [27,50]. Thus, histochemical analysis revealed ROS accumulation in tomato leaves 24 h after application of *B. cinerea* spores. A defense response accompanied by an improvement in photosystem II (PSII), possibly triggered by ROS upon short-time exposure, was observed. However, the relatively increased time of exposure to these molecules made them harmful to PSII functionality [27,51]. In addition, H_2O_2 levels in strawberry leaves were found to correlate positively with disease severity, and to be influenced by both leaf age and light quality [52].

Through photosynthesis, chloroplasts play a central role as redox sensors of environmental conditions by eliciting acclimatory or stress-defense responses [53,54]. In chloroplasts, chaperone systems refold proteins after stress, while proteases degrade misfolded and aggregated proteins that cannot be refolded [55]. A study on *Arabidopsis thaliana* demonstrated a major role of Hsp70 chaperones and Clp proteases in the folding and degradation of misfolded or damaged proteins under variable stress conditions in the chloroplast [55]. Caseinolytic proteases (Clps) function as molecular chaperones and confer thermotolerance to plants. The results of a differential gene expression analysis of Clps in wheat suggest a potential role in cold, salt and biotic stresses, and confirm the previously reported role in thermotolerance [56]. ClpATPases class I (ClpB/HSP100 and ClpC) function in assembly and disassembly with protein complexes, acting together with the HSP70/DnaK chaperone system to remodel denatured protein aggregates [57]. On the other hand, plant ClpC proteins act as stromal molecular chaperones in importing and protecting unfolded newly synthesized proteins, which are responsible for maintaining homeostasis [58–61].

In the present work two ATP-dependent Clp, two ClpC proteases and three Hsp70 (one as Chaperone DnaK) were found to be considerably increased in the susceptible genotype in response to inoculation. This result suggests that the chloroplast may respond to *B. fabae* inoculation by triggering a mechanism to repair damaged proteins. The previous proteins were highly represented at the constitutive level in resistant uninoculated plants, which may represent a temporary advantage in response to the pathogen.

4.2. Homeostatic Control as a Dynamic Regulation Mechanism for Energy and Redox Status in Response to Botrytis fabae

Enhanced photosystem II (PSII) functionality at the early stages of pathogen infection may be responsible for the increased sugar production required to strengthen the response by inducing defense genes [27,62]. However, *B. cinerea* has been reported to use large amounts of soluble sugars to grow on tomato leaves [63].

Nonphotochemical chlorophyll fluorescence quenching (NPQ) is the key photoprotective process used by plants to dissipate excess light energy as heat and preserve photosynthesis as a result [64–67]. A substantial increase in NPQ was observed in tomato leaflets up to 6 h after application of a *B. cinerea* spore suspension; the increase, however, was followed by a decrease down to control levels [27]. This outcome suggests an imbalance between energy supply and demand, resulting in increased ROS production similarly as in photoinhibition, causing damage in chloroplast and eventual cell death (necrosis) [68].

On physiological grounds, Clp and FtsH proteases are believed to play major roles in chloroplast protein homeostasis. Thus, FtsH (filamentation temperature sensitive H) proteases are membrane-bound ATP-dependent zinc metalloproteases involved in the biogenesis of thylakoid membranes and quality control in the PSII repair cycle [69]. ROS production and PSII photodamage are linked to the high turnover rate of the D1 reaction center protein, which is degraded and replaced with *de novo* synthesized protein in the so-called "PSII repair cycle" [70]. FtsH proteases are among the many components mediating coordinated turnover in D1. In addition, there is evidence that programmed inhibition of the PSII repair cycle through specific downregulation of protease activity may provide plants with a mechanism to elicit ROS production and cell death upon infection [71].

In parallel to the recognition of ROS as key signaling molecules, antioxidant enzymes and ROS scavenging, scientists have accepted their potential involvement in fine-tuning defense reactions. In chloroplasts, the antioxidants ascorbate and glutathione contribute chemically to ROS quenching. In addition, H_2O_2 can be detoxified by ascorbate peroxidases (APX), peroxiredoxine (PRX) or glutathione peroxidase (GPX), reviewed in [72], as confirmed by a study on *Gentiana triflora* which suggested that PRXQ plays a role in mediating responses against the necrotrophic fungus *B. cinerea* [73].

The proteomic analysis conducted in this work revealed that three proteins identified as chloroplastic cell division protease FtsH, and a zinc dependent protease/FTSH protease 8, were highly represented in the susceptible genotype in response to *B. fabae* inoculation. The same proteins were better represented constitutively in the resistant genotype in the first sampling. In addition, a monodehydroascorbate reductase I, an L-ascorbate peroxidase and a thioredoxin peroxidase were also better represented in both genotypes in response to inoculation in the first sampling. Zymogram analysis confirmed the results of the MS analysis, where differences in the activity bands at the molecular weights of the proteins identified by 2DE-MS were observed. In addition, inoculated and uninoculated plants of the resistant genotype exhibited some activity bands at very high molecular weights (140–250 kDa) that were not clearly observed in the susceptible genotype. This result can be ascribed to differences in the experimental conditions, which were native in the zymograms and denaturing in the 2DE-MS analysis.

Consistent with the results obtained in this work, a recent study on tomato plants revealed H_2O_2 production and enhanced photosystem II functionality 30 min after *B. cinerea* inoculation. The effect, which lasted 4 h, was suggestive of a tolerant response; however, increasing the length of exposure led to plant damage [27] by fully inhibiting PSII functionality at the application spot and nearby. This was probably a time-dependent hormetic response, suggesting a positive biological response whose effect might be reversed upon extended exposure [27,67].

5. Conclusions

Based on the proteins identified in this study (Clp and Hsp70, together with FstH proteases and ROS proteins) and their increased levels upon inoculation with *B. fabae*, a signaling response mechanism based on ROS production in the chloroplast may be elicited by the fungus. This mechanism appears to be harmful to PSII in the susceptible genotype by effect of its being associated with lengthy exposures to high ROS levels. The differential response of the two genotypes can be ascribed to a metabolic imbalance in the susceptible genotype not observed in the resistant genotype and confirming that the latter retains normal metabolism under stress. On the other hand, there is evidence that the two genotypes differ in chloroplast detoxification system, the resistant genotype exhibiting a more efficient PSII repair mechanism at the early stages of infection. Further research is required in any case to ascertain whether the ROS dose or exposure time (hormesis) is associated with the differential *V. faba* phenotypes.

Supplementary Materials: The following are available online at https://www.mdpi.com/article/10.3390/agronomy11112247/s1, Figure S1: Coomassie-stained 2DE gels images of all experimental conditions throughout the experiment: susceptible (S) and resistant (R) genotypes; control (C) and *B. fabae* inoculated (I); 6 hpi (T1) and 12 hpi (T2) and three repetitions (R1-R3), Figure S2: Chocolate spot disease severity (DS) values in the genotypes Baraca and BPL710, 2 and 6 days after *B. fabae* inoculation, Figure S3: Location of 102 identified protein spots on a virtual 2DE gel. (a) Representative Coomassie stained 2DE gel of the susceptible (left) and the resistant (right) genotypes, (b) Figure S4: Zymographic detection of proteases in faba bean leaves separated by SDS-PAGE bearing gelatin under nondenaturing conditions, control (A) and *B. fabae*-inoculated (B). Susceptible (S) and resistant genotype (R); 6 hpi (1) and 12 hpi (2). Numbers following dashes designate the particular replicates, Figure S5: Quantification of the area of protease activity bands detected by zymogram analysis corresponding to the molecular weights 40, 70, 90, 100, 140 and 250 kDa, Table S1: Dataset of the 224 protein spots detected by PDQuest image analysis.

Author Contributions: Conceptualization, M.Á.C., Á.M.V.-F. and D.R.; methodology, M.Á.C., Á.M.V.-F. and T.H.-L.; software, M.Á.C.; validation, T.H.-L. and M.Á.C.; formal analysis, M.Á.C.; investigation, M.Á.C., Á.M.V.-F. and T.H.-L.; writing—original draft preparation, M.Á.C.; writing—review and editing, M.Á.C. and Á.M.V.-F.; supervision, M.Á.C., Á.M.V.-F. and D.R.; funding acquisition, D.R. All authors have read and agreed to the published version of the manuscript.

Funding: This research was funded by Spain's Agencia Estatal de Investigación (AEI) (Grant PRIMA-DiVicia-PCI2020-111974).

Institutional Review Board Statement: Not applicable.

Informed Consent Statement: Not applicable.

Data Availability Statement: Raw data supporting reported results can be found in Supplementary Materials.

Acknowledgments: M.Á.C. is grateful to the Spanish Ministry of Science, Innovation and Universities for award of a Ramón y Cajal contract (RYC-2017-23706), and so is T.H.-L. to the University of Cordoba (Spain) for the contract under Project RYC-2017-23706.

Conflicts of Interest: The authors declare no conflict of interest.

References

1. Khazaei, H.; O'Sullivan, D.M.; Stoddard, F.L.; Adhikari, K.N.; Paull, J.G.; Schulman, A.H.; Andersen, S.U.; Vandenberg, A. Recent advances in faba bean genetic and genomic tools for crop improvement. *Legum. Sci.* **2021**, *3*, e75. [CrossRef]
2. Rispail, N.; Kaló, P.; Kiss, G.B.; Ellis, T.N.; Gallardo, K.; Thompson, R.D.; Prats, E.; Larrainzar, E.; Ladrera, R.; González, E.M.; et al. Model legumes contribute to faba bean breeding. *Field Crop. Res.* **2010**, *115*, 253–269. [CrossRef]
3. Cernay, C.; Ben-Ari, T.; Pelzer, E.; Meynard, J.-M.; Makowski, D. Estimating variability in grain legume yields across Europe and the Americas. *Sci. Rep.* **2015**, *5*, 11171. [CrossRef]
4. Kharrat, M.; Le Guen, J.; Tivoli, B. Genetics of resistance to 3 isolates of *Ascochyta fabae* on Faba bean (*Vicia faba* L.) in controlled conditions. *Euphytica* **2006**, *151*, 49–61. [CrossRef]
5. Tivoli, B.; Baranger, A.; Avila, C.; Banniza, S.; Barbetti, M.; Chen, W.; Davidson, J.; Lindeck, K.; Kharrat, M.; Rubiales, D.; et al. Screening techniques and sources of resistance to foliar diseases caused by major necrotrophic fungi in grain legumes. *Euphytica* **2006**, *147*, 223–253. [CrossRef]
6. Stoddard, F.; Nicholas, A.; Rubiales, D.; Thomas, J.; Villegas-Fernández, A. Integrated pest management in faba bean. *Field Crop. Res.* **2010**, *115*, 308–318. [CrossRef]
7. Harrison, J.G. Effects of environmental factors on growth of lesions on field bean leaves infected by *Botrytis fabae*. *Ann. Appl. Biol.* **1980**, *95*, 53–61. [CrossRef]
8. Harrison, J.G. The biology of *Botrytis* spp. on *Vicia* beans and chocolate spot disease—A review. *Plant Pathol.* **1988**, *37*, 168–201. [CrossRef]
9. Murray, G.M.; Brennan, J.P. *The Current and Potential Costs from Diseases of Pulse Crops in Australia*; Grains Research and Development Corporation: Canberra, Australia, 2012.
10. Villegas-Fernández, A.; Sillero, J.; Emeran, A.; Winkler, J.; Raffiot, B.; Tay, J.; Flores, F.; Rubiales, D. Identification and multi-environment validation of resistance to *Botrytis fabae* in *Vicia faba*. *Field Crop. Res.* **2009**, *114*, 84–90. [CrossRef]
11. Sillero, J.C.; Villegas-Fernández, A.M.; Thomas, J.; Rojas-Molina, M.M.; Emeran, A.A.; Fernández-Aparicio, M.; Rubiales, D. Faba bean breeding for disease resistance. *Field Crop. Res.* **2010**, *115*, 297–307. [CrossRef]
12. Bouhassan, A.; Sadiki, M.; Tivoli, B. Evaluation of a collection of faba bean (*Vicia faba* L.) genotypes originating from the Maghreb for resistance to chocolate spot (*Botrytis fabae*) by assessment in the field and laboratory. *Euphytica* **2004**, *135*, 55–62. [CrossRef]
13. Williamson, B.; Tudzynski, B.; Tudzynski, P.; van Kan, J.A. *Botrytis cinerea*: The cause of grey mould disease. *Mol. Plant Pathol.* **2007**, *8*, 561–580. [CrossRef] [PubMed]
14. Cimmino, A.; Villegas-Fernández, A.M.; Andolfi, A.; Melck, D.; Rubiales, D.; Evidente, A. Botrytone, a New Naphthalenone Pentaketide Produced by *Botrytis fabae*, the Causal Agent of Chocolate Spot Disease on *Vicia faba*. *J. Agric. Food Chem.* **2011**, *59*, 9201–9206. [CrossRef] [PubMed]
15. AbuQamar, S.F.; Moustafa, K.; Tran, L.S.P. 'Omics' and plant responses to *Botrytis cinerea*. *Front. Plant. Sci.* **2016**, *7*, 1658. [CrossRef] [PubMed]
16. Ren, H.; Bai, M.; Sun, J.; Liu, J.; Ren, M.; Dong, Y.; Wang, N.; Ning, G.; Wang, C. RcMYB84 and RcMYB123 mediate jasmonate-induced defense responses against *Botrytis cinerea* in rose (*Rosa chinensis*). *Plant. J.* **2020**, *103*, 1839–1849. [CrossRef] [PubMed]
17. AbuQamar, S.; Ajeb, S.; Sham, A.; Enan, M.R.; Iratni, R. A mutation in the expansin-like A2 gene enhances resistance to necrotrophic fungi and hypersensitivity to abiotic stress in *Arabidopsis thaliana*. *Mol. Plant. Pathol.* **2013**, *14*, 813–827. [CrossRef] [PubMed]
18. AbuQamar, S. Expansins: Cell Wall Remodeling Proteins with a Potential Function in Plant Defense. *J. Plant Biochem. Physiol.* **2014**, *2*, 118. [CrossRef]
19. VanEtten, H.D.; Mansfield, J.W.; Bailey, J.A.; Farmer, E.E. Two Classes of Plant Antibiotics: Phytoalexins versus 'Phytoanticipins'. *Plant Cell* **1994**, *6*, 1191. [CrossRef]
20. VAN Loon, L.; VAN Strien, E. The families of pathogenesis-related proteins, their activities, and comparative analysis of PR-1 type proteins. *Physiol. Mol. Plant Pathol.* **1999**, *55*, 85–97. [CrossRef]
21. van Loon, L.; Rep, M.; Pieterse, C. Significance of Inducible Defense-related Proteins in Infected Plants. *Annu. Rev. Phytopathol.* **2006**, *44*, 135–162. [CrossRef] [PubMed]
22. Dangl, J.L.; Jones, J.D.G. Plant pathogens and integrated defence responses to infection. *Nature* **2001**, *411*, 826–833. [CrossRef] [PubMed]
23. Colmenares, A.J.; Aleu, J.; Duran-Patron, R.M.; Collado, I.G.; Hernandez-Galán, R. The putative role of botrydial and related metabolites in the infection mechanism of *Botrytis cinerea*. *J. Chem. Ecol.* **2002**, *28*, 997–1005. [CrossRef] [PubMed]

24. Patykowski, J.; Urbanek, H. Activity of Enzymes Related to H_2O_2 Generation and Metabolism in Leaf Apoplastic Fraction of Tomato Leaves Infected with *Botrytis cinerea*. *J. Phytopathol.* **2003**, *151*, 153–161. [CrossRef]

25. Rolke, Y.; Liu, S.; Quidde, T.; Williamson, B.; Schouten, A.; Weltring, K.-M.; Siewers, V.; Tenberge, K.B.; Tudzynski, B.; Tudzynski, P. Functional analysis of H2O2-generating systems in *Botrytis cinerea*: The major Cu-Zn-superoxide dismutase (BCSOD1) contributes to virulence on French bean, whereas a glucose oxidase (BCGOD1) is dispensable. *Mol. Plant Pathol.* **2004**, *5*, 17–27. [CrossRef]

26. Hua, L.; Yong, C.; Zhanquan, Z.; Boqiang, L.; Guozheng, Q.; Shiping, T. Pathogenic mechanisms and control strategies of *Botrytis cinerea* causing post-harvest decay in fruits and vegetables. *Food Qual. Saf.* **2018**, *2*, 111–119. [CrossRef]

27. Stamelou, M.-L.; Sperdouli, I.; Pyrri, I.; Adamakis, I.-D.; Moustakas, M. Hormetic Responses of Photosystem II in Tomato to *Botrytis cinerea*. *Plants* **2021**, *10*, 521. [CrossRef]

28. El-Komy, M.H. Comparative Analysis of Defense Responses in Chocolate Spot-Resistant and -Susceptible Faba Bean (*Vicia faba*) Cultivars Following Infection by the Necrotrophic Fungus *Botrytis fabae*. *Plant Pathol. J.* **2014**, *30*, 355–366. [CrossRef]

29. Villegas-Fernández, Á.M.; Krajinski, F.; Schlereth, A.; Madrid, E.; Rubiales, D. Characterization of Transcription Factors Following Expression Profiling of *Medicago truncatula—Botrytis* spp. Interactions. *Plant Mol. Biol. Rep.* **2014**, *32*, 1030–1040. [CrossRef]

30. Birkenbihl, R.P.; Diezel, C.; Somssich, I.E. Arabidopsis WRKY33 Is a Key Transcriptional Regulator of Hormonal and Metabolic Responses toward *Botrytis cinerea* Infection. *Plant Physiol.* **2012**, *159*, 266–285. [CrossRef]

31. Windram, O.; Madhou, P.; McHattie, S.; Hill, C.; Hickman, R.; Cooke, E.; Jenkins, D.J.; Penfold, C.A.; Baxter, L.; Breeze, E.; et al. *Arabidopsis* Defense against *Botrytis cinerea*: Chronology and Regulation Deciphered by High-Resolution Temporal Transcriptomic Analysis. *Plant Cell* **2012**, *24*, 3530–3557. [CrossRef]

32. Marra, R.; Ambrosino, P.; Carbone, V.; Vinale, F.; Woo, S.L.; Ruocco, M.; Ciliento, R.; Lanzuise, S.; Ferraioli, S.; Soriente, I.; et al. Study of the three-way interaction between *Trichoderma atroviride*, plant and fungal pathogens by using a proteomic approach. *Curr. Genet.* **2006**, *50*, 307–321. [CrossRef] [PubMed]

33. Shah, P.; Powell, A.L.; Orlando, R.; Bergmann, C.; Gutierrez-Sanchez, G. Proteomic Analysis of Ripening Tomato Fruit Infected by *Botrytis cinerea*. *J. Proteome Res.* **2012**, *11*, 2178–2192. [CrossRef] [PubMed]

34. Mokhtar, M.M.; Hussein, E.H.A.; El-Assal, S.E.-D.S.; Atia, M.A.M. VfODB: A comprehensive database of ESTs, EST-SSRs, mtSSRs, microRNA-target markers and genetic maps in *Vicia faba*. *AoB Plants* **2020**, *12*, plaa064. [CrossRef]

35. Villegas-Fernández, A.M.; Sillero, J.C.; Rubiales, D. Screening faba bean for chocolate spot resistance: Evaluation methods and effects of age of host tissue and temperature. *Eur. J. Plant Pathol.* **2011**, *132*, 443–453. [CrossRef]

36. Wang, W.; Vignani, R.; Scali, M.; Cresti, M. A universal and rapid protocol for protein extraction from recalcitrant plant tissues for proteomic analysis. *Electrophoresis* **2006**, *27*, 2782–2786. [CrossRef] [PubMed]

37. Castillejo, M.A.; Bani, M.; Rubiales, D. Understanding pea resistance mechanisms in response to *Fusarium oxysporum* through proteomic analysis. *Phytochemistry* **2015**, *115*, 44–58. [CrossRef]

38. Mathesius, U.; Keijzers, G.; Natera, S.H.A.; Weinman, J.J.; Djordjevic, M.A.; Rolfe, B.G. Establishment of a root proteome reference map for the model legume *Medicago truncatula* using the expressed sequence tag database for peptide mass fingerprinting. *Proteomics* **2001**, *1*, 1424–1440. [CrossRef]

39. Sharov, A.A.; Dudekula, D.; Ko, M.S.H. A web-based tool for principal component and significance analysis of microarray data. *Bioinformatics* **2005**, *21*, 2548–2549. [CrossRef]

40. Shevchenko, A.; Wilm, M.; Vorm, O.; Mann, M. Mass Spectrometric Sequencing of Proteins from Silver-Stained Polyacrylamide Gels. *Anal. Chem.* **1996**, *68*, 850–858. [CrossRef]

41. Bradford, M.M. A rapid and sensitive method for the quantitation of microgram quantities of protein utilizing the principle of protein-dye binding. *Anal. Biochem.* **1976**, *7*, 248–254. [CrossRef]

42. Heussen, C.; Dowdle, E.B. Electrophoretic analysis of plasminogen activators in polyacrylamide gels containing sodium dodecyl sulfate and copolymerized substrates. *Anal. Biochem.* **1980**, *102*, 196–202. [CrossRef]

43. Neuhoff, V.; Arold, N.; Taube, D.; Ehrhardt, W. Improved staining of proteins in polyacrylamide gels including isoelectric focusing gels with clear background at nanogram sensitivity using Coomassie Brilliant Blue G-250 and R-250. *Electrophoresis* **1988**, *9*, 255–262. [CrossRef] [PubMed]

44. Torres, M.A.; Dangl, J.L. Functions of the respiratory burst oxidase in biotic interactions, abiotic stress and development. *Curr. Opin. Plant Biol.* **2005**, *8*, 397–403. [CrossRef]

45. Karpinski, S.; Reynolds, H.; Karpinska, B.; Wingsle, G.; Creissen, G.; Mullineaux, P. Systemic Signaling and Acclimation in Response to Excess Excitation Energy in *Arabidopsis*. *Science* **1999**, *284*, 654–657. [CrossRef]

46. Vandenabeele, S.; Vanderauwera, S.; Vuylsteke, M.; Rombauts, S.; Langebartels, C.; Seidlitz, H.K.; Zabeau, M.; Van Montagu, M.; Inzé, D.; Van Breusegem, F. Catalase deficiency drastically affects gene expression induced by high light in *Arabidopsis thaliana*. *Plant J.* **2004**, *39*, 45–58. [CrossRef]

47. Laloi, C.; Stachowiak, M.; Pers-Kamczyc, E.; Warzych, E.; Murgia, I.; Apel, K. Cross-talk between singlet oxygen- and hydrogen peroxide-dependent signaling of stress responses in *Arabidopsis thaliana*. *Proc. Natl. Acad. Sci. USA* **2007**, *104*, 672–677. [CrossRef] [PubMed]

48. Joo, J.H.; Wang, S.; Chen, J.G.; Jones, A.M.; Fedoroff, N.V. Different signaling and cell death roles of heterotrimeric G protein alpha and beta subunits in the *Arabidopsis* oxidative stress response to ozone. *Plant Cell* **2005**, *17*, 957–970. [CrossRef]

49. Belhaj, K.; Lin, B.; Mauch, F. The chloroplast protein RPH1 plays a role in the immune response of *Arabidopsis* to *Phytophthora brassicae*. *Plant J.* **2009**, *58*, 287–298. [CrossRef]

50. Kuźniak, E.; Świercz, U.; Chojak-Koźniewska, J.; Sekulska-Nalewajko, J.; Gocławski, J. Automated image analysis for quantification of histochemical detection of reactive oxygen species and necrotic infection symptoms in plant leaves. *J. Plant Interact.* **2014**, *9*, 167–174. [CrossRef]

51. Nishiyama, Y.; Yamamoto, H.; Allakhverdiev, S.; Inaba, M.; Yokota, A.; Murata, N. Oxidative stress inhibits the repair of photodamage to the photosynthetic machinery. *EMBO J.* **2001**, *20*, 5587–5594. [CrossRef]

52. Meng, L.; Höfte, M.; Van Labeke, M.-C. Leaf age and light quality influence the basal resistance against *Botrytis cinerea* in strawberry leaves. *Environ. Exp. Bot.* **2019**, *157*, 35–45. [CrossRef]

53. Li, Z.; Wakao, S.; Fischer, B.B.; Niyogi, K.K. Sensing and Responding to Excess Light. *Annu. Rev. Plant Biol.* **2009**, *60*, 239–260. [CrossRef]

54. Gawroński, P.; Burdiak, P.; Scharff, L.B.; Mielecki, J.; Górecka, M.; Zaborowska, M.; Leister, D.; Waszczak, C.; Karpiński, S. CIA2 and CIA2-LIKE are required for optimal photosynthesis and stress responses in *Arabidopsis thaliana*. *Plant J.* **2021**, *105*, 619–638. [CrossRef]

55. Pulido, P.; Llamas, E.; Rodriguez-Concepcion, M. Both Hsp70 chaperone and Clp protease plastidial systems are required for protection against oxidative stress. *Plant Signal. Behav.* **2017**, *12*, e1290039. [CrossRef] [PubMed]

56. Muthusamy, S.K.; Dalal, M.; Chinnusamy, V.; Bansal, K.C. Differential Regulation of Genes Coding for Organelle and Cytosolic ClpATPases under Biotic and Abiotic Stresses in Wheat. *Front. Plant Sci.* **2016**, *7*, 929. [CrossRef]

57. Doyle, S.M.; Wickner, S. Hsp104 and ClpB: Protein disaggregating machines. *Trends Biochem. Sci.* **2009**, *34*, 40–48. [CrossRef]

58. Park, S.; Rodermel, S.R. Mutations in ClpC2/Hsp100 suppress the requirement for FtsH in thylakoid membrane biogenesis. *Proc. Natl. Acad. Sci. USA* **2004**, *101*, 12765–12770. [CrossRef] [PubMed]

59. Sjögren, L.L.; MacDonald, T.M.; Sutinen, S.; Clarke, A.K. Inactivation of the clpC1 Gene Encoding a Chloroplast Hsp100 Molecular Chaperone Causes Growth Retardation, Leaf Chlorosis, Lower Photosynthetic Activity, and a Specific Reduction in Photosystem Content. *Plant Physiol.* **2004**, *136*, 4114–4126. [CrossRef] [PubMed]

60. Adam, Z.; Rudella, A.; van Wijk, K.J. Recent advances in the study of Clp, FtsH and other proteases located in chloroplasts. *Curr. Opin. Plant Biol.* **2006**, *9*, 234–240. [CrossRef] [PubMed]

61. Wu, H.; Ji, Y.; Du, J.; Kong, D.; Liang, H.; Ling, H.-Q. ClpC1, an ATP-dependent Clp protease in plastids, is involved in iron homeostasis in *Arabidopsis* leaves. *Ann. Bot.* **2010**, *105*, 823–833. [CrossRef] [PubMed]

62. Moghaddam, M.R.B.; Van Den Ende, W. Sugars and plant innate immunity. *J. Exp. Bot.* **2012**, *63*, 3989–3998. [CrossRef]

63. Courbier, S.; Grevink, S.; Sluijs, E.; Bonhomme, P.; Kajala, K.; Van Wees, S.C.M.; Pierik, R. Far-red light promotes *Botrytis cinerea* disease development in tomato leaves via jasmonate-dependent modulation of soluble sugars. *Plant Cell Environ.* **2020**, *43*, 2769–2781. [CrossRef]

64. Moustaka, J.; Moustakas, M. Photoprotective mechanism of the non-target organism *Arabidopsis thaliana* to paraquat exposure. *Pestic. Biochem. Physiol.* **2014**, *111*, 1–6. [CrossRef] [PubMed]

65. Moustaka, J.; Tanou, G.; Adamakis, I.-D.; Eleftheriou, E.P.; Moustakas, M. Leaf Age-Dependent Photoprotective and Antioxidative Response Mechanisms to Paraquat-Induced Oxidative Stress in *Arabidopsis thaliana*. *Int. J. Mol. Sci.* **2015**, *16*, 13989–14006. [CrossRef] [PubMed]

66. Ruban, A.V. Nonphotochemical Chlorophyll Fluorescence Quenching: Mechanism and Effectiveness in Protecting Plants from Photodamage. *Plant Physiol.* **2016**, *170*, 1903–1916. [CrossRef] [PubMed]

67. Agathokleous, E.; Kitao, M.; Harayama, H. On the Nonmonotonic, Hormetic Photoprotective Response of Plants to Stress. *Dose-Response* **2019**, *17*, 1559325819838420. [CrossRef]

68. Sperdouli, I.; Moustakas, M. A better energy allocation of absorbed light in photosystem II and less photooxidative damage contribute to acclimation of *Arabidopsis thaliana* young leaves to water deficit. *J. Plant Physiol.* **2014**, *171*, 587–593. [CrossRef]

69. Kato, Y.; Sakamoto, W. FtsH Protease in the Thylakoid Membrane: Physiological Functions and the Regulation of Protease Activity. *Front. Plant Sci.* **2018**, *9*, 855. [CrossRef]

70. Mulo, P.; Sirpiö, S.; Suorsa, M.; Aro, E.-M. Auxiliary proteins involved in the assembly and sustenance of photosystem II. *Photosynth. Res.* **2008**, *98*, 489–501. [CrossRef]

71. Seo, S.; Okamoto, M.; Iwai, T.; Iwano, M.; Fukui, K.; Isogai, A.; Nakajima, N.; Ohashi, Y. Reduced Levels of Chloroplast FtsH Protein in Tobacco Mosaic Virus–Infected Tobacco Leaves Accelerate the Hypersensitive Reaction. *Plant Cell* **2000**, *12*, 917–932.

72. Kangasjärvi, S.; Neukermans, J.; Li, S.; Aro, E.-M.; Noctor, G. Photosynthesis, photorespiration, and light signalling in defence responses. *J. Exp. Bot.* **2012**, *63*, 1619–1636. [CrossRef] [PubMed]

73. Kiba, A.; Nishihara, M.; Tsukatani, N.; Nakatsuka, T.; Kato, Y.; Yamamura, S. A Peroxiredoxin Q Homolog from Gentians is Involved in Both Resistance against Fungal Disease and Oxidative Stress. *Plant Cell Physiol.* **2005**, *46*, 1007–1015. [CrossRef] [PubMed]

Article

TMT-Based Quantitative Proteomic Analysis Reveals the Response of Tomato (*Solanum lycopersicum* L.) Seedlings to Ebb-and-Flow Subirrigation

Kelei Wang [1,2], Muhammad Moaaz Ali [1,3], Tianxin Guo [1], Shiwen Su [2,4], Xianzhi Chen [2,4], Jian Xu [2,4,*] and Faxing Chen [1,*]

[1] College of Horticulture, Fujian Agriculture and Forestry University, Fuzhou 350002, China
[2] Wenzhou Vocational College of Science and Technology, Wenzhou 325014, China
[3] College of Plant Protection, Fujian Agriculture and Forestry University, Fuzhou 350002, China
[4] Wenzhou Protected Vegetable Engineering Technology Center, Wenzhou 325014, China
* Correspondence: xujian@wzvcst.edu.cn (J.X.); cfaxing@126.com (F.C.)

Abstract: Ebb-and-flow subirrigation (EFI) is a water-saving and environmentally friendly irrigation method that can effectively improve water use efficiency and promote plant growth. In this study, we elucidated the effects of ebb-and-flow subirrigation on the protein levels in tomato roots in comparison with top sprinkle irrigation (TSI) and used an integrated approach involving tandem mass tag (TMT) labeling, high-performance liquid chromatography (HPLC) fractionation, and mass-spectrometry (MS)-based analysis. A total of 8510 quantifiable proteins and 513 differentially accumulated proteins (DAPs) were identified, of which the expressions of 283 DAPs were up-regulated, and 230 DAPs were down-regulated in the EFI vs. TSI treatment comparison. According to proteomic data, we performed a systematic bioinformatics analysis of all the identified proteins and DAPs. The DAPs were most significantly associated with the terms 'metabolic process', 'anchored component of membrane', 'oxidoreductase activity', 'phenylpropanoid biosynthesis', and 'biosynthesis of secondary metabolites' according to Gene Ontology (GO) and Kyoto Encyclopedia of Genes and Genomes enrichment (KEGG) analysis. The 272 DAPs were classified into 12 subcellular components according to their subcellular localization. Furthermore, the activities of SOD, POD, CAT, GR, and APX in tomato roots were remarkably increased under EFI, while the MDA content was decreased compared with TSI. Correlation analysis among activities of enzymes and their related DAPs showed that 30 DAPs might be responsible for the regulation of these enzymes. The results showed that ebb-and-flow subirrigation could induce a series of DAPs responses in tomato roots to be adapted to the new mode of water supply.

Keywords: irrigation; root-softening; antioxidant; ebb-and-flow; tandem mass tag; HPLC-MS

Citation: Wang, K.; Ali, M.M.; Guo, T.; Su, S.; Chen, X.; Xu, J.; Chen, F. TMT-Based Quantitative Proteomic Analysis Reveals the Response of Tomato (*Solanum lycopersicum* L.) Seedlings to Ebb-and-Flow Subirrigation. *Agronomy* **2022**, *12*, 1880. https://doi.org/10.3390/agronomy12081880

Academic Editors: Roxana Yockteng, Andrés J. Cortés and María Ángeles Castillejo

Received: 7 July 2022
Accepted: 7 August 2022
Published: 10 August 2022

1. Introduction

Tomato (*Solanum lycopersicum* L.) belongs to the Solanaceae family and is a greenhouse vegetable crop native to Peruvian and Mexican regions [1,2]. It is a rich source of natural antioxidants, including flavonoids; carotenoids (mainly lycopene and β-carotene); and vitamins A, B, and C [3–7]. Vegetable production in the greenhouse demands high fertilizer and water inputs in order to achieve better yield and superior-quality produce [8–11]. The top sprinkle irrigation system (TSI) is a common irrigation strategy for greenhouse vegetable production that is not regarded as ecologically friendly since significant amounts of water and nutrients are usually wasted and may runoff/leach, damaging surface and groundwater systems [12–14]. Besides, the seedlings around the hole plate grow poorly due to water shortage, resulting in inconsistent seedling height [15–17].

An evolved form of the continuous floating system, the ebb-and-flow irrigation (EFI) system was originally developed to grow tobacco (*Nicotiana tabacum* L.) plants in order to increase field survival and reduce transplant shock. It is now being used to grow a large number of commercial vegetables in China, Japan, the United States, and other developing countries [18]. It has many advantages, such as root moisture optimization, water saving, and fertilizer saving compared to top sprinkler irrigation [19–21]. Metal wires $\approx$ 0.20 m are used to suspend the seedling trays above the concrete floor. Every two to three days, the irrigation water is brought up to the level of the trays, held there for 15 to 45 min, and then returned to the main reservoir until the next irrigation. In this system, water or nutrient solution is transported to the plant root through the bottom of the cultivation container by capillary action of the cultivation medium, which can effectively avoid the edge or umbrella effect and improve plant uniformity [18,22,23]. A number of studies in subirrigation, which have been primarily carried out with ornamental species, have demonstrated that the concentration of the nutrient solution can be reduced by up to 50% when compared to nutrient solutions for top sprinkle irrigation, with no adverse effects on plant growth or quality [10,24]. Subirrigation systems improved the uniformity and quality of bell pepper (*Capsicum annum* L.) and tomato (*S. lycopersicum* L.) if grown with minimal nutrient and drought stress [25–27].

Different irrigation methods cause differences in root growth [28]. However, waterlogging and anaerobic respiration over an extended period of time result in the buildup of aldehydes, as well as an increase in reactive oxygen species (ROS), which finally leads to cell death and plant senescence [29,30]. Plant hormones may accumulate or degrade quickly if the gaseous exchange is hampered, and this can alter plant waterlogging tolerance [31,32]. When plants are under waterlogging or drought stress, the activities of protective enzymes in plants change dynamically [33,34]. In a recent study, the activities of superoxide dismutase (SOD), peroxidase (POD), catalase (CAT), and other protective enzymes in the roots of sesame seedlings increased at the initial stage of flooding and decreased significantly with the extension of the water stress time [35,36].

A novel tandem-mass-spectrometry (MS)-based tandem mass tag (TMT) labeling strategy has been used in quantitative proteomics in recent years [37–39]. Therefore, we used a TMT-based quantitative proteomics approach to identify differentially accumulated proteins (DAPs) in tomato roots under ebb-and-flow subirrigation so as to provide a theoretical basis for the popularization and application of ebb-and-flow irrigation technology.

2. Materials and Methods

2.1. Plant Source, Experimental Design, and Irrigation Treatments

The experiment was conducted at the experimental farm of Wenzhou Vocational College of Science and Technology, Wenzhou, China (28°05′39.5″ N 120°30′55.2″ E) from 15 August 2019 to 19 September 2019. The pure line tomato cultivar 'Ouxiu 201' (round-shaped fruit with regular leaves) was obtained from Institute of Vegetable Science, Wenzhou Academy of Agricultural Sciences, Wenzhou, China. Tomato seeds were sown in seedlings trays (540 mm × 280 mm) with 50 plugs (each having volume of 55 mL with upper and lower diameters of 48 and 18 mm, respectively). Seedling trays were purchased from Taizhou Longji Plastics Co., Ltd., Taizhou, China. Seedling plugs were filled with growing media containing peat, vermiculite, and perlite (3:1:1), which were obtained from Hangzhou Lin'an Jindalu Industrial Technology Co., Ltd., Hangzhou, China. The pH, EC, and organic matter of the growing media were 6.34, 0.87 ms·cm^{-1}, and 95.4%, respectively. The ebb-and-flow irrigation system (patent no. *ZL201520333950.6*) used in the experiment was developed by Wenzhou Academy of Agricultural Sciences, Wenzhou, China. It contained a nursery frame, a number of ebb-and-flow trays horizontally placed on the nursery frame, and a water or nutrient solution circulation device (Figure 1).

Tomato seedling trays were placed on ebb-and-flow irrigation (EFI) seedling raising system [40], as well as on simple beds for top sprinkle irrigation (TSI) water treatment. The top sprinkle irrigation was carried out by manually holding the watering can. There was no fertilizer applied during the experiment. Six seedling trays were used for this experiment, each being considered as a replication of different irrigation treatments, i.e., EFI and TSI.

Each irrigation treatment had 3 repetitions with 50 plants in each repeat. The experiment was a completely randomized design (CRD).

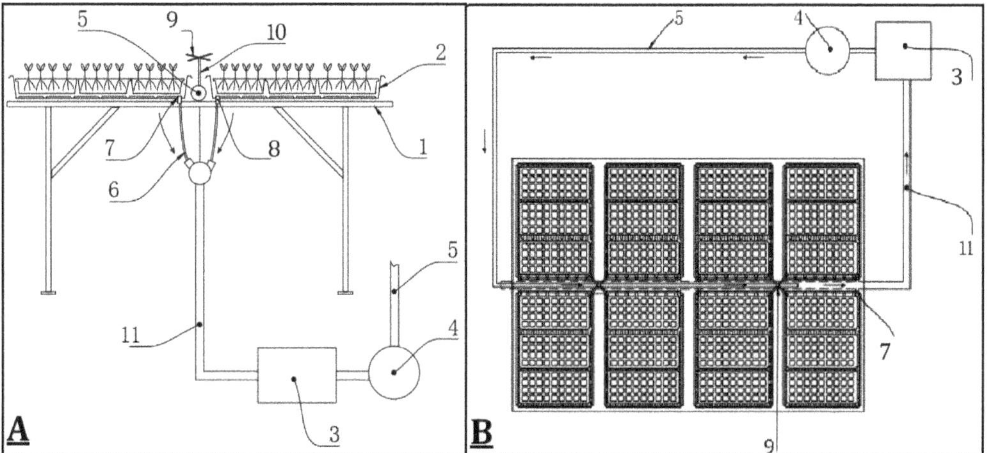

Figure 1. Lateral (**A**) and top (**B**) view of ebb-and-flow nursery raising system. 1—nursery frame; 2—ebb-and-flow tray; 3—water reservoir; 4—disinfection equipment; 5—inlet main pipe; 6—drainage branch pipe; 7—drainage hole; 8—filtering parts; 9—inlet branch pipe; 10—five-way connector; 11—drainage main pipe.

2.2. Enzymes' Activity Assay

Thirty-five days after sowing, root samples (0.5 g) were collected from ten randomly selected seedlings from each replication of each irrigation treatment and quickly frozen in liquid nitrogen. Then, 1 mL of 0.1 M phosphoric acid buffer (pH = 7) was added into frozen root samples, ground, and homogenized in an ice bath. The final volume of 5 mL was made by adding 0.1 M phosphoric acid buffer (pH = 7), and 2 mL was taken and centrifuged at 10,000 rpm for 10 min. The supernatant was obtained as the crude extract of enzyme solution [41] and stored at 4 °C for further analysis.

The activity of superoxidase dismutase (SOD; EC 1.15.1.1) was assayed using a xanthine–xanthine oxidase system [42]. The change in absorbance was read at 560 nm. Peroxidase (POD; EC 1.11.1.7) activity was determined using a previously described method [43]. The change in absorbance was read at 470 nm for 4 min. Catalase (CAT; EC 1.11.1.6) activity was assayed as described previously [44]. The reaction was initiated with the enzyme extract. The decrease in absorbance (due to decomposition of H_2O_2) at 240 nm was recorded for 1 min. Ascorbate peroxidase (APX; EC 1.11.1.11) activity was measured as previously described by Nakano and Asada [45]. The absorbance was measured at 290 nm for 1 min. The activity of glutathione reductase (GR; EC 1.6.4.2) was measured using the method earlier described by Cakmak et al. [46]. The decrease in absorbance was recorded for 1 min at 340 nm due to NADPH oxidation.

2.3. Protein Extraction, TMT Labeling, and Data Analysis

Total proteins were extracted from the EFI and TSI root tissues of tomato with three biological replicates (each containing 500 mg roots) using the cold acetone method [47]. Next, the tryptic peptides of each sample were labeled through tandem mass tags (TMT) using a TMT kit (Novogene Bio Technology Co., Ltd., Nanjing, China) according to the manufacturer's protocol, and TSI1, TSI2, TSI3, EFI1, EFI2, and EFI3 were labeled with 126, 127 N, 128 N, 129 N, 130 N, and 131 tags, respectively. Then, they were fractionated by high-pH reverse-phase HPLC using a C18 column (Waters BEH C18 4.6 × 250 mm, 5 μm) on a Rigol L3000 HPLC system. These peptides were further used for proteomic analysis. Proteome analysis for the fractionated peptides was performed using a liquid chromatography–mass spectrometry (LC-MS) system. Briefly, these peptides were performed using an EASY-nLCTM 1200 UHPLC system (Thermo Fisher Scientific Co., Ltd., Shanghai, China) coupled with a Q Exactive HF-X mass spectrometer (Thermo Fisher Scientific Co., Ltd., Shanghai, China) operating in the data-dependent acquisition (DDA) mode. A total of 1 μg sample was injected into a home-made C18 Nano-Trap column (2 cm × 75 μm, 3 μm). Peptides were separated in a home-made analytical column (15 cm × 150 μm, 1.9 μm) using a linear gradient elution. The separated peptides were analyzed by Q Exactive HF-X mass spectrometer (Thermo Fisher Scientific Co., Ltd., Shanghai, China), with ion source of Nanospray Flex™, spray voltage of 2.3 kV, and ion transport capillary temperature of 320 °C. Full scan range from 350 to 1500 m/z with resolution of 60,000 (at 200 m/z), automatic gain control (AGC) target value of 3×10^6, and a maximum ion injection time of 20 ms were used. The top 40 precursors of the highest abundant in the full scan were selected and fragmented by higher-energy collisional dissociation (HCD) and analyzed in MS, where resolution was 30,000 (at 200 m/z) for 6 plex, the automatic gain control (AGC) target value was 5×10^4, the maximum ion injection time was 54 ms, the normalized collision energy was set as 32%, the intensity threshold was 1.2×10^5, and the dynamic exclusion parameter was 20 s.

The resulting spectra from each run were searched separately against protein sequences library (Solanum_lycopersicum.SL3.0.pep.all.fasta, 34,429 sequences) database by the search engines: Proteome Discoverer 2.2 (PD 2.2, Thermo). The searched parameters were set as follows: mass tolerance for precursor ion was 10 ppm, and mass tolerance for product ion was 0.02 Da. Carbamidomethyl was specified as fixed modifications, oxidation of methionine (M) and TMT plex were specified as dynamic modification, and acetylation and TMT plex were specified as N-Terminal modification in PD 2.2. A maximum of 2 mis-cleavage sites were allowed. The principal component analysis (PCA) was used to evaluate the relationship among 6 samples. T-test was used to compare the differentially accumulated proteins (DAPs). Proteins with fold change (FC) > 1.5 or <0.67 and $p \leq 0.05$ were considered as DAPs. Notably, the protein quantification was calculated with the median ratio of its corresponding unique peptides and then normalized by taking the median of all quantified proteins.

2.4. Data Analysis

Data regarding antioxidant enzymes were subjected to Student's t-test using Microsoft Excel (ver. 2016). Correlation between antioxidant enzymes' activities and their metabolism-related proteins was determined with Pearson (n) method using IBM SPSS software (ver. 17.0) and visualized through a heatmap using TBtools software (ver. 0.6655) [48]. Principal components analysis (PCA) was performed using SIMCA-P (version 11.5) software. Gene Ontology (GO) functional analysis was conducted using the InterProScan program against the non-redundant protein database (including Pfam, PRINTS, ProDom, SMART, ProSite, PANTHER) [49], and the databases of COG (Clusters of Orthologous Groups) and KEGG (Kyoto Encyclopedia of Genes and Genomes) were used to analyze the protein families and pathways. DAPs were used for volcanic map analysis and enrichment analysis of GO and KEGG [50]. Subcellular localizations of DAPs were predicted using WoLF PSORT (https://wolfpsort.hgc.jp/ (accessed on 14 January 2022), Loctree 3 (https://rostlab.org/services/loctree3/ (accessed on 14 January 2022), TargetP (http://www.cbs.dtu.dk/services/TargetP/ (accessed on 16 January 2022), and SignalP (http://www.cbs.dtu.dk/services/SignalP/ (accessed on 16 January 2022) [51–54].

3. Results

3.1. Primary Quantitative Proteome Analysis

Based on the TMT experiment, a total of 337,532 spectra were identified from tomato roots. Moreover, 61,531 peptides, 50,462 unique peptides, and 8510 proteins were detected from 93,451 known spectra (Figure 2A). The protein molecular weight data (Figure 2B) showed that 82.28% of the total proteins were 10–80 kDa in size, and the average molecular weight was 8.25–274.25 kDa. In addition, about 91% of proteins' sequence coverage was less than 40% (Figure 2C). The length of most peptides was between 7 and 25 amino acids (aa) (Figure 2D), accounting for 97.29% of the total, of which peptides with a length of 8–12 aa were relatively large.

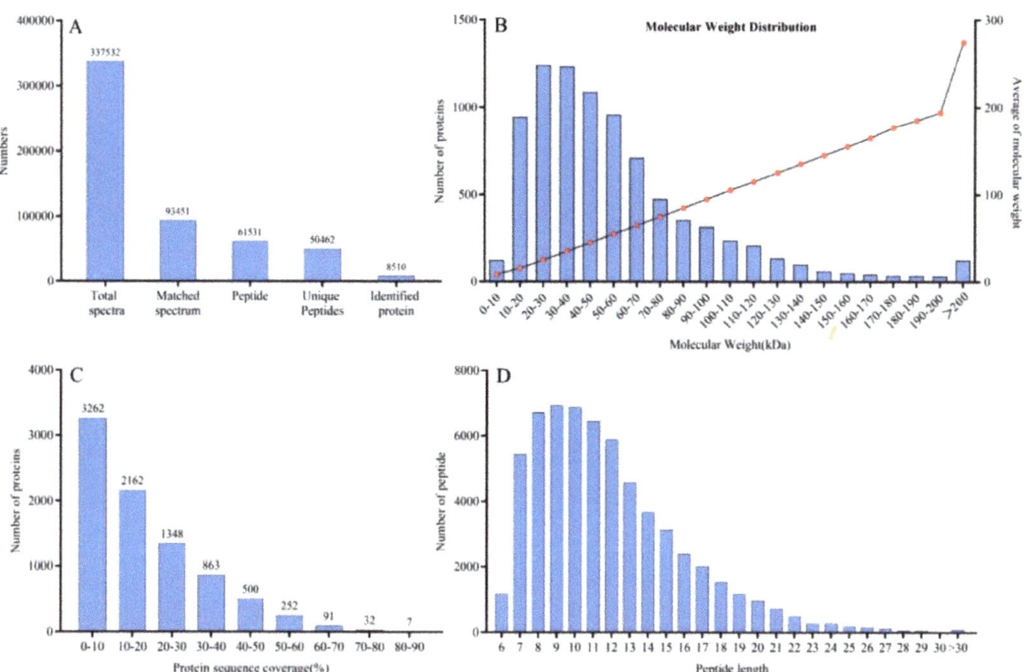

Figure 2. Quality control (QC) validation of mass spectrometer (MS) data. (**A**) Basic statistical data of MS results. (**B**) Molecular weight distribution of proteins. (**C**) Proteins' sequence coverage. (**D**) Length distribution of all identified phosphorylated peptides.

To assess the reproducibility of proteomic data, the coefficient of variation of replicates was determined. Proteins having a 20 percent coefficient of variation accounted for more than 90% of detected proteins (Figure 3A), indicating that the data were credible. EFI and TFI data were cleanly differentiated in a PCA model based on six samples (Figure 3B). PCA1 was responsible for 75.26 percent of the variability, whereas PCA2 was responsible for 13.98 percent.

3.2. Effect of EFI Treatment on the Global Proteome of Tomato Seedlings

The identified proteins and DAPs of tomato roots under ebb-and-flow subirrigation were grouped into three GO categories (biological process, cellular component, and molecular functions) (Figure 4A, Table S1). We can see that in the biological process (BP), 586 identified proteins (including 160 DAPs) were involved in the oxidation-reduction process, 357 proteins (12 DAPs) were involved in protein phosphorylation, and 288 proteins (25 DAPs) were involved in metabolic process. In the cellular component, 278 identified proteins (including 9 DAPs) were involved in integral components of the membrane, 251 proteins (18 DAPs) were involved in membrane synthesis, and 169 proteins (2 DAPs) were involved in the nucleus. In the molecular functions category, 903 identified proteins (including 30 DAPs) were involved in protein binding, 765 proteins (30 DAPs) were involved in ATP binding, whereas 356 identified proteins (12 DAPs) were involved in protein kinase activity.

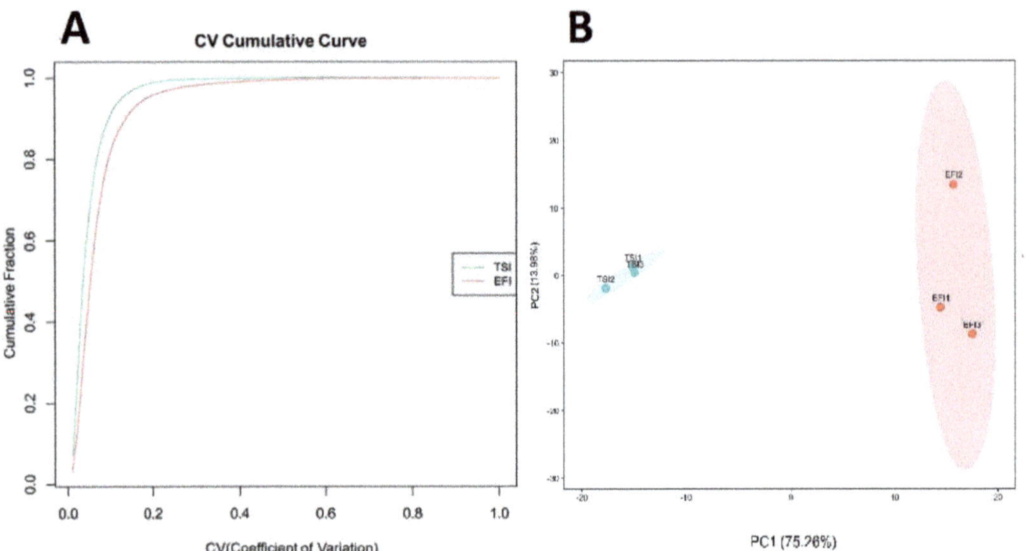

Figure 3. Scatter plot of CV (coefficient of variation) distribution and PCA (principal component analysis) of all samples using quantified proteins. (**A**) CV cumulative curve of two treatments. (**B**) Two-dimensional scatter plot of PCA distribution of all samples.

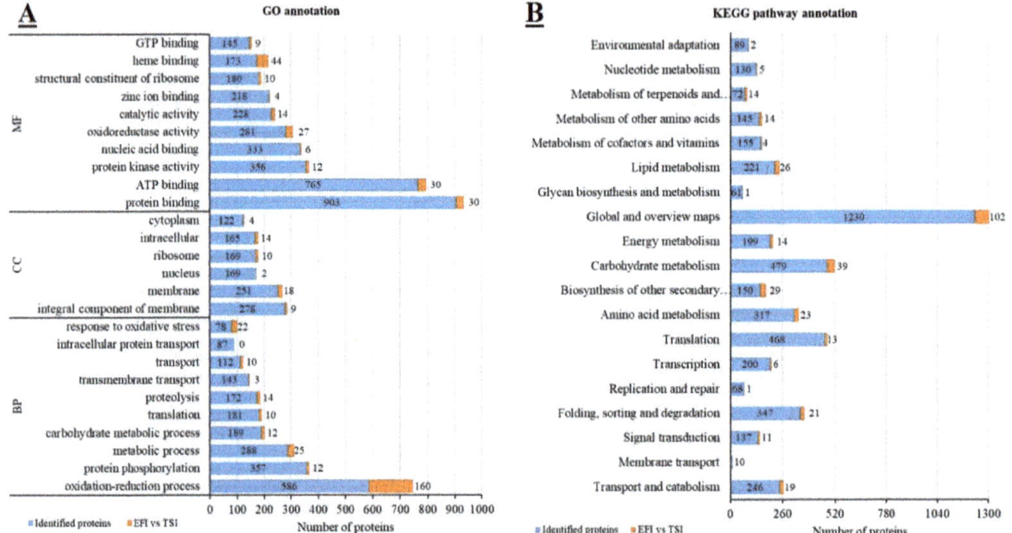

Figure 4. The GO (**A**) and KEGG (**B**) annotation of all identified proteins and DAPs (EFI vs. TSI). All proteins were classified by GO terms based on three categories: molecular function (MF), cellular component (CC), and biological process (BP).

According to KEGG annotation, 4724 identified proteins were grouped into 5 KO categories (i.e., cellular processes, environmental information processing, genetic information processing, metabolism, and organismal systems) (Figure 4B, Table S2). A total of 246 identified proteins (including 16 DAPs) were involved in cellular processes, 147 proteins (11 DAPs) were involved in environmental information processing, 1083 proteins (41 DAPs) were involved in genetic information processing, 3159 proteins (271 DAPs) were involved in metabolism, and 89 proteins (2 DAPs) were involved in organismal systems.

According to the COG database, 5258 identified proteins were categorized into 25 categories (Figure 5A). The largest group was of general function prediction only (608); followed by translation, ribosomal structure, and biogenesis (559); post-translational modification, protein turnover, and chaperones (522); signal transduction mechanisms (476); carbohydrate transport and metabolism (456); amino acid transport and metabolism (308); lipid transport and metabolism (306); secondary metabolites' biosynthesis, transport, and catabolism (270); and energy production and conversion, followed by post-translation (255).

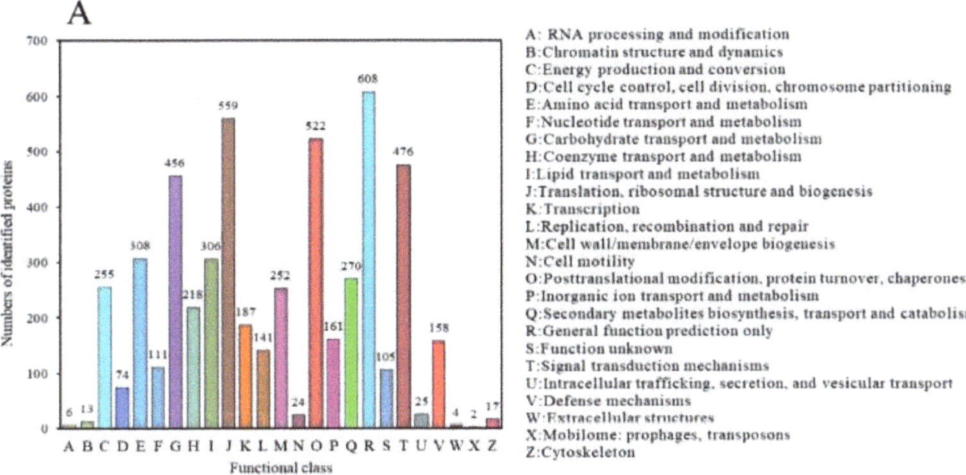

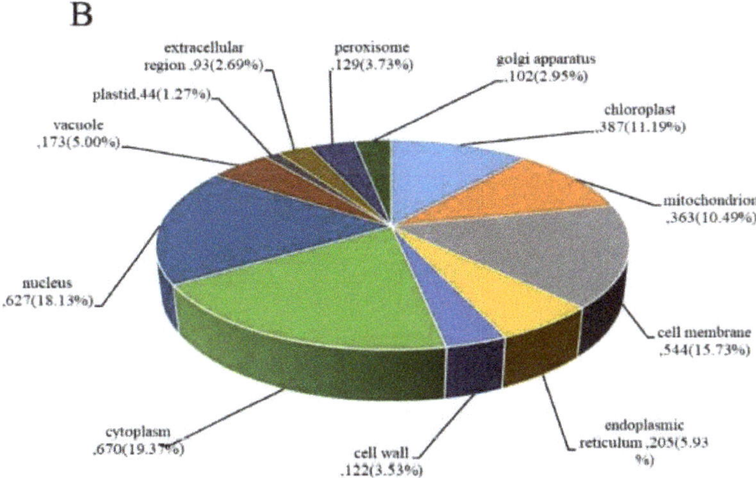

Figure 5. Categorization of all identified proteins with respect to COG annotation (**A**) and subcellular localization (**B**).

The subcellular location prediction and classification statistics for all identified proteins are shown in Figure 5B, and all identified proteins were classified into 12 subcellular components according to their subcellular localizations, including 670 proteins in the cytoplasm (19.36%), 627 in the nucleus (18.12%), 544 in the cell membrane (15.73%), 387 in the chloroplast (11.19%), and 363 in the mitochondrion (10.49%).

3.3. Enrichment Analysis of DAPs

The expression profiles of the DAPs in six samples are presented through a heat map (Figure 6A). A total of 513 DAPs were identified with a fold-changes (FC) > 1.5 or <0.67 and $p \leq 0.05$ (Table S1). Among the DAPs, 283 were up-regulated, and 280 were down-regulated (Figure 6B).

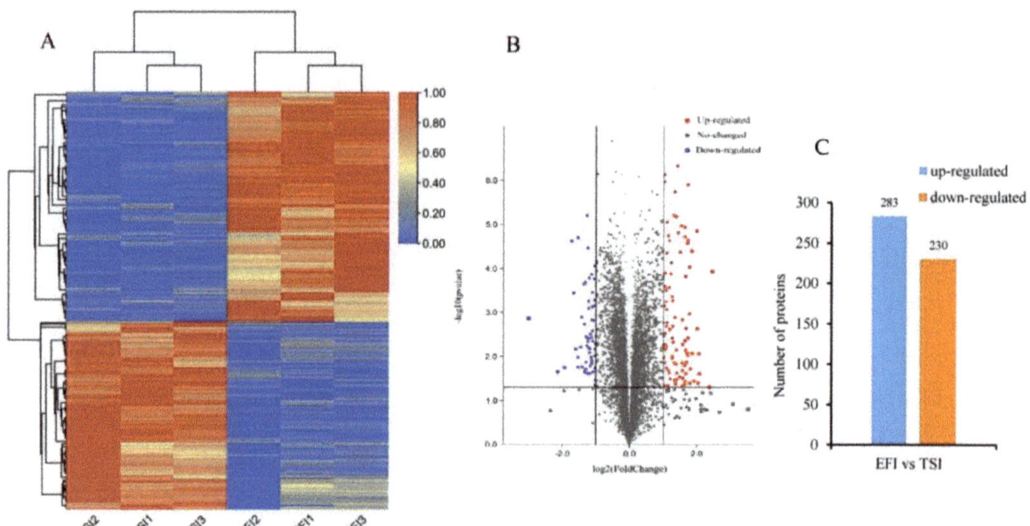

Figure 6. Effect of ebb-and-flow subirrigation strategy on proteome expressions of DAPs in tomato roots. (**A**) Expression profiles of the DAPs response to ebb-and-flow subirrigation. (**B**) Volcano plot of DAPs. (**C**) The numbers of up- and down-regulated proteins in the EFI treated plants compared to the TSI.

We completed an enrichment study of DAPs utilizing GO enrichment, KEGG pathways, and subcellular localization to see whether they were highly enriched in particular functional categories. The GO term meeting this requirement was determined as the GO term highly enriched in different proteins, using $p \leq 0.05$ as the threshold. With 222 GO terms, the 355 DAPs were divided into 3 groups: biological process (BP), cellular component (CC), and molecular function (MF). GO enrichment analysis showed that the metabolic process, single-organism process, and single-organism metabolic process were the most dominant BP terms with 195, 145, and 118 DAPs, respectively. Similarly, microtubule and the anchored component of the membrane were the major CC terms with 3 and 2 DAPs, respectively, while ion binding, oxidoreductase activity, and metal ion binding were the dominant MF terms with 58, 57, and 55 DAPs, respectively (Figure 7A).

The 180 DAPs were divided into different KEGG pathways, with 13 of them being enriched ($p \leq 0.05$). KEGG pathway enrichment revealed that 93, 71, 23, and 19 DAPs were linked to metabolic pathways, secondary metabolite production, phenylpropanoid biosynthesis, and protein processing in the endoplasmic reticulum, respectively (Figure 7B).

The 272 DAPs were classified into 12 subcellular components according to their subcellular localizations, including 56 proteins in the cytoplasm (20.59%), 49 in the cell membrane (18.01%), 25 in the cell wall (9.19%), 21 in the vacuole (7.72%), and 26 in the endoplasmic reticulum (9.56%) (Figure 7C).

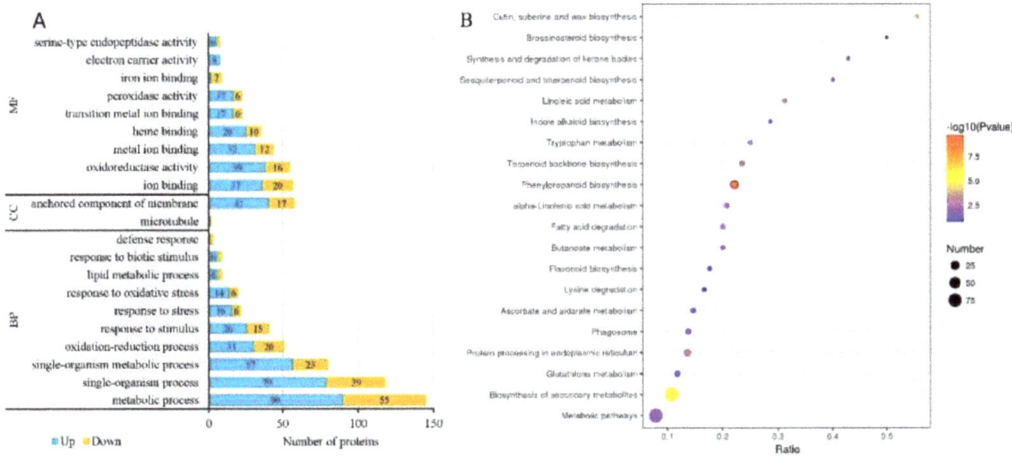

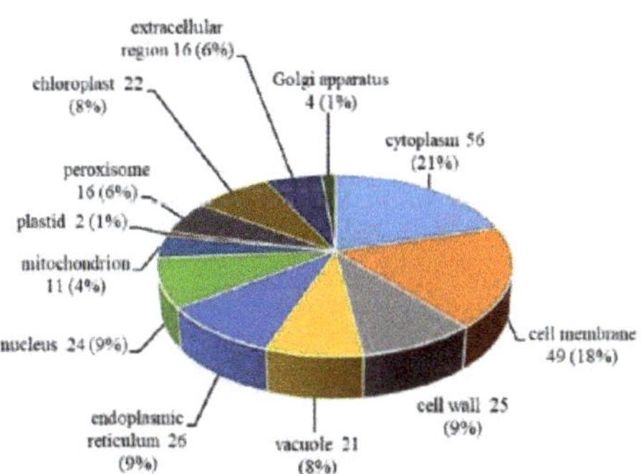

Figure 7. GO enrichment analysis (**A**) and KEGG enrichment analysis (**B**) and subcellular localizations (**C**) of DAPs in tomato roots under ebb-and-flow subirrigation.

3.4. Activities of Antioxidant Enzymes in Tomato Roots under Ebb-and-Flow Subirrigation

The activities of SOD, POD, CAT, APX, and GR in tomato roots under ebb-and-flow irrigation were significantly higher than those under top sprinkler irrigation, increasing by 12.99%, 16.98%, 18.25%, 46.51%, and 28.18%, respectively, while the content of MDA decreased significantly by 6.39% (Figure 8).

3.5. Correlation between Antioxidant Enzymes' Activities and Their Metabolism-Related Proteins

The correlation between antioxidant enzymes' activities (i.e., SOD, POD, CAT, GR, APX) and MDA content and their metabolism-related proteins in the roots of tomato seedlings grown under ebb-and-flow irrigation and top sprinkle irrigation was analyzed (Figure 9). The SOD and CAT activity was significantly ($p \leq 0.05$) and positively correlated with 11 SlPOD (Solyc10g076240.2.1, Solyc04g071900.3.1, Solyc01g105070.3.1, Solyc09g072700.3.1, Solyc09g007520.3.1, Solyc01g006300.3.1, Solyc11g018805.1.1, Solyc05g050870.3.1, Solyc01g015080.3.1, Solyc01g101050.3.1, and Solyc05g050890.2.1), 1 SlCAT Solyc12g094620.2.1, and 2 glutathione S-transferase proteins (Solyc09g011550.3.1 and Solyc08g066850.3.1). The POD activity was significantly ($p \leq 0.05$) and positively associated with 7 proteins related to peroxidase biosynthesis (Solyc10g076240.2.1, Solyc02g092580.3.1, Solyc01g105070.3.1,

Solyc05g046010.3.1, Solyc11g018805.1.1, Solyc01g101050.3.1, and Solyc05g050890.2.1), while 1 protein related to glutathione S-transferase synthesis (Solyc09g011550.3.1).

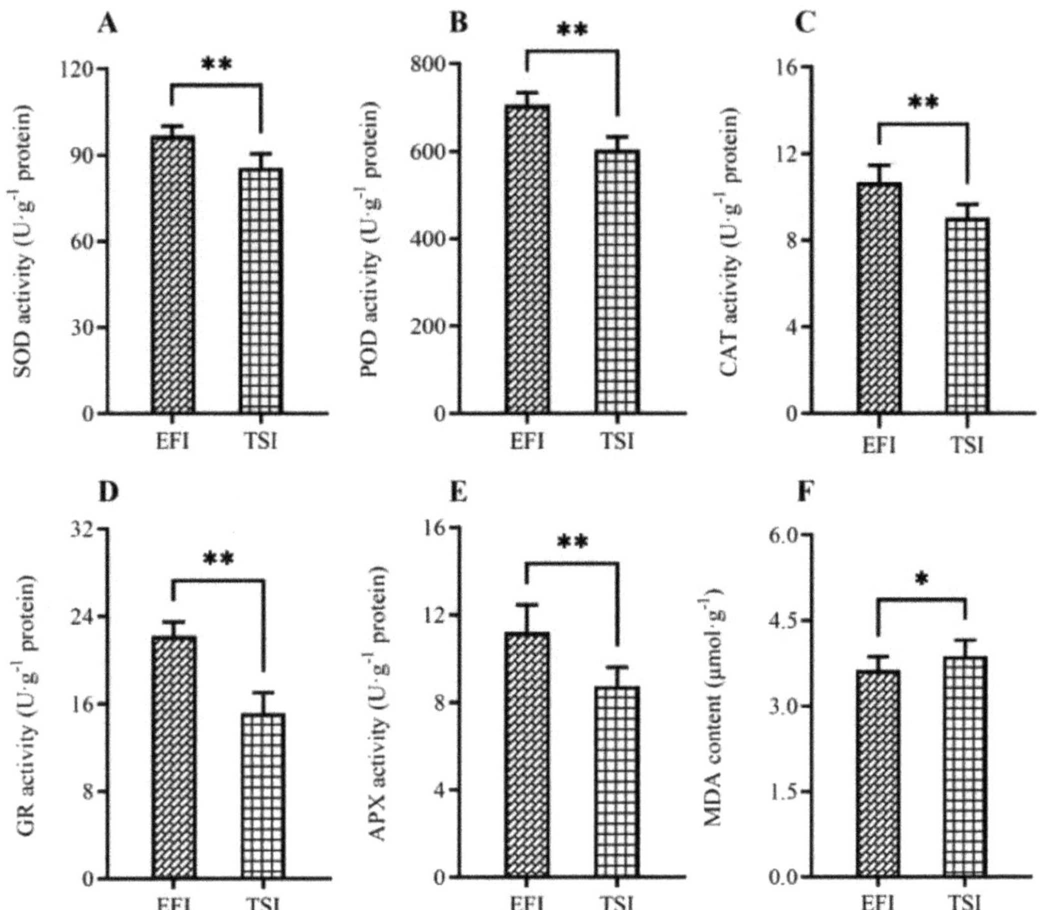

Figure 8. Effect of ebb-and-flow (EFI) irrigation and top sprinkle irrigation (TSI) on the activities of SOD (**A**), POD (**B**), CAT (**C**), GR (**D**), APX (**E**), and MDA content (**F**) of tomato roots. * and ** represent significance at $p \leq 0.05$ and $p \leq 0.01$, respectively.

Similarly, the GR activity was also significantly ($p \leq 0.05$) correlated with 13 proteins related to peroxidase and glutathione S-transferase biosynthesis (i.e., Solyc10g076240.2.1, Solyc04g071900.3.1, Solyc01g105070.3.1, Solyc05g046010.3.1, Solyc09g072700.3.1, Solyc09g007520.3.1, Solyc01g006300.3.1, Solyc11g018805.1.1, Solyc05g050870.3.1, Solyc01g101050.3.1, Solyc05g050890.2.1, Solyc09g011550.3.1, and Solyc08g066850.3.1). Thus, the activities of antioxidant enzymes were positively correlated with the expressions of their metabolism-related proteins.

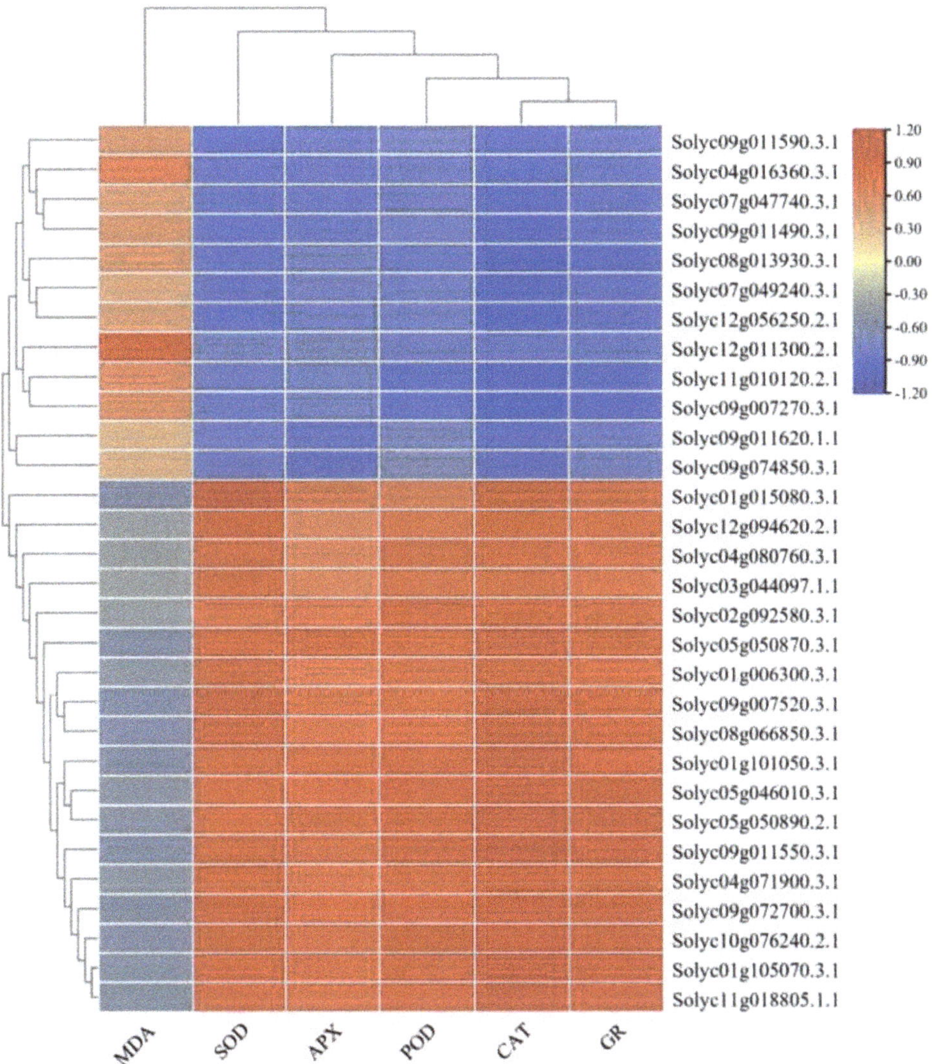

Figure 9. The heat map showing Pearson (*n*) correlation between antioxidant enzymes' activities and their metabolism-related proteins. Color codes: red—higher correlation; blue—lower correlation.

4. Discussion

In order to deeply understand the mechanism of ebb-and-flow subirrigation behind promoting seedling growth, the proteomic changes of tomato roots under two irrigation treatments were studied through the TMT-based quantitative proteomics method. A total of 513 DAPs were identified, of which 283 were up-regulated and 230 were down-regulated (Figure 7).

4.1. DAPs Participated in Carbohydrates and Energy Metabolism

Carbohydrates are essential for plants to maintain their life activities and are the main energy source for plants under abiotic stress [55]. In this experiment, it was found that ebb-and-flow irrigation treatment changed the proteomic expressions of 27 DAPs related to energy metabolism. These DAPs were enriched into six major carbohydrate metabolism-related KEGG pathways (Table S3), namely, amino sugar and nucleotide sugar metabolism (ko00520), pyruvate metabolism (ko00620), glycolysis/gluconeogenesis (ko00010), galactose metabolism (ko00052), starch and sucrose metabolism (ko00500), and pentose and glucuronate interconversions (ko00040). There were eight DAPs involved in the glycolysis/gluconeogenesis pathway, of which six were up-regulated under EFI treatment, including aldehyde dehydrogenase (NAD+), alcohol dehydrogenase class-p, pyruvate decarboxylase, and L-lactate dehydrogenase. The results showed that ebb-and-flow irrigation increased the expression of carbohydrate-metabolism-related proteins in tomato roots, enhanced carbohydrate metabolism, promoted energy production in root cells, and ultimately promoted root growth [56,57]. Studies have shown that SS (sucrose synthase) in different plants was not only involved in sucrose accumulation [58,59] but also closely related to the growth status of plants under biological and abiotic stress [60]. Moreover, enzymes related to starch or sucrose metabolism accumulate more under drought conditions [61]. In this experiment, the proteins related to sucrose synthase and beta glucosidase were up-regulated under ebb-and-flow subirrigation.

4.2. DAPs Participated in Stress Resistance and Defense Response

A large number of reactive oxygen species (ROS) are produced in the plant body due to various stress conditions [62]. Plants have evolved to form a series of complex defense systems to deal with the impact of the adverse environment so as to protect cells from excessive ROS. Under ebb-and-flow irrigation, 30 DAPs related to stress and defense changed, including 17 up-regulated and 13 down-regulated DAPs (Table S4). These DAPs are mainly enriched in KEGG pathways, including phenylpropane biosynthesis (ko00940), ascorbic acid and aldarate metabolism (ko00053), glutathione metabolism (ko00480), and peroxisome (ko04146). The accumulation of POD and GST (glutathione S-transferase) in tomato roots was also increased (Figure 9), indicating that the increase in enzymatic activity of the antioxidant system under ebb-and-flow subirrigation was an important protective method of removing or reducing ROS accumulation. Relevant studies showed that under drought stress, the accumulation of GST in alfalfa 'Longzhong' roots also increased, protecting cells from oxidative stress [63].

Previous studies showed that the proteomic expression of CAT- and POD-metabolism-related proteins changed significantly under drought, high temperatures, and other stresses [64]. In this experiment, it was found that multiple DAPs encoding CAT and POD (i.e., Solyc12g0 94620.2.1, Solyc04g071900.3.1, Solyc01g105070.3.1, Solyc05g046010.3.1, Solyc03g044097.1.1, Solyc09g072700.3.1, Solyc09g007520.3.1, Solyc01g006300.3.1, and Solyc11g018805.1.1) were up-regulated, which was consistent with the higher activity of CAT and POD in tomato roots under ebb-and-flow irrigation (Figure 9). The DAPs related to the APX were also up-regulated. APX reduces the accumulation of hydrogen peroxide in plants through the ascorbic acid metabolic pathway [65]. Studies have shown that the levels of related proteins encoding APX change under heat stress, light stress, and water stress [66]. In this study, DAPs encoding APX and its analogs (Solyc11g062440.2.1, Solyc09g007270.3.1) were up-regulated so as to improve their activity in active oxygen metabolism. Under drought stress, the expression of proteins involved in phenylpropane metabolism also changes, such as cinnamyl alcohol dehydrogenase (CAD), peroxidase, and s-adenosine-l-methionine involved in lignin synthesis [67]. In this experiment, DAPs involved in phenylpropanoid biosynthesis (Solyc05g050870.3.1, Solyc01g015080.3.1, Solyc04g080760.3.1) and glutathione metabolism (Solyc09g01150.3.1, Solyc09g011620.1.1, Solyc12g011300.2.1, and Solyc09g074850.3.1) were also up-regulated. In general, stress-related proteins can eliminate ROS in time by promoting the activity of the antioxidant system so as to maintain the high vitality of tomato seedlings.

4.3. DAPs Participated in Amino Acid Metabolism

Amino acids are important proteins for plant growth, development, and response to various stresses [68,69]. Studies have shown that the abundance and expression of proline-, aspartate-, tryptophan-, and lysine-related proteins are severely influenced under drought or heat stress [70]. In the current study, 22 DAPs participated in the metabolic pathway of a variety of amino acids (Table S5), including tryptophan metabolism (ko00380); valine, leucine, and isoleucine degradation (ko00280); lysine degradation (ko00310); arginine and proline metabolism (ko00330); histidine metabolism (ko00340); beta-alanine metabolism (ko00410); tyrosine metabolism (ko00350); cysteine and methionine metabolism (ko00270); alanine, aspartate, and glutamate metabolism (ko00250); glycine, serine, and threonine metabolism (ko00260); and the biosynthesis of amino acids (ko01230). Aldehyde dehydrogenase (NAD +) was the main enzyme involved in lysine catabolism. The expressions of aldehyde-dehydrogenase-metabolism-related proteins were down-regulated in tomato roots under ebb-and-flow subirrigation. Similarly, asparagine synthase (glutamine-hydrolysing) was the main enzyme involved in asparagine synthesis, which was up-regulated in TSI, indicating that drought caused by TSI promotes the formation of asparagine from aspartic acid and glutamine to adapt to the drought stress. The up-regulated expression of threonine dehydratase and hydroxy-methyl-glutaryl-COA synthase in EFI may increase the biosynthesis of valine, leucine, and isoleucine, which can promote tomato root system development and secondary metabolism.

4.4. DAPs Participated in Plant Hormones and Secondary Metabolism

Plant hormones widely participate in the physiological process of plant growth and development [71] and play an important role in plant response to various environmental and biological stresses [72]. Studies have shown that GH3 family proteins influence the activity of amide synthase, which catalyzes the combination of free indoleacetic acid (IAA) and amino acids so as to inactivate it [73]. The IAA is released through the hydrolysis of IAA amide hydrolase to maintain the homeostasis of IAA in cells [74,75]. Abscisic acid (ABA) plays an important role in regulating seed germination [76], plant growth [77], and the response to stress [78]. In the present study, 79 DAPs were enriched in a large number of secondary-metabolism-related pathways, mainly including phenylpropanoid biosynthesis (ko00940); the biosynthesis of secondary metabolites (ko01110); cutin, suberine, and wax biosynthesis (ko00073); terpene main chain biosynthesis (ko00900), etc. (Table S6). It was identified that the auxin-responsive GH3 gene family showed down-regulated expression in EFI compared with TSI. Moreover, the synthesis of IAA was not inhibited, which promoted the normal growth of tomato roots. Additionally, it was identified that the related proteins of the ABA receptor PYR/PTL family showed down-regulated expressions under EFI, indicating that the synthesis of ABA was inhibited under the ebb-and-flow treatment so that the normal growth of tomato roots could not be inhibited.

5. Conclusions

In the present study, a TMT-based proteomic method was used to investigate changes in protein levels in roots of tomato seedlings grown under ebb-and-flow subirrigation and top sprinkle irrigation. In total, 8510 proteins and 513 DAPs were identified in tomato roots. The bioinformatic analysis revealed that these DAPs were involved in energy production and conversion, metabolic processes, the anchoring component of microtubule and membrane, oxidoreductase activity, and response to stimuli. The KEGG enrichment showed that the DAPs were enriched in 51 KEGG pathways, of which phenylpropane biosynthesis and secondary metabolite biosynthesis were the most significant pathways. The important pathways containing DAPs related to stress response were further divided into five categories, i.e., carbohydrate and energy metabolism, stress resistance and defense response, amino acid metabolism, plant hormones, and secondary metabolism. Ebb-and-flow subirrigation could significantly activate the expressions of proteins related to stress defense and ROS detoxification in tomato roots, which could effectively maintain the balance of protein processing and degradation and enhance the ability of ion, electron, and protein transport across membranes and cell wall regulation so as to promote root growth.

Supplementary Materials: The following supporting information can be downloaded at: https://www.mdpi.com/article/10.3390/agronomy12081880/s1, Table S1: The detailed information about DAPs; Table S2: The KEGG annotation of identified proteins; Table S3: DAPs participated in carbohydrates and energy metabolism; Table S4: DAPs participated in stress resistance and defense response; Table S5: DAPs participated in amino acid metabolism; Table S6: DAPs participated in plant hormones and secondary metabolism.

Author Contributions: Conceptualization, K.W. and J.X.; methodology, K.W., T.G. and S.S.; software, K.W. and M.M.A.; validation, M.M.A., T.G. and F.C.; formal analysis, K.W. and M.M.A.; investigation, S.S. and X.C.; resources, J.X. and X.C.; data curation, M.M.A.; writing—original draft preparation, K.W. and M.M.A.; writing—review and editing, T.G., X.C., S.S., J.X. and F.C.; supervision, J.X. and F.C.; project administration, J.X. and K.W.; funding acquisition, J.X. and K.W. All authors have read and agreed to the published version of the manuscript.

Funding: This research was funded by 'China Agriculture Research System, grant number CARS-23' and the 'Doctoral Research Start Project of Wenzhou Vocational College of Science and Technology, grant number 201903'. 'Teacher professional development project of domestic visiting scholars in colleges and universities in 2021, grant number FX2021211'.

Institutional Review Board Statement: Not applicable.

Informed Consent Statement: Not applicable.

Data Availability Statement: The mass spectrometry proteomics data were deposited into the ProteomeXchange Consortium (http://proteomecentral.proteomexchange.org (accessed on 4 April 2022)) via the iProX partner repository [79] with the dataset identifier PXD033438.

Conflicts of Interest: The authors declare no conflict of interest.

References

1. Bai, Y.; Lindhout, P. Domestication and Breeding of Tomatoes: What have We Gained and What Can We Gain in the Future? *Ann. Bot.* **2007**, *100*, 1085–1094. [CrossRef] [PubMed]
2. Yousef, A.F.; Ali, M.M.; Rizwan, H.M.; Gad, A.G.; Liang, D.; Binqi, L.; Kalaji, H.M.; Wróbel, J.; Xu, Y.; Chen, F. Light quality and quantity affect graft union formation of tomato plants. *Sci. Rep.* **2021**, *11*, 9870. [CrossRef] [PubMed]
3. Guo, T.; Gull, S.; Ali, M.M.; Yousef, A.F.; Ercisli, S.; Kalaji, H.M.; Telesiński, A.; Auriga, A.; Wróbel, J.; Radwan, N.S.; et al. Heat stress mitigation in tomato (*Solanum lycopersicum* L.) through foliar application of gibberellic acid. *Sci. Rep.* **2022**, *12*, 11324. [CrossRef] [PubMed]
4. Al-Muhtaseb, A.H.; Al-Harahsheh, M.; Hararah, M.; Magee, T.R.A. Drying characteristics and quality change of unutilized-protein rich-tomato pomace with and without osmotic pre-treatment. *Ind. Crops Prod.* **2010**, *31*, 171–177. [CrossRef]
5. Ali, M.M.; Waleed Shafique, M.; Gull, S.; Afzal Naveed, W.; Javed, T.; Yousef, A.F.; Mauro, R.P. Alleviation of Heat Stress in Tomato by Exogenous Application of Sulfur. *Horticulturae* **2021**, *7*, 21. [CrossRef]
6. Yousef, A.F.; Xu, Y.; Chen, F.; Lin, K.; Zhang, X.; Guiamba, H.; Ibrahim, M.M.; Rizwan, H.M.; Ali, M.M. The influence of LEDs light quality on the growth pigments biochemical and chlorophyll fluorescence characteristics of tomato seedlings (*Solanum lycopersicum* L.). *Fresenius Environ. Bull.* **2021**, *30*, 3575–3588.
7. Yousef, A.F.; Ali, M.M.; Rizwan, H.M.; Tadda, S.A.; Xu, Y.; Kalaji, H.M.; Yang, H.; Ahmed, M.A.A.; Wro, J.; Chen, F. Photosynthetic apparatus performance of tomato seedlings grown under various combinations of LED illumination. *PLoS ONE* **2021**, *16*, e0249373. [CrossRef]
8. Ali, M.M.; Javed, T.; Mauro, R.P.; Shabbir, R.; Afzal, I.; Yousef, A.F. Effect of Seed Priming with Potassium Nitrate on the Performance of Tomato. *Agriculture* **2020**, *10*, 498. [CrossRef]
9. Richards, D.L.; Reed, D.W. New Guinea Impatiens Growth Response and Nutrient Release from Controlled-release Fertilizer in a Recirculating Subirrigation and Top-watering System. *HortScience* **2004**, *39*, 280–286. [CrossRef]
10. Zheng, Y.; Graham, T.; Richard, S.; Dixon, M. Potted Gerbera Production in a Subirrigation System Using Low-concentration Nutrient Solutions. *HortScience* **2004**, *39*, 1283–1286. [CrossRef]
11. Zheng, Y.; Graham, T.; Richard, S.; Dixon, M. Can Low Nutrient Strategies be used for Pot Gerbera Production in Closed-Loop Subirrigation? *Acta Hortic.* **2005**, *691*, 365–372. [CrossRef]
12. Ferrarezi, R.S.; Weaver, G.M.; van Iersel, M.W.; Testezlaf, R. Subirrigation: Historical Overview, Challenges, and Future Prospects. *Horttechnology* **2015**, *25*, 262–276. [CrossRef]

13. Araus, J.L.; Rezzouk, F.Z.; Thushar, S.; Shahid, M.; Elouafi, I.A.; Bort, J.; Serret, M.D. Effect of irrigation salinity and ecotype on the growth, physiological indicators and seed yield and quality of Salicornia europaea. *Plant Sci.* **2021**, *304*, 110819. [CrossRef]

14. Fereres, E.; Soriano, M.A. Deficit irrigation for reducing agricultural water use. *J. Exp. Bot.* **2006**, *58*, 147–159. [CrossRef]

15. Morison, J.I.; Baker, N.; Mullineaux, P.; Davies, W. Improving water use in crop production. *Philos. Trans. R. Soc. B Biol. Sci.* **2008**, *363*, 639–658. [CrossRef]

16. Nadeem, M.; Li, J.; Yahya, M.; Sher, A.; Ma, C.; Wang, X.; Qiu, L. Research Progress and Perspective on Drought Stress in Legumes: A Review. *Int. J. Mol. Sci.* **2019**, *20*, 2541. [CrossRef]

17. Parkash, V.; Singh, S.; Deb, S.K.; Ritchie, G.L.; Wallace, R.W. Effect of deficit irrigation on physiology, plant growth, and fruit yield of cucumber cultivars. *Plant Stress* **2021**, *1*, 100004. [CrossRef]

18. Leskovar, D.I. Root and Shoot Modification by Irrigation. *Horttechnology* **1998**, *8*, 510–514. [CrossRef]

19. Elmer, W.H.; Gent, M.P.N.; McAvoy, R.J. Partial saturation under ebb and flow irrigation suppresses Pythium root rot of ornamentals. *Crop Prot.* **2012**, *33*, 29–33. [CrossRef]

20. James, E.; van Iersel, M. Ebb and Flow Production of Petunias and Begonias as Affected by Fertilizers with Different Phosphorus Content. *HortScience* **2001**, *36*, 282–285. [CrossRef]

21. Buwalda, F.; Baas, R.; van Weel, P.A. A soilless ebb-and-flow system for all-year-round chrysanthemums. *Acta Hortic.* **1994**, *361*, 123–132. [CrossRef]

22. Naghedifar, S.M.; Ziaei, A.N.; Ansari, H. Numerical analysis of sensor-based flood-floor ebb-and-flow subirrigation system with saline water. *Arch. Agron. Soil Sci.* **2021**, *67*, 1285–1299. [CrossRef]

23. Poole, R.T.; Conover, C.A. Fertilizer Levels and Medium Affect Foliage Plant Growth in an Ebb and Flow Irrigation System. *J. Environ. Hortic.* **1992**, *10*, 81–86. [CrossRef]

24. Rouphael, Y.; Cardarelli, M.; Rea, E.; Colla, G. The influence of irrigation system and nutrient solution concentration on potted geranium production under various conditions of radiation and temperature. *Sci. Hortic.* **2008**, *118*, 328–337. [CrossRef]

25. Leskovar, D.I.; Boales, A.K. Plant Establishment Systems Affect Yield of Jalapeno Pepper. *Acta Hortic.* **1995**, *412*, 275–280. [CrossRef]

26. Leskovar, D.I.; Cantliffe, D.J. Comparison of Plant Establishment Method, Transplant, or Direct Seeding on Growth and Yield of Bell Pepper. *J. Am. Soc. Hortic. Sci.* **1993**, *118*, 17–22. [CrossRef]

27. Leskovar, D.I.; Cantliffe, D.J.; Stoffella, P.J. Transplant Production Systems Influence Growth and Yield of Fresh-market Tomatoes. *J. Am. Soc. Hortic. Sci.* **1994**, *119*, 662–668. [CrossRef]

28. Mahgoub, N.A.; Ibrahim, A.M.; Ali, O.M. Effect of different irrigation systems on root growth of maize and cowpea plants in sandy soil. *Eur. J. Soil Sci.* **2017**, *6*, 374–379. [CrossRef]

29. Xuewen, X.; Huihui, W.; Xiaohua, Q.; Qiang, X.; Xuehao, C. Waterlogging-induced increase in fermentation and related gene expression in the root of cucumber (*Cucumis sativus* L.). *Sci. Hortic.* **2014**, *179*, 388–395. [CrossRef]

30. Zhang, P.; Lyu, D.; Jia, L.; He, J.; Qin, S. Physiological and de novo transcriptome analysis of the fermentation mechanism of *Cerasus sachalinensis* roots in response to short-term waterlogging. *BMC Genom.* **2017**, *18*, 649. [CrossRef]

31. Hattori, Y.; Nagai, K.; Furukawa, S.; Song, X.-J.; Kawano, R.; Sakakibara, H.; Wu, J.; Matsumoto, T.; Yoshimura, A.; Kitano, H.; et al. The ethylene response factors SNORKEL1 and SNORKEL2 allow rice to adapt to deep water. *Nature* **2009**, *460*, 1026–1030. [CrossRef] [PubMed]

32. Kuroha, T.; Nagai, K.; Gamuyao, R.; Wang, D.R.; Furuta, T.; Nakamori, M.; Kitaoka, T.; Adachi, K.; Minami, A.; Mori, Y.; et al. Ethylene-gibberellin signaling underlies adaptation of rice to periodic flooding. *Science* **2018**, *361*, 181–186. [CrossRef] [PubMed]

33. Pan, J.; Sharif, R.; Xu, X.; Chen, X. Mechanisms of Waterlogging Tolerance in Plants: Research Progress and Prospects. *Front. Plant Sci.* **2021**, *11*, 627331. [CrossRef] [PubMed]

34. Fukao, T.; Barrera-Figueroa, B.E.; Juntawong, P.; Peña-Castro, J.M. Submergence and Waterlogging Stress in Plants: A Review Highlighting Research Opportunities and Understudied Aspects. *Front. Plant Sci.* **2019**, *10*, 340. [CrossRef] [PubMed]

35. Anee, T.I.; Nahar, K.; Rahman, A.; Mahmud, J.A.; Bhuiyan, T.F.; Alam, M.U.; Fujita, M.; Hasanuzzaman, M. Hasanuzzaman Oxidative Damage and Antioxidant Defense in Sesamum indicum after Different Waterlogging Durations. *Plants* **2019**, *8*, 196. [CrossRef]

36. Habibullah, M.; Sarkar, S.; Islam, M.M.; Ahmed, K.U.; Rahman, M.Z.; Awad, M.F.; ElSayed, A.I.; Mansour, E.; Hossain, M.S. Assessing the Response of Diverse Sesame Genotypes to Waterlogging Durations at Different Plant Growth Stages. *Plants* **2021**, *10*, 2294. [CrossRef]

37. Liu, Y.; Cao, D.; Ma, L.; Jin, X.; Yang, P.; Ye, F.; Liu, P.; Gong, Z.; Wei, C. TMT-based quantitative proteomics analysis reveals the response of tea plant (*Camellia sinensis*) to fluoride. *J. Proteom.* **2018**, *176*, 71–81. [CrossRef]

38. Hao, J.; Guo, H.; Shi, X.; Wang, Y.; Wan, Q.; Song, Y.-B.; Zhang, L.; Dong, M.; Shen, C. Comparative proteomic analyses of two Taxus species (Taxus× media and Taxus mairei) reveals variations in the metabolisms associated with paclitaxel and other metabolites. *Plant Cell Physiol.* **2017**, *58*, 1878–1890. [CrossRef]

39. Xu, D.; Yuan, H.; Tong, Y.; Zhao, L.; Qiu, L.; Guo, W.; Shen, C.; Liu, H.; Yan, D.; Zheng, B. Comparative Proteomic Analysis of the Graft Unions in Hickory (*Carya cathayensis*) Provides Insights into Response Mechanisms to Grafting Process. *Front. Plant Sci.* **2017**, *8*, 676. [CrossRef]

40. Kelei, W.; Youhe, Z.; Jianlei, S.; Zong'an, H.; Longjing, Z.; Jian, X. Application of dynamic water level management of ebb and flow irrigation for cucumber seedlings. *Acta Agric. Zhejiangensis* **2017**, *29*, 408–413.

41. Chao, Y.-Y.; Chen, C.-Y.; Huang, W.-D.; Kao, C.H. Salicylic acid-mediated hydrogen peroxide accumulation and protection against Cd toxicity in rice leaves. *Plant Soil* **2010**, *329*, 327–337. [CrossRef]
42. El-Shabrawi, H.; Kumar, B.; Kaul, T.; Reddy, M.K.; Singla-Pareek, S.L.; Sopory, S.K. Redox homeostasis, antioxidant defense, and methylglyoxal detoxification as markers for salt tolerance in Pokkali rice. *Protoplasma* **2010**, *245*, 85–96. [CrossRef]
43. Hasanuzzaman, M.; Hossain, M.A.; Fujita, M. Nitric oxide modulates antioxidant defense and the methylglyoxal detoxification system and reduces salinity-induced damage of wheat seedlings. *Plant Biotechnol. Rep.* **2011**, *5*, 353–365. [CrossRef]
44. Zhou, Y.; Ming, D.; Cui, J.; Chen, X.; Wen, Z.; Zhang, J.; Liu, H. Exogenous GSH protects tomatoes against salt stress by modulating photosystem II efficiency, absorbed light allocation and H_2O_2-scavenging system in chloroplasts. *J. Integr. Agric.* **2018**, *17*, 2257–2272. [CrossRef]
45. Nakano, Y.; Asada, K. Hydrogen Peroxide is Scavenged by Ascorbate-specific Peroxidase in Spinach Chloroplasts. *Plant Cell Physiol.* **1981**, *22*, 867–880. [CrossRef]
46. Cakmak, I.; Strbac, D.; Marschner, H. Activities of Hydrogen Peroxide-Scavenging Enzymes in Germinating Wheat Seeds. *J. Exp. Bot.* **1993**, *44*, 127–132. [CrossRef]
47. Hao, P.; Zhu, J.; Gu, A.; Lv, D.; Ge, P.; Chen, G.; Li, X.; Yan, Y. An integrative proteome analysis of different seedling organs in tolerant and sensitive wheat cultivars under drought stress and recovery. *Proteomics* **2015**, *15*, 1544–1563. [CrossRef]
48. Chen, C.; Chen, H.; Zhang, Y.; Thomas, H.R.; Frank, M.H.; He, Y.; Xia, R. TBtools: An Integrative Toolkit Developed for Interactive Analyses of Big Biological Data. *Mol. Plant* **2020**, *13*, 1194–1202. [CrossRef]
49. Jones, P.; Binns, D.; Chang, H.-Y.; Fraser, M.; Li, W.; McAnulla, C.; McWilliam, H.; Maslen, J.; Mitchell, A.; Nuka, G.; et al. InterProScan 5: Genome-scale protein function classification. *Bioinformatics* **2014**, *30*, 1236–1240. [CrossRef]
50. Huang, D.W.; Sherman, B.T.; Lempicki, R.A. Bioinformatics enrichment tools: Paths toward the comprehensive functional analysis of large gene lists. *Nucleic Acids Res.* **2009**, *37*, 1–13. [CrossRef]
51. Horton, P.; Park, K.-J.; Obayashi, T.; Fujita, N.; Harada, H.; Adams-Collier, C.J.; Nakai, K. WoLF PSORT: Protein localization predictor. *Nucleic Acids Res.* **2007**, *35*, W585–W587. [CrossRef] [PubMed]
52. Goldberg, T.; Hecht, M.; Hamp, T.; Karl, T.; Yachdav, G.; Ahmed, N.; Altermann, U.; Angerer, P.; Ansorge, S.; Balasz, K.; et al. LocTree3 prediction of localization. *Nucleic Acids Res.* **2014**, *42*, W350–W355. [CrossRef] [PubMed]
53. Emanuelsson, O.; Nielsen, H.; Brunak, S.; von Heijne, G. Predicting Subcellular Localization of Proteins Based on their N-terminal Amino Acid Sequence. *J. Mol. Biol.* **2000**, *300*, 1005–1016. [CrossRef] [PubMed]
54. Petersen, T.N.; Brunak, S.; von Heijne, G.; Nielsen, H. SignalP 4.0: Discriminating signal peptides from transmembrane regions. *Nat. Methods* **2011**, *8*, 785–786. [CrossRef] [PubMed]
55. Ghosh, D.; Xu, J. Abiotic stress responses in plant roots: A proteomics perspective. *Front. Plant Sci.* **2014**, *5*, 6. [CrossRef]
56. Chen, T.; Zhang, L.; Shang, H.; Liu, S.; Peng, J.; Gong, W.; Shi, Y.; Zhang, S.; Li, J.; Gong, J.; et al. iTRAQ-Based Quantitative Proteomic Analysis of Cotton Roots and Leaves Reveals Pathways Associated with Salt Stress. *PLoS ONE* **2016**, *11*, e0148487. [CrossRef] [PubMed]
57. Yao, K.; Wu, Y.Y. Phosphofructokinase and glucose-6-phosphate dehydrogenase in response to drought and bicarbonate stress at transcriptional and functional levels in mulberry. *Russ. J. Plant Physiol.* **2016**, *63*, 235–242. [CrossRef]
58. Yu, X.; Ali, M.M.; Li, B.; Fang, T.; Chen, F. Transcriptome data-based identification of candidate genes involved in metabolism and accumulation of soluble sugars during fruit development in 'Huangguan' plum. *J. Food Biochem.* **2021**, *45*, e13878. [CrossRef]
59. Pan, T.; Ali, M.M.; Gong, J.; She, W.; Pan, D.; Guo, Z.; Yu, Y.; Chen, F. Fruit Physiology and Sugar-Acid Profile of 24 Pomelo (*Citrus grandis* (L.) Osbeck) Cultivars Grown in Subtropical Region of China. *Agronomy* **2021**, *11*, 2393. [CrossRef]
60. Park, J.-Y.; Canam, T.; Kang, K.-Y.; Ellis, D.D.; Mansfield, S.D. Over-expression of an arabidopsis family A sucrose phosphate synthase (SPS) gene alters plant growth and fibre development. *Transgenic Res.* **2008**, *17*, 181–192. [CrossRef]
61. Tian, H.; Ma, L.; Zhao, C.; Hao, H.; Gong, B.; Yu, X.; Wang, X. Antisense repression of sucrose phosphate synthase in transgenic muskmelon alters plant growth and fruit development. *Biochem. Biophys. Res. Commun.* **2010**, *393*, 365–370. [CrossRef]
62. Wang, K.; Ali, M.M.; Pan, K.; Su, S.; Xu, J.; Chen, F. Ebb-and-Flow Subirrigation Improves Seedling Growth and Root Morphology of Tomato by Influencing Root-Softening Enzymes and Transcript Profiling of Related Genes. *Agronomy* **2022**, *12*, 494. [CrossRef]
63. Zhang, C.; Shi, S. Physiological and Proteomic Responses of Contrasting Alfalfa (*Medicago sativa* L.) Varieties to PEG-Induced Osmotic Stress. *Front. Plant Sci.* **2018**, *9*, 242. [CrossRef]
64. Du, C.; Chai, L.; Wang, Z.; Fan, H. Response of proteome and morphological structure to short-term drought and subsequent recovery in Cucumis sativus leaves. *Physiol. Plant.* **2019**, *167*, 676–689. [CrossRef]
65. Mathews, M.C.; Summers, C.B.; Felton, G.W. Ascorbate peroxidase: A novel antioxidant enzyme in insects. *Arch. Insect Biochem. Physiol.* **1997**, *34*, 57–68. [CrossRef]
66. Choi, S.; Jeong, S.; Jeong, W.; Kwon, S.; Chow, W.; Park, Y.-I. Chloroplast Cu/Zn-superoxide dismutase is a highly sensitive site in cucumber leaves chilled in the light. *Planta* **2002**, *216*, 315–324. [CrossRef]
67. Tu, Y.; Rochfort, S.; Liu, Z.; Ran, Y.; Griffith, M.; Badenhorst, P.; Louie, G.V.; Bowman, M.E.; Smith, K.F.; Noel, J.P.; et al. Functional Analyses of Caffeic Acid O-Methyltransferase and Cinnamoyl-CoA-Reductase Genes from Perennial Ryegrass (*Lolium perenne*). *Plant Cell* **2010**, *22*, 3357–3373. [CrossRef]
68. Hu, W.; Wang, B.; Ali, M.M.; Chen, X.; Zhang, J.; Zheng, S.; Chen, F. Free Amino Acids Profile and Expression Analysis of Core Genes Involved in Branched-Chain Amino Acids Metabolism during Fruit Development of Longan (*Dimocarpus longan* Lour.) Cultivars with Different Aroma Types. *Biology* **2021**, *10*, 807. [CrossRef]

69. Zhi, C.; Ali, M.M.; Zhang, J.; Shi, M.; Ma, S.; Chen, F. Effect of Paper and Aluminum Bagging on Fruit Quality of Loquat (*Eriobotrya japonica* Lindl.). *Plants* **2021**, *10*, 2704. [CrossRef]
70. Cao, L.; Lu, X.; Zhang, P.; Wang, G.; Wei, L.; Wang, T. Systematic Analysis of Differentially Expressed Maize ZmbZIP Genes between Drought and Rewatering Transcriptome Reveals bZIP Family Members Involved in Abiotic Stress Responses. *Int. J. Mol. Sci.* **2019**, *20*, 4103. [CrossRef]
71. Ali, M.M.; Anwar, R.; Malik, A.U.; Khan, A.S.; Ahmad, S.; Hussain, Z.; Hasan, M.U.; Nasir, M.; Chen, F. Plant Growth and Fruit Quality Response of Strawberry is Improved After Exogenous Application of 24-Epibrassinolide. *J. Plant Growth Regul.* **2022**, *41*, 1786–1799. [CrossRef]
72. Ali, M.M.; Yousef, A.F.; Li, B.; Chen, F. Effect of Environmental Factors on Growth and Development of Fruits. *Trop. Plant Biol.* **2021**, *14*, 226–238. [CrossRef]
73. Staswick, P.E.; Serban, B.; Rowe, M.; Tiryaki, I.; Maldonado, M.T.; Maldonado, M.C.; Suza, W. Characterization of an Arabidopsis Enzyme Family That Conjugates Amino Acids to Indole-3-Acetic Acid. *Plant Cell* **2005**, *17*, 616–627. [CrossRef]
74. Chen, Y.; Shen, H.; Wang, M.; Li, Q.; He, Z. Salicyloyl-aspartate synthesized by the acetyl-amido synthetase GH3.5 is a potential activator of plant immunity in Arabidopsis. *Acta Biochim. Biophys. Sin.* **2013**, *45*, 827–836. [CrossRef]
75. Yang, Y.; Yue, R.; Sun, T.; Zhang, L.; Chen, W.; Zeng, H.; Wang, H.; Shen, C. Genome-wide identification, expression analysis of GH3 family genes in Medicago truncatula under stress-related hormones and Sinorhizobium meliloti infection. *Appl. Microbiol. Biotechnol.* **2015**, *99*, 841–854. [CrossRef]
76. Finkelstein, R.R.; Lynch, T.J. The Arabidopsis Abscisic Acid Response Gene ABI5 Encodes a Basic Leucine Zipper Transcription Factor. *Plant Cell* **2000**, *12*, 599–609. [CrossRef]
77. Zhao, W.; Guan, C.; Feng, J.; Liang, Y.; Zhan, N.; Zuo, J.; Ren, B. The Arabidopsis CROWDED NUCLEI genes regulate seed germination by modulating degradation of ABI5 protein. *J. Integr. Plant Biol.* **2016**, *58*, 669–678. [CrossRef]
78. Mittal, A.; Gampala, S.S.L.; Ritchie, G.L.; Payton, P.; Burke, J.J.; Rock, C.D. Related to ABA-Insensitive3(ABI3)/Viviparous1 and AtABI5 transcription factor coexpression in cotton enhances drought stress adaptation. *Plant Biotechnol. J.* **2014**, *12*, 578–589. [CrossRef]
79. Ma, J.; Chen, T.; Wu, S.; Yang, C.; Bai, M.; Shu, K.; Li, K.; Zhang, G.; Jin, Z.; He, F.; et al. iProX: An integrated proteome resource. *Nucleic Acids Res.* **2019**, *47*, D1211–D1217. [CrossRef]

 *agronomy*

Article

Multi-Environment Yield Components in Advanced Common Bean (*Phaseolus vulgaris* L.) × Tepary Bean (*P. acutifolius* A. Gray) Interspecific Lines for Heat and Drought Tolerance

Esteban Burbano-Erazo [1,†], Rommel Igor León-Pacheco [2], Carina Cecilia Cordero-Cordero [1], Felipe López-Hernández [3], Andrés J. Cortés [3,*,‡,§] and Adriana Patricia Tofiño-Rivera [1,*,§]

1   Corporación Colombiana de Investigación Agropecuaria (AGROSAVIA)-CI Motilonia, Codazzi 478020, Colombia; eburbano@agrosavia.co (E.B.-E.); ccordero@agrosavia.co (C.C.C.-C.)
2   Corporación Colombiana de Investigación Agropecuaria (AGROSAVIA)-CI Caribia, Sevilla 478020, Colombia; rleon@agrosavia.co
3   Corporación Colombiana de Investigación Agropecuaria (AGROSAVIA)-CI La Selva, Rionegro 054048, Colombia; llopez@agrosavia.co
*   Correspondence: acortes@agrosavia.co (A.J.C.); atofino@agrosavia.co (A.P.T.-R.)
†   Current Address: Institute for Plant Molecular and Cell Biology–IBMCP, Campus Universitat Politècnica de València, 46022 Valencia, Spain.
‡   Secondary Address: Departamento de Ciencias Forestales, Facultad de Ciencias Agrarias, Universidad Nacional de Colombia, Sede Medellín, Medellín 050034, Colombia.
§   These authors share senior and corresponding authorship.

Citation: Burbano-Erazo, E.; León-Pacheco, R.I.; Cordero-Cordero, C.C.; López-Hernández, F.; Cortés, A.J.; Tofiño-Rivera, A.P. Multi-Environment Yield Components in Advanced Common Bean (*Phaseolus vulgaris* L.) × Tepary Bean (*P. acutifolius* A. Gray) Interspecific Lines for Heat and Drought Tolerance. *Agronomy* 2021, 11, 1978. https://doi.org/10.3390/agronomy11101978

Academic Editor: Jacqueline Batley

Received: 26 July 2021
Accepted: 22 September 2021
Published: 30 September 2021

Publisher's Note: MDPI stays neutral with regard to jurisdictional claims in published maps and institutional affiliations.

Abstract: Heat and drought are major stresses that significantly reduce seed yield of the common bean (*Phaseolus vulgaris* L.). In turn, this affects the profitability of the crop in climatic-vulnerable tropical arid regions, which happen to be the poorest and in most need of legume proteins. Therefore, it is imperative to broaden the sources of heat and drought resistance in the common bean by examining closely related species from warmer and drier environments (i.e., Tepary bean, *P. acutifolius* A. Gray), while harnessing such variation, typically polygenic, throughout advanced interspecific crossing schemes. As part of this study, interspecific congruity backcrosses for high temperature and drought tolerance conditions were characterized across four localities in coastal Colombia. Genotypes with high values of CO_2 assimilation (>24 µmol CO_2 m^{-2} s^{-1}), promising yield scores (>19 g/plant), and high seed mineral content (*Fe* > 100 mg/kg) were identified at the warmest locality, Motilonia. At the driest locality, Caribia, one intercrossed genotype (i.e., 85) and the *P. acutifolius* G40001 control exhibited sufficient yield for commercial production (17.76 g/plant and 12.76 g/plant, respectively). Meanwhile, at southernmost Turipaná and Carmen de Bolívar localities, two clusters of genotypes exhibited high mean yield scores with 33.31 g/plant and 17.89 g/plant, respectively, and one genotype had an increased *Fe* content (109.7 mg/kg). Overall, a multi-environment AMMI analysis revealed that genotypes 13, 27, 82, and 84 were environmentally stable with higher yield scores compared to the Tepary control G40001. Ultimately, this study allows us to conclude that advanced common bean × Tepary bean interspecific congruity backcrosses are capable of pyramiding sufficient polygenic tolerance responses for the extreme weather conditions of coastal Colombia, which are likely to worsen due to climate change. Furthermore, some particular recombination events (i.e., genotype 68) show that there may be potential to couple breeding for heat and drought tolerance with *Fe* mineral biofortification, despite a prevalent trade-off, as a way to fight malnutrition of marginalized communities in tropical regions.

Keywords: polygenic adaptation; abiotic stress tolerance; congruity backcrosses; germplasm characterization; plant genetic resources; multi-local analysis; AMMI model; ecophysiology; biofortification; Caribbean coast of northern South America

1. Introduction

The common bean (*Phaseolus vulgaris* L.) is an important food resource and economic income source for farmers around the world [1]. Due to its multiple nutritional components, this species also represents a relevant crop for food security of the poorest. Biofortification efforts have leveraged this potential by offering cultivars with increased *Fe* content as an alternative to reduce mineral deficiencies [2]. However, an antagonistic effect of the genetic improvement process for drought (and potentially heat) stress tolerance with mineral content has been reported [3], likely jeopardizing long-term intrinsic mineral content properties [4]. In order to minimize this trade-off, germplasm's screening must reinforce recombinant adaptive phenotypes in concert with high mineral content. Promoting a dual strategy like this promises to diminish the negative impacts of climatic variability on common bean cropping systems, while boosting their mineral content. This would eventually contribute to reduce the production losses in underfed semiarid regions affected by hydric deficit and high temperatures [5].

Drought and heat wave events are becoming more frequent and extreme since the last few decades, particularly in the world's tropics [6]. In South America, the predictions of temperature increase account for 1.5 °C with several heat waves already recorded [7]. This evidences the need to speed up breeding alternatives to reduce the negative impact of climatic events on agriculture and food security, mainly in regions with strong periods of drought during the year, such as the Caribbean coast of northern South America.

In particular, common bean yield losses due to drought and heat waves in the region may be close to 70% [4]. However, breeding efforts face major bottlenecks to obtain genotypes with good adaptability to adverse abiotic conditions [8] due to the polygenic nature of the tolerant phenotypes. A strategy to develop cultivars with these attributes requires efficiently harnessing genetic resources from other genepools of closely related species typically adapted to heat and drought [9] The Tepary bean (*P. acutifolius* A. Gray) is an annual autogamous bean from northwest Mexico [10,11], domesticated near the arid border with the USA, and with a strong preference for hot [12] and dry environments [13,14]. Thanks to this natural adaptation, the Tepary bean is the most heat tolerant among the five cultivated species of the genus *Phaseolus*. Despite this, the Tepary bean nowadays has limited relevance as a modern crop when compared with the more susceptible, yet commercially accepted, common bean. In order to bridge this gap, we envision using the Tepary bean as donor of alleles to boost drought and heat tolerance in common bean.

Previous works have set the basis to explore the potential of the Tepary bean as an exotic donor by overcoming the phylogenetic distance between the two species, and the naturally low levels of interspecific introgression [15]. For instance, the common bean has been backcrossed with Tepary donors with a relatively good rate of viability [16,17]. Yet, these efforts were unable to pyramid target alleles for heat/drought tolerance, partly due to its complex quantitative genetic inheritance [14]. A promise avenue to overcome this limitation is to rely on advanced interspecific congruity backcross generations [18], as a base population to select for heat tolerance [15]. This is because backcrossing alternately to both parents, as the congruity scheme does, can more efficiently leverage selection of complex polygenic adaptive phenotypes, which is typically more suited for Mendelian traits of simpler genetic architecture [19].

Therefore, the goal of this study was to assess heat and drought tolerance in interspecific congruity backcrosses between common and Tepary beans. First, we aimed quantifying yield components across 86 advanced interspecific lines at four localities in coastal Colombia, where heat and drought waves are frequent and extreme. Second, we envisioned condensing all trials in a multi-locality genotype × environment (C × E) model to examine genotype stability of the best-adapted genotypes. We also addressed the physiological basis of heat stress adaptation in the locality with the most extreme heat events (Motilonia) as a first step to deepen our understanding on the tolerance phenotype.

2. Materials and Methods

2.1. Interspecific Congruity-Backcross Lines

A total of 86 interspecific congruity-backcross lines between the common bean (*P. vulgaris*) and Tepary bean (*P. acutifolius*) were considered as part of this study, in addition of the G40001 *P. acutifolius* control. They correspond to the third generation (and beyond) of congruity backcrosses (detailed pedigree in Table S1). Therefore, it is expected that they partly behave as stable lines with still some potential for further segregation. These genotypes were bred to high temperature and drought conditions by the bean program of the Alliance Bioversity–CIAT (International Center for Tropical Agriculture), and transferred to AGROSAVIA after ATM subscription. All lines were examined to address genotypic differences in phenological and agronomic performance at four localities in coastal Colombia.

2.2. Multi-Locality Field Trials

The study was conducted in four localities of the Colombian Caribbean region corresponding to the following AGROSAVIA's stations: Motilonia Research Station (Codazzi, province of Cesar), Caribia Research Station (Sevilla, province of Magdalena), Carmen de Bolívar Research Station (Carmen de Bolívar, province of Bolívar), and Turipaná Research Station (Cereté, province of Córdoba). Motilonia and Caribia belong to the dry Caribbean sub-region, respectively being the warmest and driest localities, while Carmen de Bolívar and Córdoba were representatives of the humid Caribbean sub-region. Soil fertility (i.e., in terms of P, Ca, Mg, and K in mg/kg. N and Fe content in the soil are to be included as part of future screenings), and other properties of the study areas are described in Table S2, as recorded during the production cycle of 2020. The bioclimatic variation at these sub-regions offers contrasts in soil type, precipitation, season durability, maximum and minimum temperature difference, and light quality, all of which are known to affect beans' agronomical performance. Differences between dry and humid Caribbean sub-regions oscillated (1) between 23 °C and 25 °C (average of 23.7 °C) and 33 °C and 36.3 °C (average of 33 °C) for average minimum and maximum temperatures, between (2) 70% and 80% (average of 80%) for relative humidity, and between (3) 482 mm and 700 mm (average of 591 mm) for precipitation during the rainy season, respectively. Hence, environmental conditions at the localities were differential. Table S3 depicts temperature and precipitation variables for each locality during the crop cycle of July–October 2020.

2.3. Experimental Design

The study employed a completely randomized block design (CRBD) with three repetitions at each locality. The experimental unit per treatment (genotypes) was a plot of four m^2 with a conventional spatial arrangement of one row spaced at 0.8 m and 0.25 m between plants (13 plants per genotype). A total of 72 genotypes germinated and succeed to growth until scoring in Motilonia, 16 in Caribia, 61 in Turipaná, and 87 in Carmen de Bolívar. Missing genotypes were indicative of mal-adaptation at each locality.

2.4. Phenotyping

The following yield traits, standard for the common bean [20], were measured at the end of the cycle at each locality, NS: average number of seeds per pod, NP: number of pods per plant, YLP: yield per plant (g/plant), PB: filled pod weight (g), SB: seed weight (g), VB: vegetative biomass (g), WS: weight of 100 seeds, and DF: days to flowering. Content of minerals Fe and Zn was quantified for plants with a sufficient number of seeds (minimum 100 g) following the Elemental Analysis by Energy Dispersive X-ray fluorescence (EDXRF) protocol of the Nutritional Quality Laboratory at Alliance Bioversity–CIAT (Palmira, Colombia). Due to the COVID-19 contingency, mineral quantification was only possible at Motilonia, Carmen de Bolívar, and Caribia. For Turipaná, a very high pressure of soil pathogens was also observed, so the amount of seed required for the analysis was insufficient.

Detailed ecophysiological measures were possible at Motilonia for the following variables, (1) E: transpiration, (2) A: net photosynthetic rate, (3) Ca: CO_2 (in the chamber), (4) Ci: internal leaf CO_2, (5) gs: stomata conductance, (6) LT: internal leaf temperature, and (7) VPD: vapor pressure deficit. These measures were gathered using the LI-6800 portable photosynthesis system from 8:00–11:00 am. Additional compound parameters were computed as follows, (8) iWUE: instantaneous water use efficiency was measured as the relation A/E, (9) inWUE: intrinsic efficiency of water use computed as the relation A/gs, and (10) iCE: instantaneous carboxylation efficiency was measured as the relation A/Ci [21]. Measurements were made on three plants per genotype on a fully expanded leaf from the upper third part of the plant. Measurements were carried out between the pre-flowering and flowering of the genotypes. The LI-6800 equipment was calibrated for the local conditions through light and CO_2 curves, obtaining the appropriate range of photosynthetically active radiation (PARi) of 1500 μmol of photons $m^{-2}s^{-1}$, and a reference CO_2 of 400 μmol CO_2 mol^{-1}. Calibration curves were done on the *P. acutifolius* control genotype G40001.

2.5. Statistical Analyses

Analyses of variance (ANOVA) were performed for the examined traits individually at each locality after checking for normality and considering genotypes as factors in the statistical software SAS enterprise Guide 8.3. A hierarchical grouping analysis was also carried out to determine the diversity in the studied genotypes for yield traits using the *hclust* function with Ward's agglomeration method in the R software (R Core Team). The *Factoextra* [22] and *NbClust* [23] packages were further used to determine the optimum number of clusters (Figure S1).

In order to assess differences among clusters, Tukey's comparison of means was performed (p-value < 0.05) after checking for normality using R's library *agricolae*. Additionally, principal component (PCA) and Pearson's correlation analyses were carried out using R's libraries *factoextra* and *FactoMineR* [24]. Missing data were imputed when it did not exceed 10% using the *mice* library [25] from the same software.

Finally, in order to explore the extent of stability and phenotypic adaptability, previous results were further contrasted via the estimation of phenotypic stability for nine overlapping lines among localities within an AMMI analytical framework. The AMMI model used the routine proposed by [26]. It combined the analysis of variance (ANOVA) to estimate the main effects of G and E, with the principal components analysis (PCA) as a proxy for the G × E interaction, as follows (Equation (1)):

$$Y_{ij} = \mu + G_i + E_j + \Sigma_{\lambda k} \, \alpha_{ik} \, \delta_{jk} + R_{ij} + \varepsilon \tag{1}$$

where Y_{ij} corresponded to the value of the ith genotype in the jth environment, μ was the overall mean, G_i was the deviation of the ith genotype from the overall mean, E_j was the deviation of the jth environment from the overall mean, λ_k was the singular value for PC axis k, α_{ik} and δ_{jk}, respectively, were the principal component's scores for axis k of the ith genotype and jth environment, and R_{ij} and ε were the residual and error terms, respectively.

Stable genotypes for each environment were selected by the AMMI biplot, in which scores of the first principal component (CP1) were plotted against the phenotypic means of genotypes and environments. Specifically, CP1 scores close to zero in the y-axis were considered as stable genotypes. Specific genotype–environment adaptability was inferrered from the x-axis, where the means of genotypes and environments were jointly plotted.

This combined analysis was carried out in order to select the best interspecific lines adapted to the target climates at the Caribbean coastal sub-regions of Colombia given two different resolutions: (1) broad adaptation, applicable for genotypes stable across multiple environments, and (2) narrow adaptation, in which specific genotypes outperformed the rest in a single locality but not across all environments.

3. Results

3.1. Phenotypic and Physiological Variability of Interspecific Lines Per Locality

Interspecific lines characterized across four localities (i.e., Motilonia, Caribia, Turipaná, and Carmen de Bolívar) showed a statistical differential response, mainly for parameters associated with yield (Table 1). These trends are detailed per locality in the next subsections. Since Motilonia and Carmen's replicates also exhibited differences, uncontrolled variability is discussed in light of local-scale heterogeneity.

Table 1. One-site analysis of variance for the common bean (*P. vulgaris*) $\times$ Tepary bean (*P. acutifolius*) interspecific congruity backcross lines.

Yield Trait	Motilonia (72 Genotypes)		Caribia (16 Genotypes)		Turipaná (61 Genotypes)		Carmen (87 Genotypes)	
	Rep	**Gen**	**Rep**	**Gen**	**Rep**	**Gen**	**Rep**	**Gen**
NS	0.4801	<0.0001	0.1	0.52	0.28	0.0177	<0.0001	<0.0001
NP	< 0.0001	<0.0001	0.14	0.38	0.4	0.001	<0.0001	0.005
YLP	< 0.0001	<0.0001	0.04	0.46	0.07	0.04	<0.0001	<0.0001
PB	< 0.0001	<0.0001	0.09	0.38	NA	NA	0.052	0.242
SB	< 0.0001	<0.0001	0.46	NA	NA	NA	<0.0001	<0.0001
VB	0.8546	<0.0001	0.13	0.16	0.03	0.06	<0.0001	<0.0001
WS	NA	NA	0.08	0.4	0.2398	0.42	0.2497	0.001
DF	0.7666	<0.0001	NA	NA	0.08	0.262	0.442	0.122

p-value for repetition (Rep) and genotype (Gen). NS: Average number of seeds per pod, NP: number of pods per plant, YLP: yield per plant (g/plant), PB: pod biomass (g), SB: seed biomass (g), VB: vegetative biomass (g), WS: weight of 100 seeds, and DF: days to flowering.

3.2. Motilonia (Warmest Dry Caribbean Savanna Sub-Region)

At the Motilonia locality, genotypes had a variable ecophysiological behavior since all ecophysiological parameters shown statistically significant differences, indicating that at least one genotype is different (Table 2).

Table 2. Analysis of variance for ecophysiological parameters among the common bean $\times$ Tepary bean interspecific congruity-backcross lines at Motilonia (warmest dry Caribbean sub-region).

Parameter	Units	Repetitions Effect	Genotype Effect
A	μmol CO_2 m^{-2} s^{-1}	0.8126	<0.0001
Ca	μmol mol^{-1}	0.3461	<0.0001
Ci	μmol CO_2 mol^{-1}	0.4298	<0.0001
gs	mol H_2O m^{-2} s^{-1}	0.4597	<0.0001
VPD	kPa	0.2497	<0.0001
E	mmol H_2O m^{-2} s^{-1}	0.9013	<0.0001
Tleaf	°C	0.2881	<0.0001
iWUE	NA	0.6131	<0.001
inWUE	NA	0.184	<0.001
iCE	NA	0.4277	<0.001

p-value for repetition and genotype factors. E: transpiration, A: net photosynthetic rate, Ca: Ambient CO_2 (in the chamber), Ci: internal leaf CO_2, gs: stomatal conductance, E: transpiration, Tleaf: internal leaf temperature, VPD: vapor pressure deficit, iWUE: instantaneous water use efficiency, inWUE: intrinsic efficiency of water use, and iCE: instantaneous carboxylation efficiency.

Three clusters were identified according to their productive response (Figure 1a), with cluster 1 being the one with the highest average yield, followed by cluster 2, and finally cluster 3 with the lowest values (Figure 1b). Population variance was mainly explained by dimension 1 (37.5%), 2 (20.4%) and 3 (15.3%). Variables with greater contribution at explaining variability were mostly ecophysiological and performance parameters (Figure 1c, darker vectors). Performance variables like NP, YLP, PB, NS, SB, VB, gs, A, and iCE presented an inverse relationship with variables such as VPD, Ca, LT, and WS.

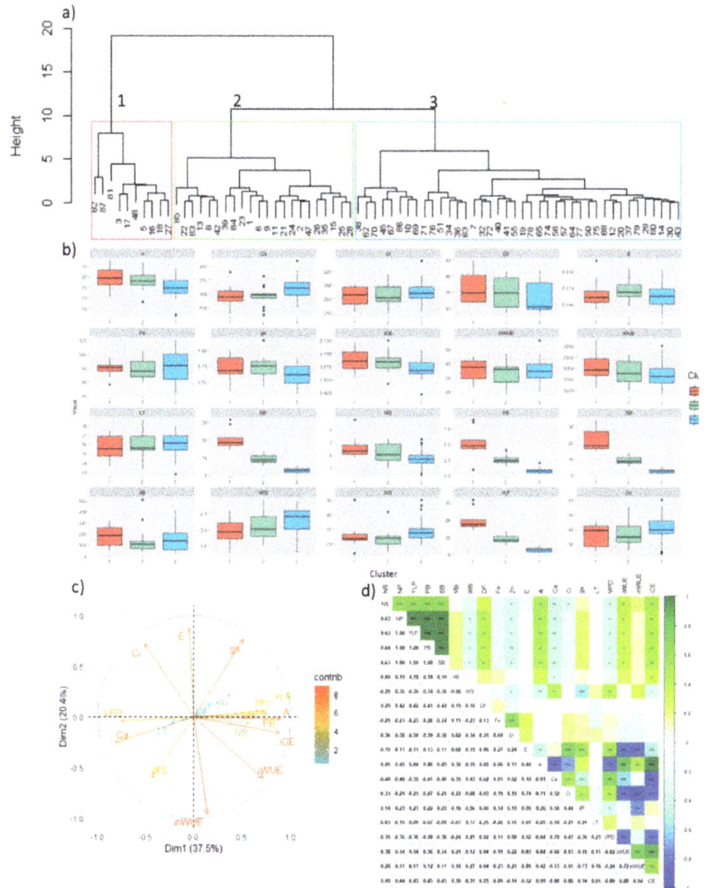

Figure 1. Yield and physiological responses of interspecific congruity-backcross lines at Motilonia (warmest dry Caribbean sub-region). (**a**) Clustering of genotypes for yield parameters. (**b**) Yield and physiological behavior of genotypes. (**c**) Principal component analysis of genotypes. (**d**) Pearson's correlation analysis for key traits. NS: average number of seeds per pod, NP: number of pods per plant, YLP: yield per plant (g/plant), PB: pod biomass (g), SB: seed biomass (g), VB: vegetative biomass (g), DF: days to flowering, WS: weight of 100 seeds (g), *Fe*: iron content (mg/kg), *Zn*: Zinc content (mg/kg), E: transpiration, A: Net photosynthetic rate, Ca: Ambient CO_2 (in the chamber), Ci: internal leaf CO_2, gs: stomatal conductance, LT: internal leaf temperature, VPD: vapor pressure deficit, iWUE: instantaneous water use efficiency, inWUE: intrinsic efficiency of water use, and iCE: instantaneous carboxylation efficiency. *: $p > 0.05$, **: $p > 0.01$; ***: $p > 0.001$. G40001 control as 87.

Pearson correlation analysis corroborated statistically significant correlations among the performance-related parameters (i.e., NS, NP, YLP, PB, SB, VB, and WS), as well as with some ecophysiological parameters (i.e., A, iWUE, gs, and iCE). In short, with higher A, iWUE, gs, and iCE, performance also increased (Figure 1d). In terms of physiological parameters, some correlations were particularly relevant, such as the inverse correlation between E and iWUE (as expected because the E term is the denominator). This result agreed with the ecophysiological behavior of the bean plant since greater transpiration, generated by greater gs, implied having less efficient use of water. Last, a direct correlation between *Fe* and *Zn* was found. *Zn* had a negative correlation with yield parameters (Figure 1d).

Regarding the physiological parameter of greatest interest, A, it was found that some genotypes carried out an efficient assimilation of CO_2 with a lower amount of Ci, involving low VPD and greater water use efficiency. It was also evident through a significant inverse correlation that the higher gs allowed to cool and reduced the temperature of the leaf, implying lower VPD and higher iCE. The inWUE, iWUE, and iCE indices were generally positively correlated (Figure 1).

Regarding differences among genotypes, it was found that the cluster with the highest yield (cluster 1) also presented the highest values for other parameters such as NS, NP, PB, SB, A, gs, and iCE. However, it presented the lowest values for WS and VPD. Similarly, the genotypes grouped in cluster 2 also showed high values for NS, A, gs and iCE, and low Ca. Hence, besides presenting high values in variables related to yield and biomass production, the ecophysiological behavior of cluster 1 also indicated a positive relationship between higher yield and CO_2 assimilation, being an attribute of interest for adaptation in the Caribbean conditions of coastal Colombian (Table 3).

Table 3. Comparisons of means for yield components and physiological traits among clusters of interspecific congruity-backcross lines in Motilonia (dry Caribe).

Trait	Unit	Cluster 1	Cluster 2	Cluster 3
NS	Seeds	2.57 [a]	2.16 [a]	1.73 [b]
NP	Pods	20.27 [a]	9.21 [b]	2.78 [c]
YLP	g/plant	19.72 [a]	9.05 [b]	2.73 [c]
PB	G	5.18 [a]	2.49 [b]	0.76 [c]
SB	G	21.36 [a]	9.05 [b]	2.68 [c]
VB	G	183.49	154.63	130.13
WS	g/100 seeds	25.46 [a,b]	23.03 [b]	28.89 [a]
DF	Days	36.3	34.96	32.92
E	mmol H_2O m^{-2} s^{-1}	0.012	0.013	0.012
A	µmol CO_2 m^{-2} s^{-1}	24.29 [a]	23.52 [a]	20.15 [b]
Ca	µmol mol^{-1}	358.51 [b]	358.41 [b]	364.04 [a]
Ci	µmol CO_2 mol^{-1}	284.55	286.45	289.25
gs	mol H_2O m^{-2} s^{-1}	0.73 [a]	0.72 [a,b]	0.60 [b]
LT	°C	35.66	35.99	36.11
VPD	kPa	1.98 [b]	2.10 [a,b]	2.29 [a]
iWUE	NA	2035.99	1819.60	1698.78
inWUE	NA	35.28	33.76	35.33
iCE	NA	0.08 [a]	0.08 [a]	0.07 [b]

footer NS: average number of seeds per pod, NP: number of pods per plant, YLP: yield per plant (g/plant), PB: pod biomass (g), SB: seed biomass (g), VB: vegetative biomass (g), WS: weight of 100 seeds, DF: days to flowering, E: transpiration, A: Net photosynthetic rate, Ca: Ambient CO_2 (in the chamber), Ci: internal leaf CO_2, gs: stomatal conductance, LT: internal leaf temperature, VPD: vapor pressure deficit, iWUE: instantaneous water use efficiency, inWUE: intrinsic efficiency of water use and iCE: instantaneous carboxylation efficiency. Different letters denote significant differences determined by Tukey's test ($\alpha \leq 0.05$).

On the contrary, cluster 3 presented the lowest values for performance traits and biomass production. In addition, regarding the physiological parameters, it was also evident that the VPD value was the highest. This indicated that, the higher the score of this parameter, the greater the impact on other physiological processes, and therefore the overall productive capacity (Table 3). In addition, cluster 3 also presented the lowest values for the variables A, gs, iWUE and iCE, and the highest for Ca and Ci.

Some of the genotypes that experienced interesting behavior at the warmest locality of Motilonia merged key physiological parameters, outstanding mineral content, and good productive responses (Table 4, despite some cases of inverse relationship between yield and *Fe* content reinforced this well-known trade-off).

Table 4. The common bean × Tepary bean interspecific congruity-backcross lines selected according to yield and *Fe* content in the locality of Motilonia (dry Caribbean sub-region). G40001 was used as control from the *P. acutifolius* genepool, and therefore is presented in the first row.

Selection Criteria	Cluster	Genotype	YLP	WS	*Fe*	*Zn*
Yield	1	G40001	31.43	13	66.8	24.3
	1	82	28.96	28.6	56.2	26.7
	1	48	20.55	50	-	-
	1	17	19.03	23.2	94.35	28.8
	1	81	17.7	27	-	-
	1	5	17.04	23.3	82.1	44.9
	1	16	16.78	23.3	83.1	41.5
	1	3	16.47	22.34	87.8	37.4
	1	18	15.07	24.7	79.2	45.4
	1	27	14.18	19.2	79.5	41.2
	2	1	13.34	25.2	80.0	30.4
	2	84	13.26	24.5	-	-
	2	85	12.14	27	61.7	31.6
	2	9	11.69	19.8	103.75	50.1
	2	6	11.69	27	61.9	25.4
	2	39	10.9	38	-	-
	2	23	10.73	23.8	66.1	34.1
	2	13	10.02	25.6	82.5	31.6
Fe content	2	9	11.69	19.8	103.75	50.1
	3	12	3.14	27.5	111.25	40.45
	2	26	6.29	19.9	103.5	44.25
	3	64	1.96	27	102.6	60.25
	3	76	1.32	25.2	118.3	56.15
	3	77	2.08	29.2	120.5	53.6
	3	80	3.79	23	107.7	52.25

footer YLP: yield (g/plant), WS: weight of 100 seeds (g), and *Fe:* iron and *Zn:* Zinc content (mg/kg).

Taking into account that seed size is a paramount commercial trait, some other genotypes were also chosen targeting this variation, specifically: genotype 35 (cluster 2) presented a yield of 5.18 g/plant, a WS of 41.72 g/100 seeds, and *Fe* and mineral content of 72.30 and 30.60 mg/kg; genotype 38 (cluster 3) had YLP of 3.07 g/plant and WS of 50 g/100 seeds; genotype 45 (cluster 3) had a YLP of 1.72 g/plant and a WS of 46 g/100 seeds; and finally genotype 36 (cluster 3) presented a YLP of 2.08 g/plant and a WS of 43.50 g/100 seeds. These multiple selection criteria also targeted *Fe* content higher than 100 mg/kg given that some recombinant genotypes may offer potential for biofortification, despite the prevailing inverse relationship between *Fe* content and yield.

3.3. Caribia (Driest Caribbean Sub-Region)

At this locality two clusters were identified for yield components (Figure 2a). Overall, the genotypes presented low yield, possibly due to the environmental conditions of the region (e.g., high relative humidity that may stimulate the presence of soil pathogens). Only two genotypes, the G40001 control from the *P. acutifolius* genepool and 85, presented high yield value. Still, other genotypes might also be considered of interest in this locality, such as 84, 86, 28, 13 and 49. However, none of the genotypes exceeded the commercially desirable weight of 40 g/100 seeds. The genotype that came closest to this value was genotype 85 with 38 g/plant, which was also the second genotype with the highest yield. None of these genotypes exceeded the value of 100 mg/kg in *Fe* content.

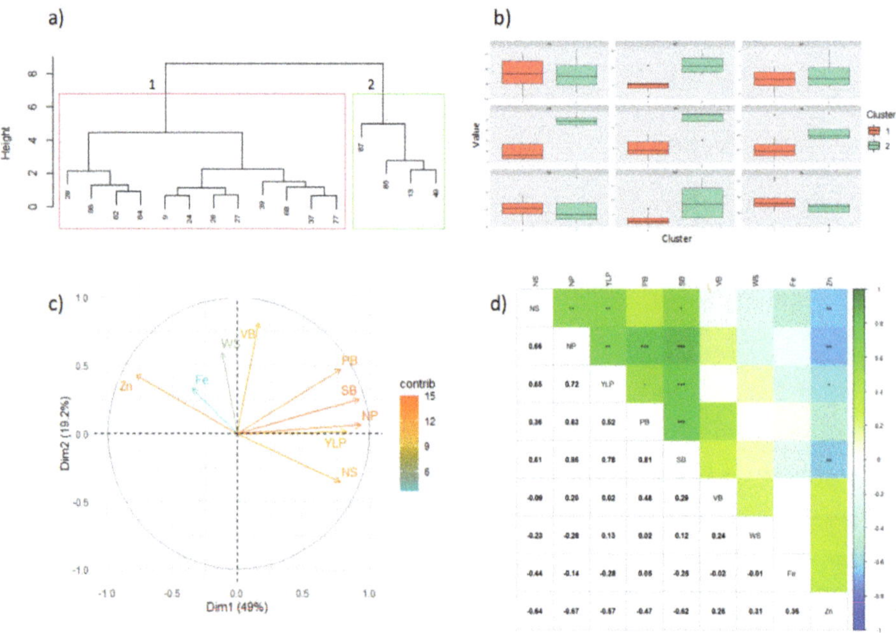

Figure 2. Yield responses of interspecific congruity-backcross lines at Caribia (driest Caribbean sub-region). (**a**) Clustering of genotypes considering yield parameters. (**b**) Yield and physiological behavior of studied genotypes. (**c**) Principal component analysis in the characterized genotypes. (**d**) Pearson's correlation analysis for the studied traits. NS: average number of seeds per pod, NP: number of pods per plant, YLP: yield per plant (g/plant), PB: pod biomass (g), SB: seed biomass (g), VB: vegetative biomass (g), WS: weight of 100 seeds (g), *Fe:* iron (mg/kg), *Zn:* Zinc (mg/kg). *: $p > 0.05$, **: $p > 0.01$: ***: $p > 0.001$. *P. acutifolius* control G40001 is abbreviated as 87.

When analyzing the relationship between the performance variables, it was found that the variability of the population was mainly explained by the first two components with 49% and 19.2%, respectively. A positive relationship was observed between the variables associated with yield, such as PB, SB, NP, YLP, and NS. However, the contents of *Fe* and *Zn* had an inverse behavior against those parameters, and to a lesser extent the variables WS and WB (Figure 2c). Pearson's correlations made it possible to corroborate these relationships between variables, showing that the yield components had a positive correlation among them. The correlations between *Fe* and *Zn* content with some yield parameters were negative, indicating an inverse behavior (Figure 2d).

In the clustering analysis, it was found that cluster 1 was made up of 75% of the genotypes, and was statistically different from cluster 2, which was the one that presented the highest values for all variables, as detailed in Table 5.

Table 5. Comparisons of means between clusters at locality of Caribia (dry Caribe).

Parameter	Unit	Cluster 1	Cluster 2
NP	Pods	10.29 [b]	22.37 [a]
YLP	g/plant	3.17 [b]	9.53 [a]
PB	G	9.37 [b]	24.32 [a]
SB	G	4.98 [b]	11.82 [a]
VB	G	21.78 [b]	45.28 [a]

footer NP: number of pods per plant, YLP: yield per plant (g/plant), PB: pod biomass (g), SB: seed biomass (g), VB: vegetative biomass (g). Different letters denote significant differences determined by Tukey's test ($\alpha \leq 0.05$).

Given the selection criteria applied to this locality, only two genotypes presented yield scores that were above the desired threshold: control genotype G40001 from the *P. acutifolius* genepool (cluster 2) with YLP of 17.76 g/plant, WS of 12.28 g/100 seeds, Fe of 63.80 mg/kg, and *Zn* of 22.40 mg/kg; and genotype 85 (cluster 2), with YLP of 12.76 g/plant, WS of 38.41 g/100 seeds, *Fe* of 71.30 mg/kg, and *Zn* of 35.20 mg/kg.

3.4. Turipaná (Humid Caribbean Sub-Region)

In the Turipaná locality, four clusters were found when considering the main performance parameters, as indicated in Figure 3. The largest cluster was the second, including up to 62.29% of the characterized interspecific genotypes, followed by clusters 4, 3, and 1, with 26.22%, 8.19%, and 3.27%, respectively (Figure 3a). The variability of the population was mostly explained by component 1 (46.7%) and component 2 (19.4%, Figure 3c). The yield (YLP) and its components (NP and NS) were positively related and provided the greatest contribution to explain the variability of the population.

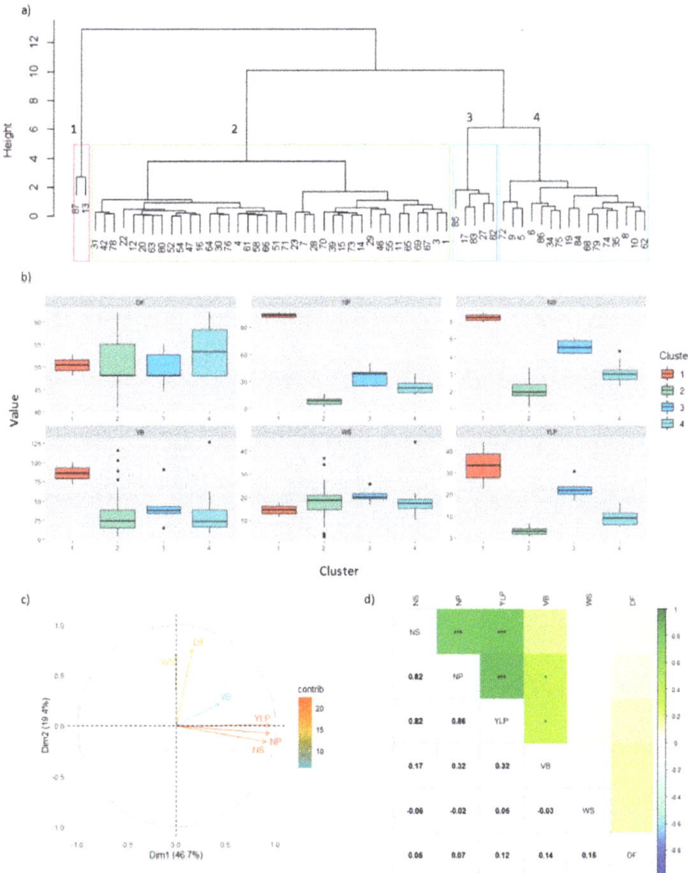

Figure 3. Yield responses of interspecific congruity-backcross lines at Turipaná (humid Caribbean sub-region). (**a**) Clustering of genotypes considering yield parameters. (**b**) Yield and physiological behavior of studied genotypes. (**c**) Principal component analysis in the characterized genotypes. (**d**) Pearson's correlation analysis for the studied traits. NS: average number of seeds per pod, NP: number of pods per plant, YLP: yield per plant (g/plant), WS: weight of 100 seeds (g), VB: vegetative biomass (g), DF: days to flowering. *: $p > 0.05$, ***: $p > 0.001$. G40001 control as 87.

Relationships among performance parameters were corroborated with Pearson's linear correlations, in which these traits presented direct and significant correlations (p-value < -0.05, Figure 3d). Finally, when comparing the clusters, it was found that the highest yield was obtained within cluster 1, followed by cluster 3, for the variables NS, NP, and YLP. On the contrary, cluster 2 presented the lowest values for these same parameters (Table 6).

Table 6. Comparisons of means for yield components among clusters of interspecific congruity-backcross lines in Turipaná (humid Caribbean sub-region).

Cluster	Units	Cluster 1	Cluster 3	Cluster 4	Cluster 2
NS	Seeds	6.20 [a]	4.55 [b]	3.07 [c]	2.09 [d]
NP	Pods	103.17 [a]	35.73 [b]	3.07 [c]	2.09 [d]
YLP	g/plant	33.31 [a]	22.59 [b]	9.22 [c]	2.81 [d]

NS: average number of seeds per pod, NP: number of pods per plant, YLP: yield per plant (g/plant). Different letters denote significant differences determined by Tukey's test ($\alpha \leq 0.05$).

In this way, the genotypes that presented the highest yield scores, and that belonged to clusters 1, 3, and 4, were chosen and ranked, as shown in Table 7.

Table 7. Interspecific congruity-backcross lines selected according to yield per plant above 10 g in Turipaná (humid Caribbean sub-region). G40001 was used as control from the *P. acutifolius* genepool.

Cluster	Genotype	NS	NP	YLP	VB	WS	DF
1	G40001	5.90	106.35	44.06	100.27	11.32	52.67
3	85	4.50	25.50	30.65	91.00	20.10	48.00
3	17	4.20	50.00	23.50	33.67	16.50	52.67
1	13	6.50	100.00	22.57	72.00	17.77	48.00
3	83	4.10	38.50	21.80	43.00	25.76	55.00
3	27	4.90	40.00	20.00	15.00	19.50	44.50
3	82	5.07	24.67	17.00	37.00	21.56	48.00
4	86	2.78	32.33	16.04	37.67	21.49	60.67
4	19	3.11	16.33	13.67	17.67	18.39	62.00
4	34	3.07	31.67	12.70	22.33	17.09	57.33
4	84	2.57	16.00	12.50	11.00	15.00	48.00
4	72	4.30	25.00	11.00	33.00	18.61	55.00
4	75	2.90	27.00	10.67	126.33	17.52	48.00
4	6	3.47	38.33	10.30	42.00	15.39	48.00

NS: average number of seeds per pod, NP: number of pods per plant, YLP: yield per plant (g/plant), VB: Vegetative biomass (g), WS: weight of 100 seeds (g), DF: days to flowering.

On the other hand, given the importance of grain weight, genotype 68 presented a good WS of 43.82 g/100 seeds, although in terms of NS had a modest score of 2.90 (besides a NP of 22.50, YLP of 9.00 g/plant, GV of 39.50, and DF of 51.50).

3.5. Carmen de Bolívar (Humid Caribbean Sub-Region)

In Carmen de Bolívar cluster 1 was the most numerous with 44.82% genotypes, cluster 2 with 36.78%, and cluster 3 with 18.39% (Figure 4a). Population variability was mainly explained by component 1 (46.4%) and 2 (17.46%). It was also found that most of the yield components were related, except for DM, DF, WS, *Fe* and *Zn*, which presented an opposite response to the rest of the parameters (Figure 4c). This was corroborated with Pearson's correlations (Figure 4d), in which a high inverse correlation was found between YLP and WS (p-value < -0.01), indicating that the genotypes with the highest yield had low grain weight (i.e., higher yield was obtained at the expense of WS).

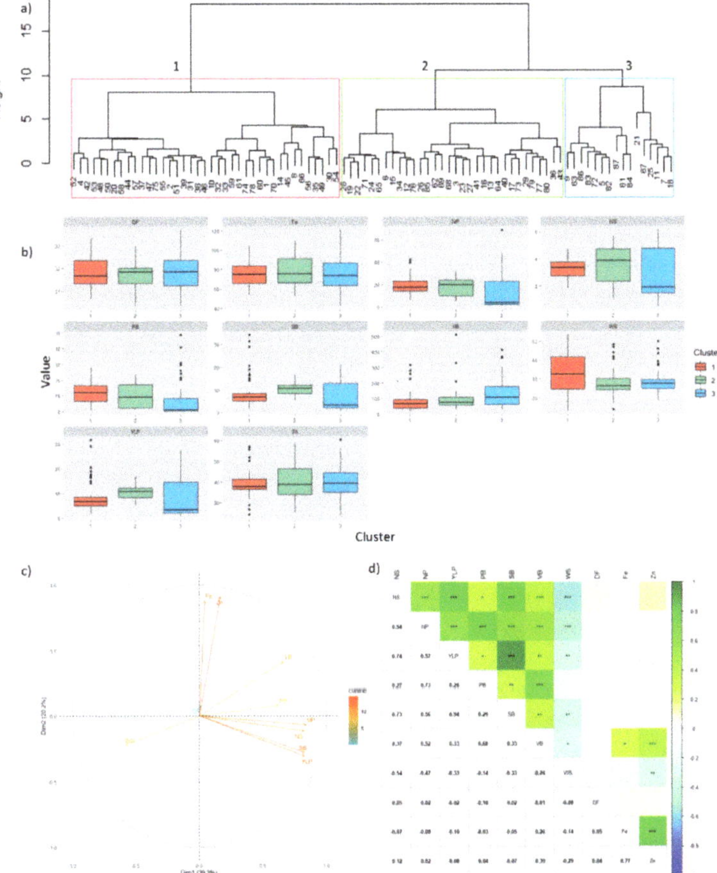

Figure 4. Yield responses of interspecific congruity-backcross lines at Carmen de Bolívar (humid Caribbean sub-region). (**a**) Clustering of genotypes considering yield parameters. (**b**) Yield and physiological behavior of genotypes. (**c**) Principal component analysis across genotypes. (**d**) Pearson's correlation among traits. NS: average number of seeds per pod, NP: number of pods per plant, YLP: yield per plant (g/plant), PB: pod biomass (g), SB: seed biomass (g), VB: vegetative biomass (g), WS: weight of 100 seeds (g), *Fe:* iron (mg/kg), *Zn:* Zinc (mg/kg). *: $p > 0.05$, **: $p > 0.01$: ***: $p > 0.001$. *P. acutifolius* control G40001 is abbreviated as 87.

In addition, other statistically significant correlations of interest were found, although with lower Pearson's correlation scores. Specifically, positive correlations between *Zn* and VB, and between *Fe* and VB, suggested that plants with higher VB might accumulate higher content of *Zn* and *Fe* in the grain.

On the contrary, a low negative correlation was found between WS and *Fe*. This trend suggested that the grains with the highest weight had the lowest *Fe* content, as expected due to the prevailing trade-off between yield and nutrition. A high and positive correlation was also found between *Fe* and *Zn*.

Cluster 3 presented the highest values for NS, NP, YLP, SB, and VB, and the lowest values for WS and DM. Regarding these latter two traits, cluster 1 was the one with the highest scores (Table 8). It was expected that genotypes with the lowest yield could reach the highest grain weight, possibly due to more photo-assimilate accumulation on less grains.

Table 8. Comparisons of means between clusters of interspecific congruity-backcross lines, as assessed at Carmen de Bolívar (humid Caribbean sub-region).

Cluster	NS	NP	YLP	SB	VB	WS	DM
1	3.46 [c]	18.84 [c]	5.98 [c]	5.99 [c]	57.60 [b]	36.18 [a]	76.11 [a]
2	4.51 [b]	23.34 [b]	11.61 [b]	11.61 [b]	67.13 [b]	30.35 [b]	72.70 [b]
3	5.25 [a]	34.39 [a]	17.89 [a]	16.91 [a]	87.86 [a]	28.11 [b]	72.37 [b]

NS: average number of seeds per pod, NP: number of pods per plant, YLP: yield per plant (g/plant), SB: seed biomass (g), VB: vegetative biomass (g), WS: weight of 100 seeds (g), DM: days to maturation. Different letters denote significant differences as by the Tukey's test ($\alpha \leq 0.05$).

The genotypes selected by yield mainly belonged to clusters 3 and 2, in decreasing order, as indicated in Table 9. Of these groups, only genotypes 77 and 64 presented high values for minerals. Some additional genotypes of interest were identified considering weight of seeds.

Table 9. The common bean × Tepary bean interspecific congruity-backcross lines selected for yield, seed weight, and seed *Fe* in the locality of Carmen de Bolívar (humid Caribbean sub-region). G40001 was used as control from the *P. acutifolius* genepool.

Criteria	Cluster	Genotype	YLP	WS	*Fe*	*Zn*
	3	G40001	27.12	29.47		
	3	21	21.12	26	71.5	36.9
	3	86	20.96	27.33	64.4	36.8
	3	82	20.15	28.33	53.1	34.2
	3	84	20.07	28.33	58	30.9
	3	81	19.77	30.33	70.3	44.4
	3	5	19.11	27.67	62	38
	3	72	18.7	28.67	83.6	41
	3	83	17.43	26	67.9	32.9
	3	9	16	26	79.2	45.5
	3	63	15.79	32	92.25	48.4
	3	67	15.76	33	81.8	51.2
	3	7	15	29	77.5	42
	3	18	14.9	24.67	66.4	33.6
	3	25	13.37	24	56.3	31.6
	3	11	11	29	73.6	43.9
	2	43	16.48	35.67	70.2	28.4
	2	62	15.13	31	76.6	42.4
	2	80	14.78	32.67	90.55	41.7
	2	77	14.75	29.42	109.7	58.4
	2	40	14.61	44	66.6	27
Yield	2	79	13.96	26.25	99.9	56.3
	2	29	13.91	40.67	63.1	29.7
	2	17	13.52	24.89	70.3	35.2
	2	85	12.72	28	53.5	35.9
	2	73	12.57	30.33	82.3	46.3
	2	26	12.23	25.33	74.9	35
	2	69	12.03	33.67	92	46.8
	2	41	11.99	48.33	64.4	34.1
	2	16	11.95	28.67	64.6	35.8
	2	13	11.9	26	75.1	45
	2	36	11.81	47.02	73.5	33.5
	2	28	11.7	26.03	56.4	35.4
	2	24	11.35	25	61.1	32.1
	2	65	11.02	29.33	93.15	58.7
	2	23	10.92	25.62	62.6	35.5
	2	19	10.72	25	72.3	40.4

Table 9. *Cont.*

Criteria	Cluster	Genotype	YLP	WS	*Fe*	*Zn*
	2	22	10.62	24.33	82.2	39
	2	64	10.34	30	101.1	50.4
	2	3	10.14	27.7	71.4	38.1
	2	71	10.09	30.67	92.65	46.4
	1	45	9.42	41	65.4	39.2
	1	56	9.03	47.87	74	38.4
	1	35	8.24	44	74.3	39.5
	1	47	8.15	46.75	74.3	37.4
WS	1	37	7.36	40.04	61.7	41.9
	1	49	7.09	39.78	74.7	33.1
	1	50	7.02	41.42	96.5	33.9
	1	48	6.76	40.89	65.8	37.9
	1	53	6.58	39.91	57.8	37.5
	1	14	7.68	23.49	101.4	50.2
Fe content	1	78	7.52	25.33	104.4	57
	2	6	8.4	28.66	99.45	41.6
	2	76	7.44	27	106.05	57.9

YLP: yield per plant (g/plant), WS: weight of 100 seeds (g), *Fe:* iron (mg/kg), *Zn:* Zinc (mg/kg).

Finally, mineral content was also accounted as selection criterion for some genotypes in order to advance future biofortification outcomes in the presence of increasing heat and drought constrains.

3.6. Phenotypic Stability

The interspecific congruity-backcross lines with the highest yield and the best stability across localities were identified in a panel of nine genotypes that grew and produced in all four environments. Remaining genotypes were interpreted as mal-adapted.

The selection criterion for phenotypic stability was based on distance from the origin to the first component (ASV) in the standard AMMI analysis, which meant that the genotypes with the lowest stability index value (YSI) were discarded (Table 10, Figure 5).

Table 10. Stability index with AMMI statistics for yield per plant in nine common bean × Tepary bean interspecific congruity-backcross lines evaluated across four localities in the Caribbean coast of Colombia. G40001 (*P. acutifolius*) was used as control.

Genotypes	ASV	YSI	rASV	rYSI	Square Root	Yield (g/plant)
13	1.08	10	5	5	3.26	12.14
27	0.94	10	4	6	3.17	11.54
28	1.1	14	6	8	2.61	7.43
68	0.80	12	3	9	2.10	5.62
82	1.43	11	8	3	3.83	17.32
84	0.74	6	2	4	3.51	13.22
85	1.69	11	9	2	3.88	17.06
G40001	0.12	2	1	1	5.13	28.16
9	1.36	14	7	7	2.76	8.89

ASV: AMMI stability value, YSI: stability index value, rASV: ranking ASV, rYSI: ranking.

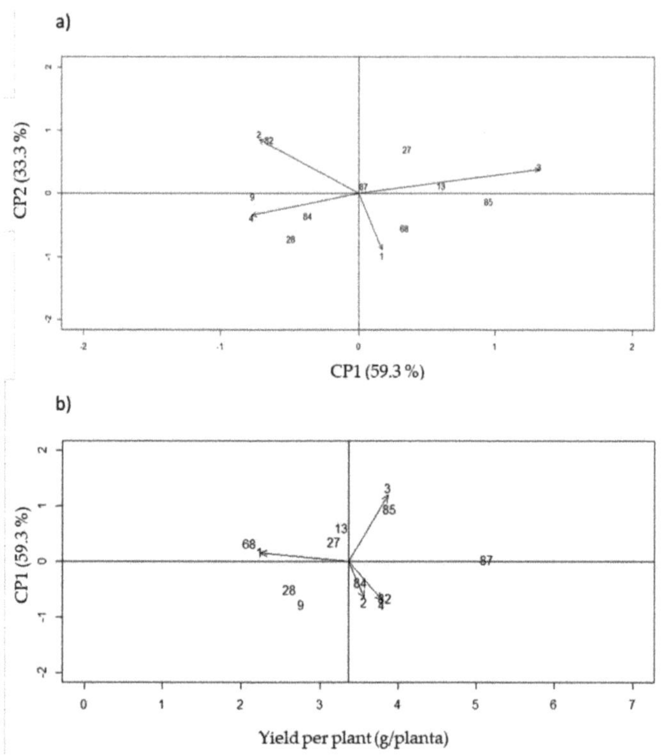

Figure 5. AMMI analysis for nine interspecific congruity-backcross lines evaluated across four localities in the Caribbean region of Colombia. (**a**) Main components (CPs) for yield per plant. (**b**) Biplot of main component 1 (CP1) against yield per plant. Arrows depicts localities 1: Caribia, 2: Motilonia, 3: Turipaná, 4: Carmen de Bolívar. *P. acutifolius* control G40001 is abbreviated as 87.

The AMMI analysis showed that the first component explained 59.3% of the variance due to G × E interaction for total yield, while the second component explained 33.3%. Therefore, the selection of a single multiplicative term of the AMMI model was sufficient to explain a large proportion of the relevant data [27]. The genotypes closest to the origin point in the AMMI biplot were those with little contribution to the interaction effects, and therefore more stable. Among those, genotypes 13, 27, 82, and 84 were more stable (i.e., broad adaptation) with higher yields, compared to the G40001 *P. acutifolius* control. Genotypes 68, 84, 85, and the *P. acutifolius* G40001 control, respectively located closer to localities Caribia, Motilonia, Turipaná and Carmen de Bolívar, suggesting narrow adaptation [28] to such specific environments in the dry and humid Caribbean sub-regions of Colombia.

Heat and drought events were likely more extreme in the former (i.e., Motilonia, Turipaná), while severe drought was presumably preponderant in the latter (i.e., Turipaná and Carmen de Bolívar). In short, this AMMI approach allowed selecting four stable genotypes across environments, and four additional genotypes with narrow adaptations to specific sites [29–32]. Furthermore, the AMMI analysis made possible a straightforward interpretation of the results using a biplot. This enabled improving the precision of the G × E interaction, since it eliminated the error around phenotypic stability parameters generated by the effect of some environments on particular genotypes [30].

4. Discussion

Across all four localities, the *P. acutifolius* control genotype G40001 presents the best performance, evidencing its ability to adapt to the different conditions in coastal Colombia. This behavior is consistent with extensive reports that mention the potential of the species *P. acutifolius* to grow and produce in warm arid and semi-arid conditions [13,14]. In addition, this study reveals that genotypes from advanced interspecific congruity backcrosses with this species exhibit promising responses to extreme conditions in Coastal Colombia, suggesting the inheritance of polygenic adaptation across interspecific boundaries. Yet, selection of elite genotypes may also be guided at each locality with the aim to identify genotypes with particular narrow adaptability.

Looking back into the pedigree of the elite interspecific lines, key donors include the SER, SEF, HTA, GGR, and SMG genotypes (i.e., general combining ability due to the additive parental breeding values), as well as the progenies from the crosses among G 40119, G 40264, G 40036, DAA 9, SAR, SAB, SEF, RAB, SMR 180, ALB 74, INB 841, SMC 199, and DAB 295 (i.e., specific combining ability due to relative non-additive performance of particular crosses). Hence, genotypes resulting from the Alliance Bioversity–CIAT bean breeding program show potential as parental donors for interspecific crosses [33–36], or even as cultivars themselves, in the face of heat and drought stresses in coastal Colombia [37]. In this way, this works offers novel perspectives to harness, via advanced interspecific schemes, the highly contrasting diversity present at beans' germplasm, while bridging major bottlenecks to breed traits of interest in the face of climate change.

Currently, the Tepary bean represents an important genetic resource to improve resistance to heat and drought stresses, which are the most limiting factors in common bean [12]. Throughout this research, we have demonstrated the potential of this species to contribute outstanding trait scores via interspecific congruity backcrosses with the common bean. This approach offers new scopes to utilize the conserved bean variation to strength the sustainability of the common bean crop in hot semiarid and arid conditions in tropical regions [38]. Such hybrid breeding strategy is necessary to leverage the over-dominance of genetic variation previously locked within species because the genomic architecture [9,39] conferring thermotolerance in the Tepary bean [40] likely differs from the one in the common bean [41]. Specifically, GWAS (Genome Wide Association Studies) and candidate gene approaches have unveiled complex genomic architectures for drought tolerance, subjected to rampant divergent selection and limited cross-species standing variation [42–45]. Similarly, the genetic basis of heat tolerance, as inferred in common bean from GEA (Genome–Environment Association) scans and candidate genes [46,47], only holds partially in the Tepary bean [9]. In short, the Tepary bean could help bridging long-standing gaps in the breeding of common bean for drought and heat tolerance, due to the interspecies mosaic in genetic determination mechanisms for these abiotic stresses. Specifically, the Tepary bean may offer promising [47], accessible, and novel genetic pathway sources of resistance [48], targeting tropical geographical regions with prolonged periods of intense heat and drought events [49].

Additionally, our research also invites consideration of the feasibility to simultaneously breeding for drought tolerance and adaptation to low soil fertility, such as via line 68 (a.k.a., SMG8). This interspecific backcross recombinant is particularly outstanding because it breaks a prevalent inverse relationship between *Fe* content and yield, pervasively observed across the common bean × Tepary bean congruity backcross families. The fact that some interspecific recombination events break intrinsic trait correlations evidences the potential to mitigate phenotypic trade-offs by taking advantage of cross-species variability. In order to achieve a more integrative view on the interaction among various trait types, oncoming trials must select promissory interspecific lines after considering further localities in the Caribbean region of Colombia experiencing combined restrictive environments (e.g., high water limitation and low soil quality).

Moreover, this study reinforces the potential of interspecific congruity backcrosses between exotic donors of adaptation and more susceptible elite genepools. Congruity schemes perform multiple alternating backcrosses [8,15,18,50], allowing for a better retention of the exotic phenotype while overcoming the hybrid sterility, genotype incompatibility, and embryo abortion typically found in simpler backcrosses [9]. As shown here, congruity backcross pedigrees behave equally well for the introgressive breeding of polygenic quantitative adaptation [51] (i.e., heat/drought stress tolerance [17]). Therefore, this scheme brings a realistic pathway to efficiently unlock polygenic adaptation, and effectively utilize hidden variation from tolerant landraces, related crop species, and well-adapted crop wild genepools [9].

Yet, we must recognize the pilot nature of the present study when it comes to the physiological basis of the heat and drought tolerance phenotypes. Even though we carried out ambitious enviromics' multi-locality trials of interspecific crosses, typically regarded as of low viability, more efforts are required to bridge the gap between yield and physiological components during stress responses. In addition, unexplained variation among replicates was evidenced for Motilonia and Carmen de Bolívar in the single-locality analysis. This is possibly due to some intrinsic factors such as intra-family segregation, spatial variability of the soil, or variation of the edaphic microbiota due to high temperature and low organic matter content (i.e., below 2%, which has been reported as a condition that can affect the site-specific response of bean yield [52]).

Local-scale spatial variability is undeniably regarded as a major driving force in various cropping systems [53–55]. Fortunately, the effect of soil heterogeneity in the statistical analysis of bean trials is starting to be recognized by the workflow that the Alliance Bioversity–CIAT has carried out as part of the bean genetic improvement program. Specifically, a novel platform called Mr. Bean is undergoing testing in order to better handle local-scale spatial information [56], and will certainly prove useful for future trials. Meanwhile, as part of future projects, we aim extending time-seriated physiological screening from the warmest Motilonia locality to more mild climates in coastal Colombia such as Turipaná and Carmen de Bolívar, where heat is predicted to become more intense and frequent in the oncoming years. Beyond this, it would also be desirable to better control for the effect of specific microbiological activity on the performance parameters. Envisioning novel spatial analysis and experimental designs would help minimizing the impact of uncontrolled intra-genotype variability, even at the G × E interface. These transdisciplinary approaches are ultimately needed to gather a more mechanistic understating of the tolerance response across the environmental continuum, specifically concerning traits' trade-offs.

5. Perspectives

After having assessed adaptation in advanced common bean × Tepary bean interspecific congruity backcrosses through four localities in coastal northwest South America, the study of G × E interaction would require to better account for the inter-annual variability. We aim addressing this perspective in the years to come by validating and extending in time the key predictions from the present study. Future field trials will have to be explicitly contrasted with the current analysis of the 2020 production cycle, which was typically heat and warm for coastal Colombia (therefore serving as a standard condition to capture adaptive variation for heat and drought).

Furthermore, the use of parental Tepary beans may also confer resistance to soil pathogens as part of future introgressive breeding schemes. This expectation is due to the ability of common bean × Tepary bean interspecific lines to survive in localities with presumably high incidence of biotic stresses, such as Caribia. After all, some common bean parental genotypes are derived from selection schemes carried out by the Alliance Bioversity–CIAT at the locality of Santander de Quilichao (province of Cauca), where the pressure of soil pathogens is high and similar to the one observed in some regions of coastal Colombia.

Other interspecific lines examine here might also exhibit future potential to introgress resistance to bean common mosaic virus (BCMV). This is the case of lines 84 and 86 (a.k.a., SEF 16 and SEF 54), both with red seed coat color, as well as the pure 118 line (a.k.a., SER118), developed by the breeding program of Escuela Agrícola Panamericana Zamorano (EAP) in Honduras. Yet, these expectations remain speculative at the studied localities in Coastal Colombia. Hence, future studies must carry out detailed characterizations of in situ biotic pressures at these localities, as a key step to deepen our understanding on the role of hybrid breeding for biotic stress resistance in bean species. These studies must be envisioned within the framework of the plant disease triangle (PDT), which postulates that any plant disease is the result of the interaction between a host's genotype, the biotic stress, and their unique environmental combination.

Another necessary next step will involve exploring the underlying genomic basis of drought and heat tolerance across the species boundary via high-throughput sequencing [9]. First, genomic diversity and signatures of selection for heat and drought tolerance may be explored by tracing back changes in the allelic and haplotype frequencies across the available interspecific congruity-backcross genealogies that made part of this study [15]. Theoretical expectations are that selection signatures of introgressive breeding will be enriched in (i) key underlying regions of the tolerance adaptive phenotype [18], as well as on (ii) low recombining regions in high linkage disequilibrium that are more prone to lineage sorting (i.e., genetic drift) [57] and linked selection due to smaller effective population size [58,59].

Second, more explicit additive models such as GWAS mapping [60], and polygenetic genomic prediction (GP) will respectively enlighten the genomic architecture of drought and heat tolerance in *Phaseolus* beans, while improving predictions of the genomic estimated breeding values (GEBVs) in yet to be established congruity-backcross hybrid seedlings (i.e., genomic-assisted backcrossing–GABC) [60]. Given the complex nature of the heat and drought tolerances, it is foreseen that the GP and GABC approaches would outperform GWAS-type modeling [61], without meaning that the latter will be incapable to capture variants with moderate effects segregating at medium frequencies, still useful to boost more traditional and scalable marker-assisted backcrossing (MABC) initiatives [62]. Merging these approaches will eventually insight into molecular evolution and pre-breeding under heat and drought stresses [12], both at the yield and ecophysiological levels [63].

6. Conclusions

This research explored the yield, ecophysiology, and mineral content of Tepary and common bean interspecific congruity-backcross lines in four localities with contrasting conditions at the Caribbean region of Colombia. Mineral content of the grain in terms of *Fe* and *Zn* was carried out completely in three localities, while ecophysiological responses were recorded in one location as a pilot proxy to intermingle yield and physiological responses at extreme heat and drought conditions. Genotypes obtained by interspecific congruity backcrosses emphasize the importance of conserving and utilizing agrobiodiversity as part of the pre-breeding for adaptation to heat and drought conditions. Inter-specific backcrosses 13, 27, 82, and 84 showed desirable adaptation, yield scores, and stability across all four localities, close to the Tepary bean's G40001 control genotype. Yet, selection for narrow adaptation is also feasible because some interspecific lines exhibited good responses in terms of yield and mineral content per locality.

In the same way, as part of the characterization of ecophysiological parameters, it was possible to identify how the cluster of genotypes with the best performance also presents the highest scores for assimilation of CO_2 and WUE. Therefore, parameters related to gas exchange may be used as indirect selection indices (i.e., proxy traits to speed-up breeding [64,65]). In order to unveil this potential, we recommend continuing with high-throughput phenotypic characterizations of target yield, ecophysiological and mineral content parameters in an extended panel of interspecific congruity-backcross lines.

Last but not least, expanded field trials across contrasting agroecological regions must procure including localities beyond dry and humid coastal northwest South America, as part of an ambitious enviromic [66] approach [67]. A multi-climate enviromic perspective offers opportunities to better optimize G × E arrangements for yield components and biofortification initiatives in the face of novel antagonistic abiotic and biotic interactions unleashed by climate change. This initiative will ultimately enable selecting for heat/drought tolerant [68,69] parents and new recombinants [70], hopefully by breaking the long-standing genetic trade-off between yield and *Fe* biofortification.

Supplementary Materials: The following are available online at https://www.mdpi.com/article/10.3390/agronomy11101978/s1, Table S1. Genealogy of 86 interspecific congruity-backcross lines, between common (*P. vulgaris*) and Tepary (*P. acutifolius*) beans, evaluated during the 2020 production cycle in four regions of coastal Colombia. Table S2. Localities' description and soil fertility scores in coastal Colombia. Table S3. Climate variables, temperature (minimum, maximum and average), and precipitation at the four localities in coastal Colombia during the crop cycle July-October 2020. Figure S1. Optimal number of clusters at (a) Motilonia, (b) Caribia [in (c) Caribia genotypes with only one repetition due to early mortality], (d) Turipaná, and (e) Carmen de Bolívar localities as determined by *Factoextra* and *NbClust* R's packages.

Author Contributions: E.B.-E., R.I.L.-P., C.C.C.-C., F.L.-H., A.J.C. and A.P.T.-R. designed sampling and experiments with assistance from S. Beebe. R.I.L.-P. and C.C.C.-C. collected data. F.L.-H. provided R code for analyses. E.B.-E. and R.I.L.-P. carried out data analysis with assistance from F.L.-H., E.B.-E., R.I.L.-P., C.C.C.-C., F.L.-H., A.J.C. and A.P.T.-R. contributed to data interpretation. All authors have read and agreed to the published version of the manuscript.

Funding: The Korea-Latin America Food and Agriculture Cooperation Initiative (KoLFACI) funded this research in alliance with the Colombian Agricultural Research Corporation (AGROSAVIA) throughout project number 1001513 entitled "Obtaining commercial and peasant market varieties of drought tolerant beans under sustainable production systems in the Colombian Caribbean".

Institutional Review Board Statement: Ethical review and approval were waived for this study because it did not involve humans or animals.

Informed Consent Statement: Not applicable because this study did not involve humans.

Data Availability Statement: Data supporting reported results can be found is the Supplementary Materials or per request to the corresponding authors.

Acknowledgments: The authors express their acknowledgments to the Korea-Latin America Food and Agriculture Cooperation Initiative (KoLFACI) for funding, the Alliance Bioversity–CIAT (International Center for Tropical Agriculture) for providing interspecific congruity-backcross lines between common (*P. vulgaris*) and Tepary (*P. acutifolius*) beans, the Colombian Agricultural Research Corporation (AGROSAVIA) for technical assistance, and Ministerio de Agricultura y Desarrollo Rural de Colombia (MADR) for administrative support. Authors also thank S. Beebe and V. Mayor for providing seed material, and the genealogies of the interspecific congruity-backcross lines. E.B.-E acknowledges insightful discussions around common bean with M.W. Blair, which took place during the summer of 2019 in Rionegro (Antioquia, Colombia) as part of the funding provided by the Fulbright's U.S. Specialist Program. F.L.-H. recognizes AGROSAVIA's Department for Research Capacity Building for subsidizing his internship in 2018, which enabled start exploring G × E models for heat stress in common bean. A.J.C. appreciates M.J. Torres-Urrego's support during the writing of this manuscript. Two reviewers and the topic editor are acknowledged for proving insightful recommendations. The invited co-editors R. Yockteng and M.A. Castillejo are also recognized for envisioning the highly relevant Special Issue on "Omics Approaches for Crop Improvement". Finally, we are in debt to MDPI and Agronomy's Editorial Office for encouraging this work's submission, and subsidizing its APC as part of this Special Issue.

Conflicts of Interest: The authors declare no conflict of interest. The funders had no role in the design of the study, in the collection, analyses, or interpretation of data, in the writing of the manuscript, or in the decision to publish the results.

References

1. Beebe, S. Common Bean Breeding in the Tropics. *Plant Breed. Rev.* **2012**, *36*, 357–426.
2. Caproni, L.; Raggi, L.; Talsma, E.F.; Wenzl, P.; Negri, V. European landrace diversity for common bean biofortification: A genome-wide association study. *Sci. Rep.* **2020**, *10*, 19775. [CrossRef]
3. He, M.; Dijkstra, F.A. Drought effect on plant nitrogen and phosphorus: A meta-analysis. *New Phytol.* **2014**, *204*, 924–931. [CrossRef] [PubMed]
4. Smith, M.R.; Veneklaas, E.; Polania, J.; Rao, I.M.; Beebe, S.E.; Merchant, A. Field drought conditions impact yield but not nutritional quality of the seed in common bean (*Phaseolus vulgaris L.*). *PLoS ONE* **2019**, *14*, e0217099. [CrossRef]
5. González-Orozco, C.E.; Porcel, M.; Velásquez, D.F.A.; Orduz-Rodríguez, J.O. Extreme climate variability weakens a major tropical agricultural hub. *Ecol. Indic.* **2020**, *111*, 106015. [CrossRef]
6. Mukherjee, S.; Ashfaq, M.; Mishra, A.K. Compound drought and heatwaves at a global scale: The role of natural climate variability-associated synoptic patterns and land-surface energy budget anomalies. *J. Geophys. Res. Atmos.* **2020**, *125*, e2019JD031943. [CrossRef]
7. Dosio, A.; Mentaschi, L.; Fischer, E.M.; Wyser, K. Extreme heat waves under 1.5 °C and 2 °C global warming. *Environ. Res. Lett.* **2018**, *13*, 054006. [CrossRef]
8. Beebe, S.; Rao, I.; Blair, M.; Acosta, J. Phenotyping common beans for adaptation to drought. *Front. Physiol.* **2013**, *4*, 35. [CrossRef]
9. Buitrago-Bitar, M.A.; Cortés, A.J.; López-Hernández, F.; Londoño-Caicedo, J.M.; Muñoz-Florez, J.E.; Muñoz, L.C.; Blair, M.W. Allelic Diversity at Abiotic Stress Responsive Genes in Relationship to Ecological Drought Indices for Cultivated Tepary Bean, Phaseolus acutifolius A. Gray, and Its Wild Relatives. *Genes* **2021**, *12*, 556. [CrossRef]
10. Mhlaba, Z.B.; Mashilo, J.; Shimelis, H.; Assefa, A.B.; Modi, A.T. Progress in genetic analysis and breeding of tepary bean (Phaseolus acutifolius A. Gray): A review. *Sci. Hortic.* **2018**, *237*, 119. [CrossRef]
11. Jiri, O.; Mafongoya, P.L.; Chivenge, P. Climate Smart Crops for food and nutritional security for semi-arid zones of Zimbabwe. *Afr. J. Food Agric. Nutr. Dev.* **2017**, *17*, 12280–12294. [CrossRef]
12. Moghaddam, S.M.; Oladzad, A.; Koh, C.; Ramsay, L.; Hart, J.P.; Mamidi, S.; McClean, P.E. The tepary bean genome provides insight into evolution and domestication under heat stress. *Nat. Commun.* **2021**, *12*, 14. [CrossRef]
13. Mwale, S.E.; Shimelis, H.; Mafongoya, P.; Mashilo, J. Breeding tepary bean (Phaseolus acutifolius) for drought adaptation: A review. *Plant Breed.* **2020**, *139*, 821–833. [CrossRef]
14. Muñoz, L.C.; Duque, M.C.; Debouck, D.G.; Blair, M.W. Taxonomy of Tepary Bean and Wild Relatives as Determined by Amplified Fragment Length Polymorphism (AFLP) Markers. *Crop. Sci.* **2006**, *46*, 1744–1754. [CrossRef]
15. Mejía-Jiménez, A.; Muñoz, C.; Jacobsen, H.J.; Roca, W.M.; Singh, S.P. Interspecific hybridization between common and tepary beans: Increased hybrid embryo growth, fertility, and efficiency of hybridization through recurrent and congruity backcrossing. *Appl. Genet.* **1994**, *88*, 324–331. [CrossRef] [PubMed]
16. Belivanis, T.; Doré, C. Interspecific hybridization of Phaseolus vulgaris L. and Phaseolus angustissimus A. Gray using in vitro embryo culture. *Plant Cell Rep.* **1986**, *5*, 329–331. [CrossRef]
17. Souter, J.R.; Gurusamy, V.; Porch, T.G.; Bett, K.E. Successful Introgression of Abiotic Stress Tolerance from Wild Tepary Bean to Common Bean. *Crop. Sci.* **2017**, *57*, 1160–1171. [CrossRef]
18. Muñoz, L.C.; Blair, M.W.; Duque, M.C.; Tohme, J.; Roca, W. Introgression in Common Bean × Tepary Bean Interspecific Congruity-Backcross Lines as Measured by AFLP Markers. *Crop. Sci.* **2004**, *44*, 637–645. [CrossRef]
19. Blair, M.W.; Cortes, A.J.; Farmer, A.D.; Huang, W.; Ambachew, D.; Penmetsa, R.V.; Carrasquilla-Garcia, N.; Assefa, T.; Cannon, S.B. Uneven recombination rate and linkage disequilibrium across a reference SNP map for common bean (*Phaseolus vulgaris L.*). *PLoS One* **2018**, *13*, e0189597. [CrossRef]
20. Blair, M.W.; Cortés, A.J.; Soler, A. Diversification and Population Structure in Common Beans (*Phaseolus vulgaris L.*). *PLoS ONE* **2012**, *7*, e49488. [CrossRef]
21. Borba, M.E.A.; Maciel, G.M.; Júnior, E.F.; Júnior, C.M.; Marquez, G.R.; Silva, I.G.; Almeida, R.S. Gas exchanges and water use efficiency in the selection of tomato genotypes tolerant to water stress. *Genet. Mol. Res.* **2017**, *16*, gmr16029685. [CrossRef] [PubMed]
22. Kassambara, A.; Mundt, F. Package 'factoextra'. Extract and visualize the results of multivariate data analyses. *RCRAN* **2017**, *76*, 1–74.
23. Charrad, M.; Ghazzali, N.; Boiteau, V.; Niknafs, A. NbClust: An R Package for Determining the Relevant Number of Clusters in a Data Set. *J. Stat. Softw.* **2014**, *61*, 36. [CrossRef]
24. Kassambara, A. Practical guide to principal component methods in R: PCA, M (CA), FAMD, MFA, HCPC, factoextra. *STHDA* **2016**, *2*, 152.
25. Van Buuren, S.; Groothuis-Oudshoorn, K. Mice: Multivariate imputation by chained equations in R. *J. Stat. Softw.* **2011**, *45*, 67. [CrossRef]
26. De Mendiburu, F. Agricolae: Statistical Procedures for Agricultural Research. *RCRAN* **2011**, *a1748*, 1–155.
27. Gauch, H.G., Jr.; Piepho, H.-P.; Annicchiarico, P. Statistical Analysis of Yield Trials by AMMI and GGE: Further Considerations. *Crop Sci.* **2008**, *48*, 866–889. [CrossRef]
28. Yan, W.; Hunt, L.A.; Sheng, Q.; Szlavnics, Z. Cultivar Evaluation and Mega-Environment Investigation Based on the GGE Biplot. *Crop Sci.* **2000**, *40*, 597–605. [CrossRef]

29. Gonçalves, J.G.R.; Chiorato, A.F.; Perina, E.F.; Carbonell, S.A.M. Estabilidade fenotípica em feijoeiro estimada por análise ammi com genótipo suplementar. *Bragantia* **2009**, *68*, 871. [CrossRef]

30. Barros, M.A.; Rocha, M.D.M.; Gomes, R.L.F.; Silva, K.J.D.; Neves, A.C.D. Adaptabilidade e estabilidade produtiva de feijão-caupi de porte semiprostrado. *Pesqui. Agropecuária Bras.* **2013**, *48*, 410. [CrossRef]

31. De Oliveira, E.J.; de Freitas, J.P.X.; de Jesus, O.N. AMMI analysis of the adaptability and yield stability of yellow passion fruit varieties. *Sci. Agric.* **2014**, *71*, 145. [CrossRef]

32. Escobar, E.A.V.; Sánchez, J.E.V.; García, D.B. Analysis of stability and adaptability of QPM hybrids of maize growing in different Colombian agroecological zones. *Acta Agronómica* **2016**, *65*, 79.

33. Polania, J.; Rao, I.M.; Cajiao, C.; Grajales, M.; Rivera, M.; Velasquez, F.; Raatz, B.; Beebe, S.E. Shoot and Root Traits Contribute to Drought Resistance in Recombinant Inbred Lines of MD 23–24 × SEA 5 of Common Bean. *Front. Plant Sci.* **2017**, *8*, 296. [CrossRef] [PubMed]

34. Suárez, J.C.; Polanía, J.A.; Contreras, A.T.; Rodríguez, L.; Machado, L.; Ordoñez, C.; Beebe, E.; Rao, I.M. Adaptation of common bean lines to high temperature conditions: Genotypic differences in phenological and agronomic performance. *Euphytica* **2020**, *216*, 22. [CrossRef]

35. Guevara-Escudero, M.; Osorio, A.N.; Cortés, A.J. Integrative Pre-Breeding for Biotic Resistance in Forest Trees. *Plants* **2021**, *10*, 2022. [CrossRef]

36. Camilo, S.; Odindo, A.O.; Kondwakwenda, A.; Sibiya, J. Root Traits Related with Drought and Phosphorus Tolerance in Common Bean (*Phaseolus vulgaris* L.). *Agronomy* **2021**, *11*, 552. [CrossRef]

37. Da Silva Sá, F.V.; Ferreira Neto, M.; de Lima, Y.B.; de Paiva, E.P.; Prata, R.C.; Lacerda, C.F.; Brito, M.E.B. Growth, gas exchange and photochemical efficiency of the cowpea bean under salt stress and phosphorus fertilization. *Comun. Sci.* **2018**, *9*, 668–679.

38. Parsons, L.R.; Howe, T.K. Effects of water stress on the water relations of *Phaseolus vulgaris* and the drought resistant Phaseolus acutifolius. *Physiol. Plant.* **1984**, *60*, 197–202. [CrossRef]

39. Jimenez-Galindo, J.C.; Alvarez-Iglesias, L.; Revilla-Temino, P.; Jacinto-Soto, R.; Garcia-Dominguez, L.E.; de La Fuente-Martinez, M.; Osorno, J.M. Screening for Drought Tolerance in Tepary and Common Bean Based on Osmotic Potential Assays. *Plant* **2018**, *6*, 24–32. [CrossRef]

40. Gujaria-Verma, N.; Ramsay, L.; Sharpe, A.G.; Sanderson, L.A.; Debouck, D.G.; Tar'an, B.; Bett, K.E. Gene-based SNP discovery in tepary bean (*Phaseolus acutifolius*) and common bean (*P. vulgaris*) for diversity analysis and comparative mapping. *BMC Genom.* **2016**, *17*, 239. [CrossRef]

41. Rosas, J.C.; Beaver, J.S.; Escoto, D.; Pérez, C.A.; Llano, A.; Hernández, J.C.; Araya, R. Registration of 'Amadeus 77' Small Red Common Bean. *Crop Sci.* **2004**, *44*, 1867–1868. [CrossRef]

42. Cortés, A.J.; Blair, M.W. Genotyping by Sequencing and Genome–Environment Associations in Wild Common Bean Predict Widespread Divergent Adaptation to Drought. *Front. Plant Sci.* **2018**, *9*, 128. [CrossRef]

43. Blair, M.W.; Cortés, A.J.; This, D. Identification of an ERECTA gene and its drought adaptation associations with wild and cultivated common bean. *Plant Sci.* **2016**, *242*, 250–259. [CrossRef] [PubMed]

44. Galeano, C.H.; Cortés, A.J.; Fernández, A.C.; Soler, Á.; Franco-Herrera, N.; Makunde, G.; Vanderleyden, J.; Blair, M.W. Gene-Based Single Nucleotide Polymorphism Markers for Genetic and Association Mapping in Common Bean. *BMC Genet.* **2012**, *13*, 48. [CrossRef] [PubMed]

45. Cortés, A.J.; Chavarro, M.C.; Madriñán, S.; This, D.; Blair, M.W. Molecular ecology and selection in the drought-related Asr gene polymorphisms in wild and cultivated common bean (*Phaseolus vulgaris* L.). *BMC Genet.* **2012**, *13*, 58. [CrossRef] [PubMed]

46. López-Hernández, F.; Cortés, A.J. Last-Generation Genome–Environment Associations Reveal the Genetic Basis of Heat Tolerance in Common Bean (*Phaseolus vulgaris* L.). *Front. Genet.* **2019**, *10*, 954. [CrossRef]

47. Mhlaba, Z.B.; Amelework, B.; Shimelis, H.; Modi, A.T.; Mashilo, J. Genetic differentiation among selected tepary bean collections revealed by morphological traits and simple sequence repeat markers. *Acta Agric. Scand. Sect. B Soil Plant Sci.* **2018**, *68*, 608–618. [CrossRef]

48. Blair, M.W.; Pantoja, W.; Carmenza Munoz, L. First use of microsatellite markers in a large collection of cultivated and wild accessions of tepary bean (*Phaseolus acutifolius* A. Gray). *Appl. Genet.* **2012**, *125*, 1137–1147. [CrossRef]

49. Crespo-Muñoz, S.; Rivera-Peña, M.; Rosero-Alpala, D.A.; Muñoz-Florez, J.E.; Rao, I.M.; Muñoz-Florez, L.C. Pollen viability of Tepary bean (*Phaseolus acutifolius* A. Gray) mutant lines under water stress conditions and inoculation with rhizobia. *Acta Agron.* **2018**, *67*, 319–325. [CrossRef]

50. Rao, I.; Polania, J.; Beebe, S.; Ricaurte, J.; Cajiao, C.; Garcia, R.; Rivera, M. Can tepary bean be a model for improvement of drought resistance in common bean. *Afr. Crop. Sci. J.* **2013**, *21*, 256–281.

51. Barghi, N.; Hermisson, J.; Schlötterer, C. Polygenic adaptation: A unifying framework to understand positive selection. *Nat. Rev. Genet.* **2020**, *21*, 769–781. [CrossRef]

52. Rivera, A.P.T.; Murgas, R.E.C.; Ríos, A.E.M.; Merini, L.J. Efecto del glifosato sobre la microbiota, calidad del suelo y cultivo de frijol biofortificado en el departamento del Cesar, Colombia. *Rev. Argent. Microbiol.* **2020**, *52*, 61–71.

53. Isik, F.; Holland, J.; Maltecca, C. *Genetic Data Analysis for Plant and Animal Breeding*; Springer International Publishing: Berlin/Heidelberg, Germany, 2017.

54. Medina, C.A.; Hawkins, C.; Liu, X.P.; Peel, M.; Yu, L.X. Genome-Wide Association and Prediction of Traits Related to Salt Tolerance in Autotetraploid Alfalfa (*Medicago sativa* L.). *Int. J. Mol. Sci.* **2020**, *21*, 3361. [CrossRef]

55. Murillo, D.A.; Gezan, S.A.; Heilman, A.M.; Walk, T.C.; Aparicio, J.S.; Horsley, R.D. FielDHub: A Shiny App for Design of Experiments in Life Sciences. *J. Open Src. Softw.* **2021**, *6*, 3122. [CrossRef]
56. Aparicio, J.; Ariza-Suarez, D.; Raatz, B. Web Application for Spatial Modelling of Field Trials. In Proceedings of the XXIX Simposio Internacional de Estadística, Barranquilla, Colombia, 15–19 July 2019.
57. Cortés, A.J.; Skeen, P.; Blair, M.W.; Chacón-Sánchez, M.I. Does the Genomic Landscape of Species Divergence in Phaseolus Beans Coerce Parallel Signatures of Adaptation and Domestication? *Front. Plant Sci.* **2018**, *9*, 1816. [CrossRef] [PubMed]
58. Wolf, J.B.W.; Ellegren, H. Making sense of genomic islands of differentiation in light of speciation. *Nat. Rev. Genet.* **2017**, *18*, 100. [CrossRef] [PubMed]
59. Ellegren, H.; Galtier, N. Determinants of genetic diversity. *Nat. Rev. Genet.* **2016**, *17*, 433. [CrossRef]
60. Cortés, A.J.; Restrepo-Montoya, M.; Bedoya-Canas, L.E. Modern Strategies to Assess and Breed Forest Tree Adaptation to Changing Climate. *Front. Plant Sci.* **2020**, *11*, 1606. [CrossRef]
61. Cortés, A.J.; López-Hernández, F. Harnessing Crop Wild Diversity for Climate Change Adaptation. *Genes* **2021**, *12*, 783. [CrossRef]
62. Pandit, E.; Pawar, S.; Barik, S.R.; Mohanty, S.P.; Meher, J.; Pradhan, S.K. Marker-Assisted Backcross Breeding for Improvement of Submergence Tolerance and Grain Yield in the Popular Rice Variety 'Maudamani'. *Agronomy* **2021**, *11*, 1263. [CrossRef]
63. Leitão, S.T.; Bicho, M.C.; Pereira, P.; Paulo, M.J.; Malosetti, M.; de Sousa Araújo, S.; van Eeuwijk, F.; Patto, M.C.V. Common bean SNP alleles and candidate genes affecting photosynthesis under contrasting water regimes. *Hortic. Res.* **2021**, *8*, 4. [CrossRef]
64. Watson, A.; Ghosh, S.; Williams, M.J.; Cuddy, W.S.; Simmonds, J.; Rey, M.D.; Md Hatta, M.A.; Hinchliffe, A.; Steed, A. Speed breeding is a powerful tool to accelerate crop research and breeding. *Nat. Plants* **2017**, *4*, 23–29. [CrossRef] [PubMed]
65. Cortés, A.J.; López-Hernández, F.; Osorio-Rodriguez, D. Predicting Thermal Adaptation by Looking Into Populations' Genomic Past. *Front. Genet.* **2020**, *11*, 1093. [CrossRef] [PubMed]
66. Resende, R.T.; Piepho, H.P.; Rosa, G.J.; Silva-Junior, O.B.; e Silva, F.F.; de Resende, M.D.V.; Grattapaglia, D. Enviromics in breeding: Applications and perspectives on envirotypic-assisted selection. *Theor. Appl. Genet.* **2021**, *134*, 95–112. [CrossRef] [PubMed]
67. Arenas, S.; Cortés, A.J.; Mastretta-Yanes, A.; Jaramillo-Correa, J.P. Evaluating the accuracy of genomic prediction for the management and conservation of relictual natural tree populations. *Tree Genet. Genomes* **2021**, *17*, 12. [CrossRef]
68. Cortés, A.J.; Monserrate, F.A.; Ramírez-Villegas, J.; Madriñán, S.; Blair, M.W. Drought Tolerance in Wild Plant Populations: The Case of Common Beans (*Phaseolus vulgaris* L.). *PLoS ONE* **2013**, *8*, e62898. [CrossRef]
69. Cortés, A.J.; This, D.; Chavarro, C.; Madriñán, S.; Blair, M.W. Nucleotide diversity patterns at the drought-related DREB2 encoding genes in wild and cultivated common bean (*Phaseolus vulgaris* L.). *Theor. Appl. Genet.* **2012**, *125*, 1069–1085. [CrossRef]
70. Blair, M.W.; Cortés, A.J.; Penmetsa, R.V.; Farmer, A.; Carrasquilla-Garcia, N.; Cook, D.R. A high-throughput SNP marker system for parental polymorphism screening, and diversity analysis in common bean (*Phaseolus vulgaris* L.). *Theor. Appl. Genet.* **2013**, *126*, 535–548. [CrossRef] [PubMed]

agronomy

Article

High-Throughput Canopy and Belowground Phenotyping of a Set of Peanut CSSLs Detects Lines with Increased Pod Weight and Foliar Disease Tolerance

Davis Gimode [1,2], Ye Chu [1], Corley C. Holbrook [3], Daniel Fonceka [4], Wesley Porter [5], Iliyana Dobreva [6], Brody Teare [7], Henry Ruiz-Guzman [7], Dirk Hays [7,8] and Peggy Ozias-Akins [1,*]

[1] Institute of Plant Breeding Genetics and Genomics, University of Georgia, Tifton, GA 31793, USA; dgimode@gmail.com (D.G.); ychu@uga.edu (Y.C.)
[2] International Crops Research Institute for the Semi-Arid Tropics, Nairobi P.O. Box 39063-00623, Kenya
[3] United States Department of Agriculture, Agricultural Research Service, Tifton, GA 31793, USA; corley.holbrook@ars.usda.gov
[4] Centre d'Etude Régional pour l'Amélioration de l'Adaptation à la Sécheresse (CERAAS), Institut Sénégalais de Recherches Agricoles (ISRA), BP 3320, Route de Khombole, Thiès 21000, Senegal; daniel.fonceka@cirad.fr
[5] Crop and Soil Sciences Department, University of Georgia, Tifton, GA 31793, USA; wporter@uga.edu
[6] Department of Geography, The Ohio State University, Colombus, OH 43210, USA; dobreva.iliyana.phd@gmail.com
[7] Molecular & Environmental Plant Sciences, Texas A&M University, College Station, TX 77843, USA; blteare@gmail.com (B.T.); henry.ruiz@tamu.edu (H.R.-G.); dirk.hays@ag.tamu.edu (D.H.)
[8] Department of Soil and Crop Sciences, Texas A&M University, College Station, TX 77843, USA
* Correspondence: pozias@uga.edu

Citation: Gimode, D.; Chu, Y.; Holbrook, C.C.; Fonceka, D.; Porter, W.; Dobreva, I.; Teare, B.; Ruiz-Guzman, H.; Hays, D.; Ozias-Akins, P. High-Throughput Canopy and Belowground Phenotyping of a Set of Peanut CSSLs Detects Lines with Increased Pod Weight and Foliar Disease Tolerance. *Agronomy* 2023, 13, 1223. https://doi.org/10.3390/agronomy13051223

Academic Editors: Roxana Yockteng, Andrés J. Cortés and María Ángeles Castillejo

Received: 21 March 2023
Revised: 17 April 2023
Accepted: 21 April 2023
Published: 26 April 2023

Abstract: We deployed field-based high-throughput phenotyping (HTP) techniques to acquire trait data for a subset of a peanut chromosome segment substitution line (CSSL) population. Sensors mounted on an unmanned aerial vehicle (UAV) were used to derive various vegetative indices as well as canopy temperatures. A combination of aerial imaging and manual scoring showed that CSSL 100, CSSL 84, CSSL 111, and CSSL 15 had remarkably low tomato spotted wilt virus (TSWV) incidence, a devastating disease in South Georgia, USA. The four lines also performed well under leaf spot pressure. The vegetative indices showed strong correlations of up to 0.94 with visual disease scores, indicating that aerial phenotyping is a reliable way of selecting under disease pressure. Since the yield components of peanut are below the soil surface, we deployed ground penetrating radar (GPR) technology to detect pods non-destructively. Moderate correlations of up to 0.5 between pod weight and data acquired from GPR signals were observed. Both the manually acquired pod data and GPR variables highlighted the three lines, CSSL 84, CSSL 100, and CSSL 111, as the best-performing lines, with pod weights comparable to the cultivated check Tifguard. Through the combined application of manual and HTP techniques, this study reinforces the premise that chromosome segments from peanut wild relatives may be a potential source of valuable agronomic traits.

Keywords: peanut; phenomics; high-throughput phenotyping; ground penetrating radar; tomato spotted wilt virus; leaf spot; pod weight

1. Introduction

A key challenge in peanut breeding is increasing the genetic diversity of the crop. Unlike their wild relatives, cultivated varieties have severely limited variation as a result of their genetic heritage and the process of domestication [1]. What is now recognized as cultivated peanut arose from the hybridization of two wild species, namely *A. duranensis* and *A. ipaensis*. Spontaneous doubling of the chromosomes of the hybrid resulted in tetraploid *A. hypogaea* [2,3]. The resultant ploidy barrier restricted the ability of cultivated peanut to exchange genetic material with the wilds. This limitation was further enforced by the process of human selection and domestication [4,5]. Consequently, peanuts are under the constant threat of biotic pressures, such as insects and diseases, as well as abiotic stresses, such as drought, which are exacerbated by the effects of climate change. In contrast, because of their diversity, wild peanut relatives are more adaptable since they have maintained genes that enable them to cope with these pressures. This makes them precious sources of diversity for cultivated peanut [6–9].

One of the ways that breeders sought to harness genetic diversity from the wild was by creating a chromosome segment substitution line (CSSL) population [1]. This was achieved by crossing the two known diploid ancestors of peanut to result in a diploid hybrid. The genome of the hybrid was doubled to form a synthetic tetraploid, essentially recreating the allotetraploid ancestor of cultivated peanut [10]. The synthetic allotetraploid was crossed with Fleur 11, a cultivated variety of peanut that is popularly grown in West Africa. Subsequent judicious selection resulted in a population comprised of 122 individual lines. Each line is genetically similar to the cultivated variety; however, the selection was made such that each line retained a small segment of the synthetic genome. Therefore, each line is distinguished by the portion of the synthetic genome it contains, while at the same time, the whole genome of the wild tetraploid is captured in the entire population [1]. The CSSL population, thus, constitutes a critical peanut genetic resource, with the potential to enable understanding of the basis of genetic variation in peanut.

High-resolution genetic characterization of wild introgressions in this population has been achieved previously, made possible by taking advantage of technological improvements and reduced costs, which have made it easier to obtain genotype data. These factors have enabled peanut researchers to achieve the phenomenal feat of releasing high-quality whole genome sequences of the tetraploid species [11], as well as the A and B genome progenitors of cultivated peanut [12]. Other genetic resources, including two high-quality SNP arrays [13–16], a reference transcriptome [17] as well as a comprehensive genomics database [18], have ensured the graduation of peanut from an orphan crop status [19].

Despite the ease of access and routine deployment of high-throughput genotyping, peanut phenotyping is mostly carried out at low throughput. Typically, this involves studying single plants in controlled environments such as greenhouses and growth chambers, or in small field plots for traits such as disease resistance. At the same time, it is often necessary to harvest plants destructively and at fixed growth stages. Greenhouse and growth chamber conditions fail to capture the true attributes of the plants since, when they are grown in the field, they behave differently at the various growth stages and as a result of competition for water, nutrients, and sunlight [20–22]. Additional drawbacks of such manual phenotyping include the fact that it is time-consuming, labor and cost intensive, and prone to human error and biases. This phenotyping challenge forms a bottleneck in the peanut breeding pipeline that curtails the full exploitation of available genetic and genomic resources for association studies.

A practical way of alleviating this bottleneck is the use of phenomics, or high-throughput phenotyping (HTP), a novel approach that is increasingly finding application in plant breeding research. It combines cutting-edge technologies such as spectroscopy, noninvasive imaging, and high-performance computing to capture phenotypic data at high resolution and throughput to address breeding problems [22]. HTP involves the use of equipment that can facilitate the collection of large quantities of high-quality data in a short period, thus availing the possibility of linking genotype data with phenotype data of equal throughput obtained in a "real world" environment. Field-based HTP, in particular, enables the accurate measurement of plant growth, architecture, and performance non-destructively and in the complexity of their true environment [22–24]. Such phenomic approaches can facilitate the effective use of genetic data to discover novel variations that could improve the quality and yield of crops [25].

Two commonly used field HTP techniques in studying crop plants involve remote sensing using visible light cameras (RGB) and multispectral cameras. These are typically borne on unmanned aerial vehicles (UAV) and can be useful for evaluating plant biotic and abiotic stress, plant vigor and phenology, soil characterization, and field mapping [26]. Examples of studies utilizing such techniques include studying the response to nitrogen and fertilizer treatment in maize [27,28], evaluation of water and nitrogen use efficiency in rice [29,30], and evaluation of pest pressure and yield estimation in soybean [31,32].

In peanut, UAV-based remote sensing has been used to acquire data on NDVI, canopy temperature, and RGB images to discriminate between various varieties based on yield under drought and late leaf spot pressure [33,34]. An attempt was made to derive canopy spectral signatures for predicting pod maturity, though without significant success [35]. Patrick et al. [36] experimented with various multispectral indices to determine the best one for detecting TSWV and settled on NDRE, which detected TSWV as early as 93 days into the season. Sarkar et al. [37] used a digital camera mounted on a UAV to successfully evaluate leaf wilting for irrigation triggering using indices derived from red–blue–green images. Manual and UAV-based HTP were compared for their utility to assess key traits in the US mini-core collection by Sarkar et al. [38]. The study showed good correlations between NDVI and RGB indices with traits such as plant height, lateral growth, leaf wilting, thrips damage, and yield, suggesting that vegetative indices could be used as surrogates for trait selection. Bagherian et al. [39] combined the use of aerial hyperspectral imaging with machine learning to predict biomass, pod count, and yield. Correlations of up to 0.73 were observed between measured biomass and its prediction.

Of equal novelty is the use of ground penetrating radar (GPR) for belowground studies. GPR works by emitting electromagnetic waves into the ground and detecting the waves that are reflected back to a detector. This enables the detection and rendering of 3D images of belowground biomass [40]. GPR has been effectively used for the non-destructive study of belowground biomass, especially for trees under various soil conditions [41], in forest systems [42], tree intercropping systems [43], and other agroforestry systems [44,45]. However, its application in studying fine belowground biomass, characteristic of agricultural crops, has been limited [46].

Most studies of the application of GPR to belowground phenotyping have used image analysis and image-thresholding analysis to extract GPR features, which are then correlated to crop characteristics such as biomass [40,46]. Dobreva et al. [47] demonstrated the limitations of this approach for the assessment of peanut yield. Specifically, it was demonstrated that the information about peanut yield is located within narrow vertical strips of the radargram. Moreover, due to the sensitivity of the GPR signal to soil heterogeneity, applying the same approach to a different geographic site or even to the same geographic site with different soil moisture conditions results in a different depth from which maximum peanut information can be detected. A frequency-based approach to agricultural GPR involves transforming the radar information from the time to the frequency domain. Agbona et al. [48] demonstrated the application of Fourier transform to belowground phenotyping of cassava. The advantage of this implementation is that a larger vertical portion of the radargram can be analyzed, which eliminates the need to determine the narrow vertical location of maximum peanut yield if image thresholding is applied. It is also expected that this approach is less sensitive to the specific soil conditions of different geographic sites. Peanut is a geocarpic plant that flowers aboveground but sets seed belowground [49], and, therefore, the most important biomass of peanut is below the ground. Consequently, the phenotyping of peanut pods requires a destructive harvest. This makes peanut an important candidate for the development of non-destructive belowground phenotyping techniques such as GPR.

Thus, the peanut CSSL population was an ideal candidate for deploying HTP techniques to bridge the genotype-phenotype gap. In this study, we incorporated HTP into the screening of a subset of the CSSLs that exhibited contrasting morphological attributes for canopy and belowground traits. Initially, the objective was to capture morphological diversity inherent in the population as a result of the differences in their genetic composition. However, as serendipity would have it, increased tomato spotted wilt disease and leaf spot pressure afforded the opportunity to evaluate the performance of the lines under disease pressure. Aerial images were collected and correlated with manual disease ratings. In addition, GPR was used to evaluate pod weight prior to harvest.

2. Materials and Methods

A subset of 26 lines from a CSSL population originally developed by [1] was used for this study. In addition, Fleur 11, the cultivated parent, as well as Tifguard and Florunner, which are morphologically distinct from the CSSLs, were also planted as adapted high-yielding leaf spot and TSWV-resistant checks for a total number of 29 lines. Two field trials were conducted with the lines planted in May 2018 at the University of Georgia Tifton Campus' Bowen and Gibbs farms, with the soil type being Tift loamy sand (fine-loamy, kaolinitic, thermic Plinthic Kandiudults). The experimental setup in each field was RCBD, with three replicates in each field except for CSSL 15, which had 2 replicates in both fields, and CSSL 100, which had 2 replicates at Gibbs due to seed shortage. Each 3 m long plot had two rows with the seeds sown with a spacing of 7 cm between seeds. Normal cultural practices, including scheduled pesticide and fungicide application and irrigation, were followed. Ratings for mid-season TSWV were recorded at approximately 80 days after planting (DAP) for both fields, as well as RGB and thermal aerial imaging. At the end of the season (110 DAP), ratings for leaf spot were carried out on both fields, while TSWV was rated on the Gibbs farm only. For TSWV, the ratings entailed assessing canopy TSWV symptoms on a scale of 0 to 10, with 0 being no symptoms observed and 10 being the presence of symptoms in the entire plot. An immunostrip assay was carried out to confirm the presence of TSWV in susceptible lines and its absence in lines that were asymptomatic. Leaf spot ratings were carried out without making a distinction between early and late leaf spot on a scale of 1, indicating no leaf spot symptoms, to 10, indicating complete defoliation and plant death. Thermal and multispectral images were collected on the same day that TSWV rating was undertaken. All cameras used for aerial imaging were mounted on a 3 DR Solo quadcopter.

GPR data were collected at 117 DAP, which was 1 day before plot inversion. The GPR imaging system was an experimental prototype consisting of an array of 7 antennas developed by IDS GeoRadar systems. Both rows in each of the 85 plots were scanned at Gibbs farm; however, only 49 of the 86 plots had both rows scanned at Bowen, with the rest having only single rows scanned. Post-harvest, manual data was collected by taking the total pod weight of each plot.

Analysis

Indices derived from RGB images were the Visible Atmospherically Resistant Index (VARI), Green Vegetation Index-R (GRV), and Green Leaf Index (GLI). Multispectral indices were the Chlorophyll Index-RE (CIRE), Difference Vegetation Index (DVI), Green Normalized Difference Vegetation Index (GNDVI), Normalized Difference Vegetation Index (NDVI), Normalized Green (NG), Optimized Soil Adjusted Vegetation Index (OSAVI), Ratio Vegetation Index (RVI), Soil Adjusted Vegetation Index (SAVI), and Triangular Vegetation Index (TVI). Canopy temperature depression (CTD, a measure of the temperature difference between the canopy and the surrounding area) was derived from the thermal images. Collected images were stitched using the photogrammetry software Pix4D (Prilly, Switzerland) and resulted in whole-field orthomosaics. ArcGIS [50] was used for further downstream analysis. Briefly, the various indices were calculated to extract data from the RGB, multispectral, and thermal orthomosaics. Boundaries were manually drawn to delineate each plot in the fields with appropriate buffering to ensure no overlap between plots. Pixels outside the plot boundaries were eliminated. Within the plots, thresholds of pixels representing soil were manually determined using the identity function and eliminated. Canopy pixels were averaged to derive quantitative data for each line. CTDs were calculated by subtracting the average plot temperature value from the air temperature which was obtained for each field from the University of Georgia Weather Network (www.weather.uga.edu, accessed on 20 March 2023).

GPR data were processed using the Python software platform GPR-Studio (Hays, pers. Comm.). Data from channel 3 were used for downstream signal processing. To begin with, the radargrams were split into plots based on electronic markers that had been incorporated into the images at the time of GPR acquisition. In an ideal HTP setting, the radargrams should be split into plots using the electronic markers alone, but in this study, each agricultural plot was examined manually, and the markers adjusted to ensure that all plots were of the same length. The radargrams were also subset vertically to remove the top and bottom portions of the radargram. Specifically, the analysis was performed on the portion of the radargram that started at pixel row 160 and for a depth of 200 pixel rows.

Three GPR processing pipelines were tested. In the first pipeline, no processing was applied other than the vertical subset of the radargram. In the second and third pipelines, subset radargrams were filtered with Kirchoff migration and a migration window of 5 and 35, respectively. Migration focuses the signal, and a migration window refers to the width around each sample that is used to perform the migration, with a small window corresponding to a smaller effect of the migration. The output of the three pipelines was subjected to discrete Fourier transformation (DFT), and similar to Agbona et al. [48], the results of the DFT were averaged for each agricultural plot. Raw data following processing with the three pipelines and application of DFT are available in Tables S1–S3. The GPR features were derived from the power of the Fourier coefficients for specific frequency bins. The features were within the frequency ranges of 1–13, with each frequency delineated to 7 sub-frequencies (such that frequency 1 contained sub-frequencies 1, 1.143, 1.286, 1.429, 1.571,1.714, and 1.857, and so on) for a total of 91 features per single row scanned. The 7 sub-frequencies were combined to derive one variable per frequency by a trapezoidal approximation of the area under the sub-frequency curve (AUFC) using the formula:

$$AUFC = \sum_{i=1}^{n-1} \frac{d_i + d_{i+1}}{2} \times (s_{i+1} - s_i) \tag{1}$$

where d_i is the value of DFT at the ith observation, s_i is sub-frequency at the ith observation, and n is the total number of observations. The AUFC variables were regarded as the quantitative data points representing belowground pod variation for the population. The variables were labeled by prefixing the pipeline and suffixing the frequency from which it was derived; for example, p1_freq_1 was the variable for the first frequency of pipeline 1. Thirteen variables were derived for each pipeline to yield a total 39 quantitative GPR variables.

All aerial, belowground, and manual quantitative data were analyzed using R [51]. The partitioning of variance was carried out using mixed model linear regression with the lmerTest package [52] in R with FDR used for multiple hypothesis correction. Trait summaries and correlations were derived from the obtained coefficients. Where appropriate, broad sense heritability was obtained by calculating the ratio of genetic variance to the total variance. For statistically significant traits ($p < 0.05$), lines that were significantly different from Fleur 11 ($p < 0.05$) were determined by running a Dunnett's multiple comparison test [53]. The relative effects of introgression on the traits were calculated by taking the difference between the coefficients of each line and Fleur 11 and getting the percentage relative to Fleur 11 ((CSSL−Fleur 11)/Fleur 11) × 100). Pearson correlation was used to investigate the relationship between the HTP data and the appropriate manual data.

3. Results

3.1. Manual Phenotyping

Ratings for TSWV were produced in the middle and at the end of the season. Due to more severe leaf spot incidence at the Bowen farm, it was not possible to accurately collect TSWV data at the end of the season. For mid-season TSWV, there were significant differences among the lines with no interaction between the fields. Late-season ratings at the Gibbs farm also revealed significant differences between the lines (Tables 1 and S4).

The line with the best rating mid-season was CSSL 100, which outperformed the cultivated check Tifguard. CSSL 111 was ranked third, while Fleur 11 ranked last (Table 2). End-season data from Gibbs showed that CSSL 100 was still the best-ranked within the population; however, Tifguard outperformed it (Figure 1 and Table 2). Overall, the best-performing CSSL lines under TSWV pressure were CSSL 100, CSSL 84, CSSL 15, and CSSL 111.

Table 1. Summary statistics for manually collected traits and spectral indices for the chromosome segment substitution line (CSSL) population.

Name [a]	Field	Significance	Mean	Min	Max	Fleur 11	SE [b]	Heritability [c]
Leaf spot late-season	Bowen	***	5.39	2.55	7.33	6.17	0.273	0.64
Leaf spot late-season	Gibbs	***	5.00	3.00	7.00	6.36	0.212	0.57
TSWV mid-season	Bowen	***	2.11	−0.12	4.67	4.67	0.243	0.57
TSWV mid-season	Gibbs	***	2.89	0.67	5.33	5.00	0.219	0.40
TSWV late-season	Gibbs	***	4.98	1.00	8.17	5.67	0.354	0.48
Pod weight (g)	Bowen	***	688.64	414.00	1320.00	498.00	45.189	0.54
Pod weight (g)	Gibbs	***	898.82	540.67	1610.00	705.33	44.817	0.74
GLI	Bowen	***	0.17	0.14	0.19	0.14	0.002	0.37
GLI	Gibbs	***	0.18	0.15	0.20	0.15	0.002	0.74
GRV	Bowen	***	0.13	0.11	0.16	0.11	0.003	0.56
GRV	Gibbs	***	0.15	0.13	0.16	0.13	0.002	0.70
VARI	Bowen	***	0.19	0.16	0.26	0.16	0.006	0.72
VARI	Gibbs	***	0.22	0.19	0.28	0.20	0.004	0.76
Mid CTD	Bowen	NS	−9.29	−10.62	−8.37	−9.74	0.076	0.00
Mid CTD	Gibbs	NS	−10.79	−11.53	−10.10	−11.28	0.076	0.03
End CTD	Bowen	***	−9.14	−11.01	−6.76	−10.30	0.178	0.22
End CTD	Gibbs	***	−10.31	−11.36	−9.01	−10.91	0.093	0.25
CIRE	Bowen	***	−0.56	−0.73	−0.47	−0.48	0.015	0.55
CIRE	Gibbs	***	−0.59	−0.70	−0.46	−0.48	0.012	0.69
DVI	Bowen	***	0.31	0.22	0.45	0.23	0.013	0.64
DVI	Gibbs	***	0.35	0.27	0.43	0.28	0.008	0.65
GNDVI	Bowen	***	0.70	0.62	0.82	0.66	0.010	0.78
GNDVI	Gibbs	***	0.73	0.68	0.81	0.69	0.006	0.62
NDVI	Bowen	***	0.71	0.59	0.86	0.63	0.014	0.66
NDVI	Gibbs	***	0.76	0.67	0.84	0.69	0.008	0.64
NG	Bowen	***	0.13	0.08	0.15	0.14	0.004	0.76
NG	Gibbs	***	0.12	0.09	0.14	0.13	0.002	0.66
OSAVI	Bowen	***	0.60	0.47	0.76	0.50	0.015	0.65
OSAVI	Gibbs	***	0.65	0.55	0.74	0.57	0.009	0.64
RVI	Bowen	***	7.37	4.05	14.88	4.62	0.579	0.76
RVI	Gibbs	***	8.72	5.63	13.26	6.04	0.384	0.73
SAVI	Bowen	***	0.49	0.37	0.65	0.40	0.016	0.64

Table 1. *Cont.*

Name [a]	Field	Significance	Mean	Min	Max	Fleur 11	SE [b]	Heritability [c]
SAVI	Gibbs	***	0.54	0.44	0.63	0.46	0.009	0.65
TVI	Bowen	***	18.62	13.08	27.34	13.73	0.814	0.61
TVI	Gibbs	***	21.33	15.94	25.99	16.84	0.504	0.65

NS: $p > 0.05$; *** $p \leq 0.001$. [a] Abbreviations for names: TSWV—tomato spotted wilt virus, GLI—Green Leaf Index, GRV—Green Vegetation Index-Red, VARI—Visible Atmospherically Resistant Index, CTD—canopy temperature depression, CIRE—Chlorophyll Index-Red Edge, DVI—Difference Vegetation Index, GNDVI—Green Normalized Difference Vegetation Index, NDVI—Normalized Difference Vegetation Index, NG—normalized green, OSAVI—Optimized Soil Adjusted Vegetation Index, RVI—Ratio Vegetation Index, SAVI—Soil Adjusted Vegetation Index, TVI—Triangular Vegetation Index; [b] standard error; [c] broad-sense heritability.

Table 2. Rankings for the chromosome segment substitution lines (CSSLs) based on manually collected data for both Bowen and Gibbs fields.

Sample	TSWV (MS)	Rank TSWV (MS)	TSWV (LS)	Rank TSWV (LS)	Leaf Spot	Rank Leaf Spot	Pod Weight (g)	Rank Pod Weight
CSSL 009	4.67	28	7.33	27	6.00	19	681.67	22
CSSL 010	3.50	21	8.17	29	6.17	20	705.67	18
CSSL 013	3.17	18	6.50	24	6.42	24	805.33	7
CSSL 014	2.67	13	6.33	22	5.25	9	747.67	9
CSSL 015	1.26	4	3.50	6	4.25	6	589.83	27
CSSL 022	3.83	27	7.83	28	6.42	25	732.00	11
CSSL 025	3.50	23	5.67	16	6.67	27	702.67	19
CSSL 027	3.67	26	6.67	25	6.50	26	681.67	21
CSSL 031	2.17	8	4.00	7	5.67	13	696.00	20
CSSL 044	2.33	12	4.83	11	5.83	15	713.33	16
CSSL 051	2.17	9	5.00	13	5.83	16	717.67	15
CSSL 053	3.00	16	4.50	9	5.75	14	512.67	29
CSSL 055	2.33	10	4.33	8	5.00	7	547.33	28
CSSL 056	3.67	25	4.67	10	6.25	22	666.33	23
CSSL 058	3.33	20	6.17	19	6.83	28	725.33	13
CSSL 060	2.33	11	6.17	18	5.33	11	828.33	6
CSSL 061	3.17	19	7.00	26	6.92	29	709.00	17
CSSL 062	2.83	14	5.33	14	5.92	18	661.33	24
CSSL 069	3.67	24	6.17	20	5.25	10	832.33	5
CSSL 084	1.33	5	3.17	4	4.17	5	1465.00	1
CSSL 100	0.41	1	2.00	2	3.33	2	994.10	4
CSSL 111	1.00	3	3.50	5	3.75	3	1299.33	2
CSSL 112	3.00	15	4.83	12	5.58	12	728.00	12
CSSL 113	3.17	17	6.50	23	5.83	17	722.33	14
CSSL 115	1.67	6	6.33	21	5.17	8	743.67	10
CSSL 121	3.50	22	5.67	15	6.17	21	645.67	25
Fleur 11	4.83	29	5.67	17	6.25	23	601.67	26
Florunner	1.83	7	2.67	3	3.83	4	802.00	8
Tifguard	0.50	2	1.00	1	3.25	1	1014.67	3

MS: mid-season; LS: late-season.

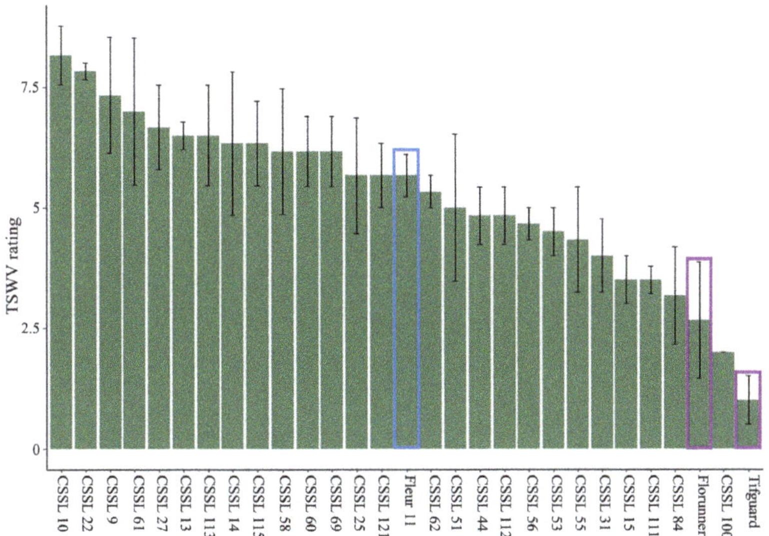

Figure 1. Manually collected end-season tomato spotted wilt virus (TSWV) scores for the chromosome segment substitution lines (CSSLs) showing multiple lines with a lower disease rating than Fleur 11, highlighted in blue. Adapted checks are highlighted in purple. The data were collected at Gibbs farm at 110 DAP.

Leaf spot was rated at the end of the season at both Gibbs and Bowen farms. The data revealed significant differences among the lines with no interaction between the fields (Tables 1 and S4). Overall, the best lines were CSSL 100 and CSSL 111, which were just below Tifguard, while CSSL 84 and CSSL 15 were third and fourth and ranked below the other check variety, Florunner (Figure 2 and Table 2).

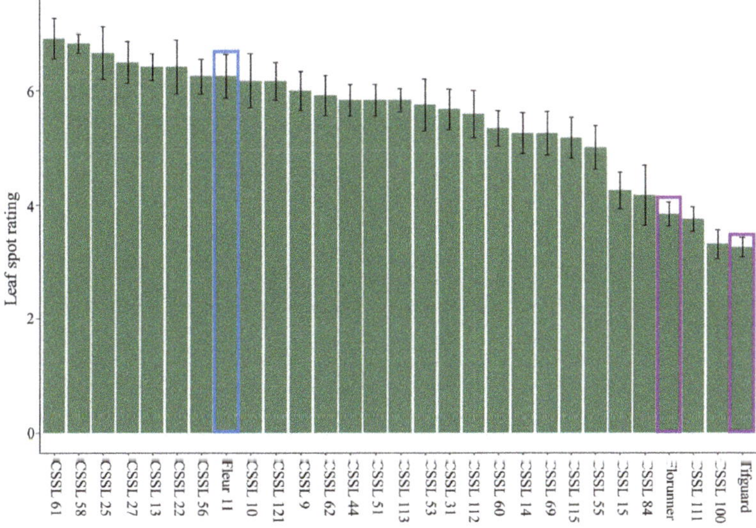

Figure 2. Manual leaf spot scores for the chromosome segment substitution lines (CSSLs) showing multiple lines with a lower disease rating than Fleur 11, highlighted in blue. Adapted checks are highlighted in purple. The data were collected from both Gibbs and Bowen farms at 110 DAP.

Pod weight data showed significant differences between the lines with some interaction between the fields (Tables 1 and S4). Overall, the best lines were CSSL 84 and CSSL 111, which performed better than the checks. CSSL 100 was the third best, ranking just below Tifguard (Figure 3 and Table 2). From Gibbs, the order of the top-ranked lines was CSSL 84, CSSL 111, and CSSL 100, followed in the fourth rank by Tifguard. At Bowen, CSSL 84 and CSSL 111 were at the top, followed by Florunner and Tifguard, with CSSL 100 falling to the seventh rank (Table S5). This could be attributed to the fewer number of seeds available for planting that resulted in two reps planted at Bowen for CSSL 100. Comparison of introgression lines showed consistently superior performance of CSSL 84 and CSSL 111 in both fields, with CSSL 100 being significantly different from Fleur 11 only at Gibbs. The analysis also highlighted CSSL 69 as having a superior pod weight relative to Fleur 11, but only at Gibbs (Table S6).

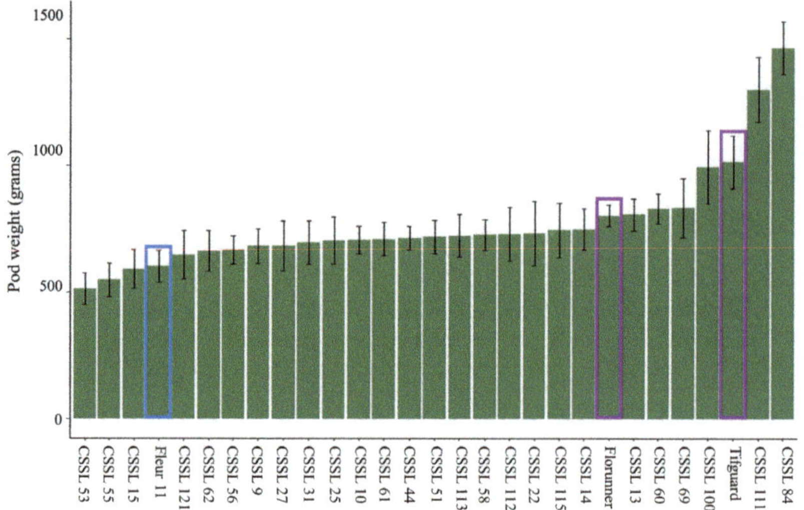

Figure 3. Pod weight for the chromosome segment substitution lines (CSSLs) showing multiple lines with a superior performance relative to Fleur 11, highlighted in blue. Adapted checks are highlighted in purple. The data were collected from both Gibbs and Bowen farms after harvest.

3.2. Aerial Phenotyping

A total of 14 traits that included 3 RGB indices, 9 multispectral indices and 2 canopy temperature traits were derived from aerial images. RGB images were acquired in the middle of the season, and the derived indices showed significant differences between the genotypes with significant line-by-field interaction for VARI and GRV (Table S4). GLI had some correlation with late-season TSWV at the Gibbs location (R of 0.53) but not significantly with any other traits. At Gibbs, VARI had a correlation of -0.78 with mid-season TSWV, which became increasingly negative to -0.92 with end-season TSWV and -0.83 with leaf spot (Table 3). At Bowen, the correlations were -0.8 for mid-season TSWV and -0.88 for leaf spot. At Gibbs, the correlation of GRV with mid-season TSWV, end-season TSWV, and leaf spot was -0.82, -0.85 and -0.79, respectively. At Bowen, the correlations were -0.75 for mid-season TSWV and -0.83 for leaf spot. The two indices also correlated well with pod weight, especially at Bowen, where the correlation was 0.82 and 0.81 for VARI and GRV, respectively (Table 3).

Table 3. Correlations between red green blue (RGB) indices and manually collected data.

Trait	Spectrum	Formula	Citation	Field	Pod Weight	TSWV (MS)	TSWV (LS)	Leaf Spot (LS)
Visible Atmospherically Resistant Index (VARI)	RGB	$(G - R)/(G + R - B)$	[54]	Bowen	0.82	−0.8	NA	−0.88
Visible Atmospherically Resistant Index (VARI)	RGB	$(G - R)/(G + R - B)$	[54]	Gibbs	0.6	−0.78	−0.92	−0.83
Green Vegetation Index-R (GRV)	RGB	$(G - R)/(G + R)$	[55]	Bowen	0.81	−0.75	NA	−0.83
Green Vegetation Index-R (GRV)	RGB	$(G - R)/(G + R)$	[55]	Gibbs	0.58	−0.82	−0.85	−0.79
Green Leaf Index (GLI)	RGB	$(G \times 2 - R - B)/(G \times 2 + R + B)$	[56]	Bowen	0.01	−0.02	NA	0.16
Green Leaf Index (GLI)	RGB	$(G \times 2 - R - B)/(G \times 2 + R + B)$	[56]	Gibbs	−0.18	0.14	0.53	0.40

MS: mid-season; LS: late-season.

Mid-season CTD showed no significant differences between the lines (Table 1). There were significant differences among the lines for end-season CTD with no sample-by-field interaction (Tables 1 and S4). There was a strong correlation between end-season CTD and leaf spot (R = −0.82) and some correlation with end-season TSWV at Gibbs (−0.63). For pod weight, the correlation was 0.82 at Bowen and 0.65 at Gibbs (Table 4).

Table 4. Correlations between multispectral indices and manually collected data.

Trait	Spectrum	Formula	Citation	Leaf Spot	TSWV [a]	Pod Weight Gibbs	Pod Weight Bowen
Chlorophyll Index-RE (CIRE)	Multispectral	$IR/(RE - 1)$	[57]	−0.95	−0.80	0.43	0.84
Difference Vegetation Index (DVI)	Multispectral	$IR - R$	[58]	−0.95	−0.85	0.45	0.84
Green Normalized Difference Vegetation Index (GNDVI)	Multispectral	$(IR - G)/(IR + G)$	[59]	−0.92	−0.94	0.49	0.81
Normalized Green (NG)	Multispectral	$G/(IR + R + G)$	[60]	0.92	0.94	0.52	0.80
Ratio Vegetation Index (RVI)	Multispectral	IR/R	[61]	−0.93	−0.90	0.55	0.84
Normalized Difference Vegetation Index (NDVI)	Multispectral	$(NIR - R)/(NIR + R)$	[62]	−0.93	−0.86	0.49	0.81
Optimized Soil Adjusted Vegetation Index (OSAVI)	Multispectral	$(1 + 0.16)(IR - R)/(IR + R + 0.16)$	[63]	−0.94	−0.85	0.48	0.82
Soil Adjusted Vegetation Index (SAVI)	Multispectral	$(1 + 0.5)(IR - R)/(IR + R + 0.5)$	[64]	−0.95	−0.85	0.47	0.83
Triangular Vegetation Index (TVI)	Multispectral	$0.5 \times (120 \times (IR - G) - 200 \times (R - G))$	[65]	−0.95	−0.83	0.45	0.83
End season canopy temperature depression	Thermal	NA	NA	−0.82	−0.63	0.65	0.82

[a] Data for Gibbs late-season.

The multispectral indices were all significantly different, without interaction, and generally showed strong correlations with the disease scores (Tables 4 and S4). For leaf spot, CIRE, DVI, SAVI, and TVI were best correlated with an R of 0.95. In the case of TSWV ratings which were only collected at Gibbs, the best correlations were derived from GNDVI and NG with an R of 0.94. The range of correlations with pod weight were 0.43–0.55 at

Gibbs and 0.8–0.84 at Bowen. Consistent with the strong correlations, the indices ranked the lines comparably with the manual data (Table S5).

Generally, the mean of most traits was comparable to the value of Fleur 11 (Table 1) and varied from the cultivated checks, Florunner and Tifguard, which had better performance than the population. This was expected, since the CSSLs have the genetic background of Fleur 11. For the agronomic traits studied, the broad sense heritabilities were comparable between fields and ranged from 0.4 for mid-season TSWV to 0.74 for pod weight at Bowen. This is indicative of significant and reliable genetic influence on trait variations, a fact also attested by the low standard errors of the trait means. The lowest heritability for significant traits was 0.2 for end-season CTDs, while the rest of the indices, with the exception of GLI at Bowen, had generally high heritabilities. Since the genetic background of the population is known to be uniform, any variation is hypothesized to be a result of introgression effects from wild chromosome segments.

Multicomparison testing was carried out to evaluate how the individual lines differ from Fleur 11 in the different traits (Table S6). It was evident that the three lines, CSSL 84, CSSL 100, and CSSL 111, were the most outstanding in all traits, with the introgressions making them clearly distinct from Fleur 11.

3.3. Belowground Phenotyping

The initial output of a GPR scan is a radargram of the reflected waveform over time, known as a B Scan. The various pipelines for signal processing resulted in the conversion of the B Scans to 13 variables of quantitative data for each pipeline that could be evaluated as proxies for manually collected pod weight data (Table S7). The GPR variables showed significant differences between the genotypes at Bowen. However, the opposite observation was made for Gibbs, as the variables did not show significant differences between the genotypes (Table 5).

Table 5. Summary of GPR variables that showed significant correlations with pod weight.

GPR Variable	Pipeline	Field	Significance of GPR Variable	Mean	Max	Min	SE	Fleur 11	Correlation	Significance of Correlation with Pod Weight
p1_freq_1	1	Bowen	*	0.37	0.459	0.254	0.009	0.35	−0.512	**
p2_freq_1	2	Bowen	*	0.37	0.464	0.259	0.009	0.35	−0.512	**
p3_freq_1	3	Bowen	*	0.34	0.438	0.222	0.010	0.32	−0.518	**
p1_freq_1	3	Gibbs	NS	0.29	0.352	0.246	0.0038	0.35	−0.391	*
p2_freq_1	2	Gibbs	NS	0.29	0.350	0.247	0.0038	0.35	−0.407	*
p3_freq_1	3	Gibbs	NS	0.23	0.266	0.206	0.0026	0.27	−0.453	*
p3_freq_3	3	Gibbs	NS	0.07	0.077	0.063	0.0007	0.07	−0.405	*
p3_freq_4	3	Gibbs	NS	0.06	0.071	0.057	0.0006	0.07	−0.420	*
p3_freq_5	3	Gibbs	NS	0.06	0.068	0.054	0.0006	0.06	−0.409	*
p3_freq_6	3	Gibbs	NS	0.06	0.066	0.053	0.0006	0.06	−0.449	*
p3_freq_7	3	Gibbs	NS	0.06	0.066	0.054	0.0005	0.06	−0.432	*
p3_freq_8	3	Gibbs	NS	0.06	0.065	0.053	0.0005	0.06	−0.440	*
p3_freq_9	3	Gibbs	NS	0.06	0.064	0.052	0.0006	0.06	−0.428	*
p3_freq_10	3	Gibbs	NS	0.06	0.064	0.051	0.0006	0.06	−0.411	*
p3_freq_11	3	Gibbs	NS	0.06	0.064	0.051	0.0006	0.06	−0.430	*
p3_freq_12	3	Gibbs	NS	0.06	0.064	0.052	0.0006	0.06	−0.409	*
p3_freq_13	3	Gibbs	NS	0.06	0.065	0.052	0.0006	0.06	−0.437	*

NS: $p > 0.05$; * $p \leq 0.05$; ** $p \leq 0.01$.

Considerable correlations were observed between the coefficients of GPR variables and those of manual pod weight of the genotypes. At Bowen, three variables had significant correlations ($p < 0.05$). The variables corresponded to frequency one of pipeline one (p1_freq_1, R = −0.512), pipeline two (p2_freq_1, R = −0.512), and pipeline three (p3_freq_1, R = −0.516) (Figure 4a–c). Similarly, at Gibbs, the variable representing the first frequency for the pipelines had a significant correlation with pod weight ($p < 0.05$) with values of R = −0.391, R = −0.407 and R = −0.453 for p1_freq_1, p2_freq_1, and p3_freq_1, respectively (Figure 4d,e). However, at Gibbs, contrary to the observation at Bowen, all frequency variables derived from pipeline three, with the exception of frequency 2, also had significant correlations with pod weight at $p < 0.05$ and R between −0.449 and −0.405 (Table 5).

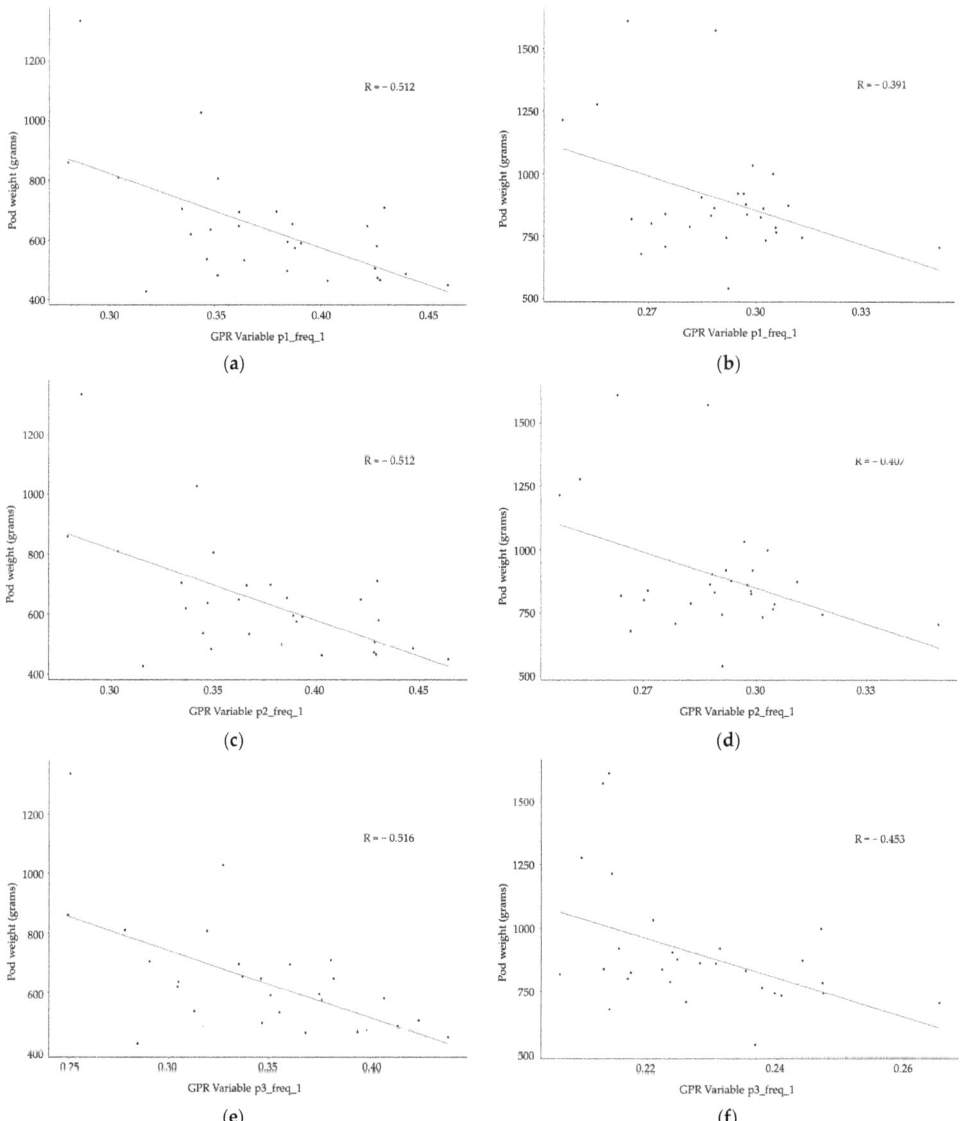

Figure 4. Correlation between pod weight and GPR variables extracted from frequency 1 using pipeline 1 (**a**), pipeline 2 (**b**), pipeline 3 (**c**), at Bowen and pipeline 1 (**d**), pipeline 2 (**e**), pipeline 3 (**f**), at Gibbs.

The ranking capacity of the variables with significant correlations were evaluated by comparing their ability to rank the top and bottom ten genotypes in comparison to manual pod weight (Figure 5 and Table S8). For Bowen, frequency one of all three pipelines captured five of the top genotypes while pipeline one and two captured six of the bottom genotypes and pipeline three captured five of the bottom genotypes. Among the top genotypes detected by all pipelines were the cultivated checks, Florunner (ranked first) and Tifguard (ranked third), as well as CSSL 84 (ranked second). CSSL 111 was ranked seventh by pipelines one and two, and eleventh by pipeline three, while CSSL 100 had an intermediate rank for all pipelines. Curiously, the lowest-performing CSSL 55 was ranked among the top ten at rank four for all pipelines. At Gibbs, frequency one of pipelines one and two captured four of the top genotypes, while pipeline three captured five. For the bottom ranks, frequency one of pipelines one and two captured five, and pipeline three captured seven of the bottom genotypes. The rest of the frequencies of pipeline three that had significant correlations performed marginally worse than frequency one, capturing four of the top genotypes (except frequency seven, which captured five) and three of the bottom genotypes (except frequency four, which captured two). This indicated that, similar to Bowen, the first frequency is sufficient for detecting pod weight. All three pipelines ranked among the top for the cultivated check Tifguard (ranked first by pipelines one and two, and seventh by pipeline 3), CSSL 100 (ranked second by all three pipelines), and CSSL 84 (ranked third by pipeline one and two and five by pipeline 3), while CSSL 111 had an intermediate rank.

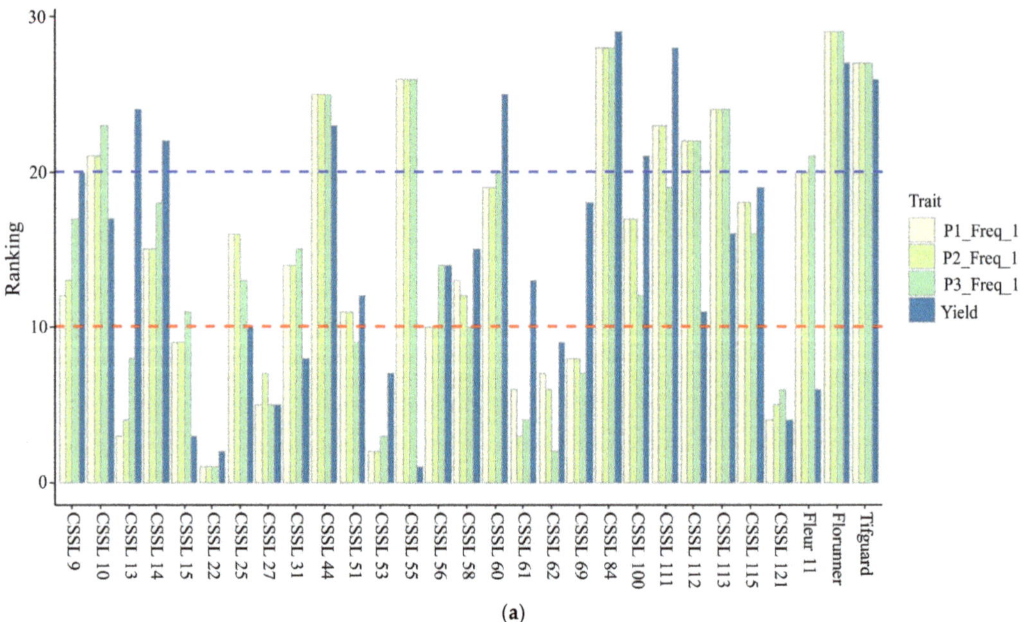

Figure 5. *Cont.*

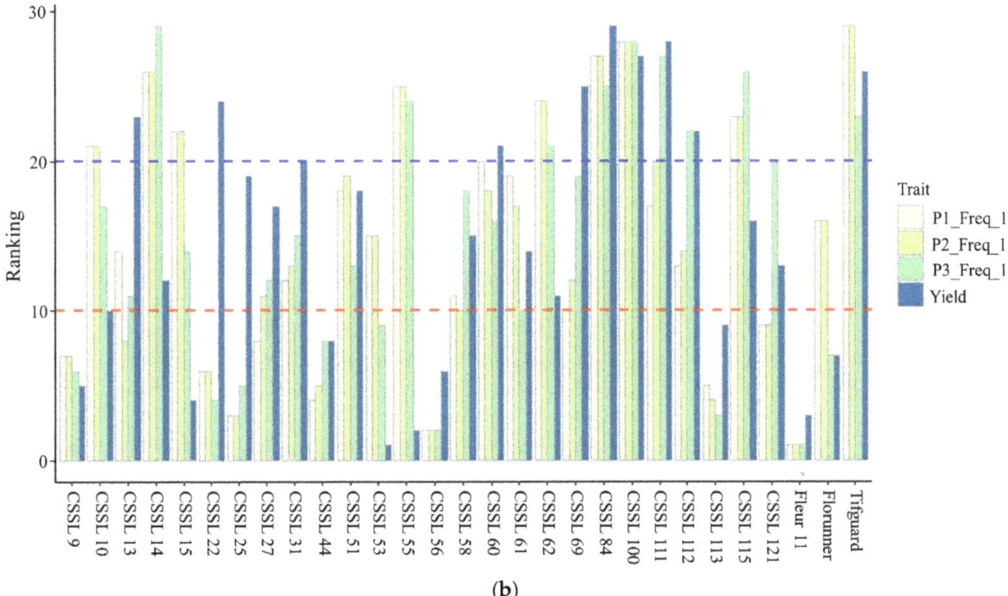

Figure 5. Comparison of the ability of the GPR variables to rank the lines relative to pod weight at Bowen (**a**) and Gibbs (**b**). Levels above the blue dashed line highlight lines ranked among the top 10 by pod weight, while levels below the red dashed line highlight lines ranked among the bottom 10 by pod weight.

4. Discussion

Since the cultivated background of the population, Fleur 11, is a Spanish variety predominantly grown in West Africa, there was little expectation that the CSSLs would have agronomic performance similar to those of runner varieties that are highly adaptable in Georgia. The genotypes Tifguard and Florunner were used as adapted runner-type checks. Tifguard is a high-yield cultivar with field resistance to TSWV [66], while Florunner is high-yielding but susceptible to TSWV [67,68]. Both genotypes are susceptible to early and late leaf spot. TSWV causes spotted wilt disease in peanut and was first reported in the US in 1974 [69]. The virus is transmitted by thrips and leads to drastic losses of yield on infected peanut plants [70,71]. In the 1980s and 1990s, severe epidemics that resulted in yield losses of up to USD 40 million were reported in Georgia alone by Bertrand [72], cited in Srinivasan et al. [73]. The use of insecticides to control thrips does not provide sufficient protection against TSWV [74]. This means that the primary strategy for combating the disease is by using cultivars with genetic resistance. Examples of released TSWV-resistant cultivars include Southern Runner [75], Georgia Green [76], Georgia-06G [77], Tifguard [66], and Georgia-09B [78].

Of equally devastating consequences to peanut production are leaf spot diseases. These are classified into early leaf spot, caused by *Passalora arachidicola* (*Hori*) U. Braun (syn. *Cercospora arachidicola*) and late leaf spot, caused by *Nothopassalora personata* (Berk. & M.A. Curtis) U. Braun, C. Nakash., Videira & Crous (syn. *Cercosporidium personatum*) [79]. The two are distinguishable by the color and location of the lesions that they form. ELS typically forms brown lesions on the adaxial (upper) side of the leaves, while LLS forms black lesions on the abaxial (under) side of the leaves [79]. While the diseases can co-occur in the field, the onset of ELS is usually earlier in the season, while for LLS, it is later in the season [80,81]. Management of both diseases is primarily by the costly, regular application of fungicides [82,83]. As with TSWV, sources of genetic resistance for both leaf spots are available, which when used in combination with optimum management practices lead to reduced costs and losses. Georganic is a cultivar with resistance to leaf spot. Georganic and Tifrunner, both of which were derived from PI 203396, also have resistance to TSWV [84,85]. The most widely utilized source of genetic resistance to leaf spot was from the wild diploid *Arachis cardenasii*, that was introduced into *A. hypogaea* via the hexaploid route to interspecific hybridization [86]. From these introgression lines, GP-NC WS 16, which is resistant to multiple diseases such as ELS, Cylindrocladium black rot (CBR), Sclerotinia blight (SB), and TSWV, was developed [87]. This line has been used as the source of resistance to several populations, including the peanut nested association mapping (NAM) population [88]. Other leaf spot-resistant cultivars derived from the *A. cardenasii* resistance source are the Indian cultivars ICGV 87165 [89] and GPBD 4 [9,90], as well as the Brazilian cultivar IAC Sempre Verde [91].

4.1. Performance of the CSSLs under TSWV and Leaf Spot Pressure

The occurrence of both TSWV and leaf spot diseases provided the opportunity to evaluate the CSSL population for the diseases. CSSL 100, CSSL 84, CSSL 15, and CSSL 111 had superior performance relative to Fleur 11 and the rest of the CSSLs. This indicated that wild chromosome segments conferred some level of disease protection for these lines. Their performance was comparable to that of the resistant check Tifguard. Paradoxically, this good performance was also comparable to that of the notoriously susceptible Florunner. However, qualitative assays of the root tips of the resistant CSSLs and Fleur 11 using immunostrips confirmed that the absence of typical canopy chlorosis was due to the absence of TSWV in the resistant lines [69]. The CSSLs, being Spanish varieties, have a short growing cycle of fewer than 120 days, in contrast to the runner checks that mature at 140 days. With more days of observation, the contrast between the two runner checks and CSSLs may have been more apparent, though the possibility of reduced disease pressure in the year of study cannot be ruled out. However, for these CSSLs, the combination of disease resistance and shorter growing cycle were sufficient to provide adequate protection against TSWV in the Georgia growing environment.

At the time of rating, both early and late leaf spot were observed in the field, with ELS being predominant. As such, the leaf spot rating did not distinguish between the two diseases. The same four lines were superior to Fleur 11 and the rest of the CSSLs, with their performance comparable to Tifguard and Florunner. Under the management practices employed, it was clear that the superior CSSLs had genetic-based suppression of leaf spot. Previously, $[A. ipaensis \times A. duranensis]^{4X}$, which is the source of introgressions for the CSSLs, was observed to have late leaf spot resistance. BC1F5 lines derived from crossing this allotetraploid to IAC-886, a derivative of the leaf spot susceptible Florunner, were shown to have resistance to LLS [92,93]. Our results suggest that, indeed, wild derived introgressions may have conferred some level of leaf spot resistance to these lines.

4.2. Evaluation of TSWV and Leaf Spot using Vegetative Indices

Strong correlations were observed between disease scores and various image-based vegetative indices. Among the RGB indices, VARI and GRV, which were derived from mid-season ratings, were able to accurately estimate TSWV and leaf spot disease at the end of the season. VARI was designed to remotely estimate the vegetation fraction of canopies with less sensitivity to differences in atmospheric conditions [54]. The index is closely associated with the leaf pigment components, chlorophyll, and carotenoids [94]. It has been used to study green biomass in maize [95], to discriminate between water stress and nitrogen stress in maize [96], and in harvest date optimization in soybean [97]. The GRV index was a modification of VARI and, thus, it had a similar performance with slightly greater sensitivity than VARI. These indices were derived from RGB images taken by a GoPro digital camera. Their informativeness and ability to select the best lines (Table S5) show that, together with the availability of low-cost UAVs, they may be a convenient way to deploy HTP for disease phenotyping by exploiting electromagnetic radiation within the visible range.

Multispectral indices are derived from the canopy reflectance of specific wavelengths in the near-infrared (NIR; 750–1350 nm) as opposed to visible (350–750 nm) regions of the electromagnetic spectrum [98]. Their informativeness is premised on the fact that healthy vegetation absorbs visible light while strongly reflecting NIR. On the contrary, vegetation in poor health absorbs most of the NIR (Mullan, 2012). Spectral indices take advantage of these properties and typically depends on the maximization of differences between the red wavelengths and NIR wavelengths to indicate plant health [98,99]. Examples include the simplest index, known as the ratio vegetative index (RVI) [61], and the most widely used NDVI [62]. While maintaining the same principles, other indices have been developed by modifying parameters to increase their sensitivity to various physiological properties of plants [100]. An example of this is the incorporation of the green channel into NDVI to form GNDVI [59]. Another example is increasing the reflectance of leaf chlorophyll by incorporating red edge (RE) in the index, as in the case of CIRE, which is a modification of the simple index, RVI [61]. Examples of other modifications include adjustments to correct for soil background in indices such as SAVI [64] and its optimized form, OSAVI [63].

Taking these relationships into consideration, it is no surprise that all the multispectral indices we used were comparable with each other and had very strong correlations with the manual disease ratings. Essentially, the indices were detecting the state of the canopy health and not necessarily discriminating between the two diseases nor determining the underlying pathophysiology. The superior CSSLs that were previously not known to have resistance were ranked high by all the indices, showing their sufficiency in phenotyping for this population. All the same, we considered GNDVI to be a representative index for multispectral phenotyping of the population in this study (Figure 6).

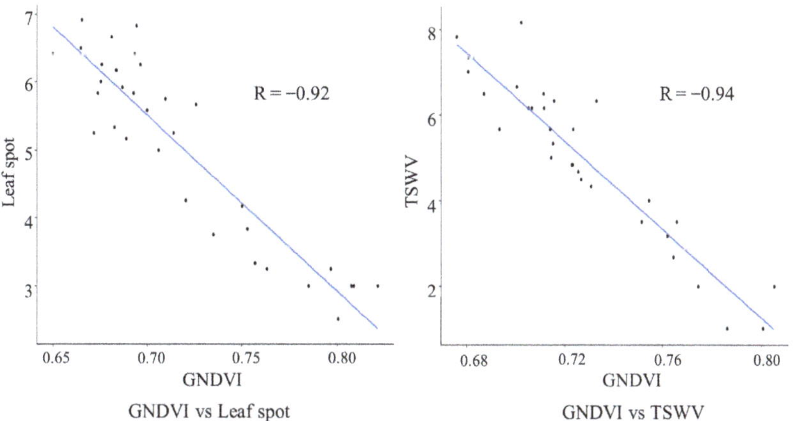

Figure 6. Correlations of green normalized difference vegetative index (GNDVI), one of the multi-spectral indices used with leaf spot rated at Gibbs and Bowen, and tomato spotted wilt virus TSWV, rated at Gibbs.

The four lines with significantly superior pod weight compared to Fleur 11 showed that introgressions from the wild could be used to improve this important economic trait. In fact, CSSL 69 has been demonstrated to have beneficial introgressions that increase pod and seed size and is a source for pyramiding 100-seed weight QTLs [101]. However, introgressions that conferred protection from the disease were also responsible for the improved performance. This is apparent, especially considering that CSSL 69 had significantly higher pod weight than Fleur 11 at Gibbs, but its disease susceptibility reduced its performance relative to the other three. At Bowen, because of the higher disease pressure, its pod weight performance was reduced to normal. It was impressive that the pod weight of the un-adapted CSSLs with a Spanish background were comparable to those of the high-yielding Tifguard and Florunner. A factor that may have reduced the pod weight of the checks was the earlier harvest time. Hence the pod-filling process may not have been complete, leading to reduced weight. Even with that taken into account, the performance of the CSSLs is no less impressive. The RGB, multispectral, and CTD indices correlated strongly with pod weight, particularly at Bowen, where the leaf spot pressure was more severe. This points more to a relationship between plant health and pod weight rather than the ability to select for pod weight based on the indices.

4.3. Use of GPR for Pre-Harvest Pod Weight Phenotyping

This is among the first research demonstrating the use of GPR as a peanut phenotyping tool, and the first evaluation of peanut belowground biomass using a frequency-based analysis. Preharvest evaluation of belowground biomass is an area of interest in peanut research. While canopy traits are important due to the pervasiveness of a plethora of foliar diseases, the geocarpic nature of peanut makes belowground phenotyping an area of particular importance. So far, any subsurface evaluation of traits such as white mold, root system architecture, and most importantly, yield and its components, has necessitated destructive digging of the plants at the end of the season. This study implemented a potentially revolutionary technique in studying peanut pod traits prior to harvest.

GPR technology has been used extensively as a noninvasive way to detect a coarse large structure belowground biomass, primarily tree roots, with considerable success [102–104]. However, its application for fine biomass structure detection, which would be ideal for agricultural research, has been limited by the fact that such structures are typically below the detection threshold of commonly used GPR frequencies, which are typically in the 500–1500 MHz range [105–107]. Still, the technology has been applied in root phenotyping of crops such as winter wheat and energy cane, though rather than describing root architecture, root cohort parameters such as biomass and density were studied [46,107]. An important case where it was used was the study of the root bulking rate of cassava, a crop whose economic potential is stored belowground. In this case, GPR detection was sufficient, with a correlation of up to 0.65 [48] and 0.79 with the ability to discriminate between varieties [40].

Like cassava, the economic yield component of peanut is situated belowground, since after flowering, the peanut develops a peg that grows downwards, penetrates the soil, and elongates sideways to form the pods. Pod formation occurs in the pod zone, a region that is approximately 4 cm below the surface and hence shallower than the root zone. This made the choice of a GPR system with a high central frequency antenna (1800 MHz) appropriate for this study, since higher frequencies promise higher radar resolution at shallow depths [107]. The resultant radargrams were processed using three different pipelines to derive quantitative data from the reflectance of GPR frequencies ranging from 1–13 GHz for association with pod weight. The features were grouped into 13 variables, each representing a frequency band, such that variable 1 represented frequencies within 1 GHz and so on.

These variables yielded useful information that could be related to the belowground biomass properties of the population, suggesting that GPR had potential utility for the non-destructive preharvest HTP of peanut. Statistical analysis of the variables showed disparity in the performance of GPR between the Gibbs and Bowen fields. This may be attributed to the more intense disease pressure at Bowen, particularly leaf spot. This may have resulted in severe penalties, as observed by the more intense inverse correlation between pod weight and late-season leaf spot at Bowen than at Gibbs (R = −0.79 and R = −0.47, respectively). Hence, the variation in GPR-detectable biomass may have been more pronounced at Bowen than at Gibbs. It is also worth considering that the soil condition differences between the two fields may have contributed to the difference in results. While standardization of the GPR processing pipeline would be ideal for the automation and applicability of the technology across different geographic sites, with respect to the current implementation, optimization of the processing pipelines for each specific site may be required.

For all three analysis pipelines used to derive quantitative GPR features, variable one, corresponding to the seven frequency bands within 1 GHz was the most informative. This was within expectation since the central frequency of the GPR system used was 1.8 GHz. The comparable performance of the three pipelines indicated that either was sufficient for GPR processing with only a slight bias for pipeline three. Combining the features by calculating the area under the frequency curve for the DFTs was also an effective way to derive variables from GPR output that could be implemented into a breeder's analysis protocol similar to other conventional data. Linear regression of these variables yielded coefficients that correlated relatively well with the coefficients of manual pod weight data. Specifically, there was a negative correlation between the GPR features based on Fourier transform and pod weight. The amplitude of the Fourier coefficients is typically related to greater or smaller GPR signal variability at the specified frequency. With this assumption, the negative correlations reported in this study imply that smaller GPR signal variability is associated with greater peanut pod weight for these frequencies. A plot with more pods belowground may mean a greater clustering of peanut pods with less soil in between the peanut pods and thus the smaller GPR signal variability, as opposed to a plot with fewer pods where more soil is present between the peanut pods resulting in greater signal variability. Since this is the first application of a frequency-based GPR analysis to peanut pod weight assessment, our interpretations are only based on the results of this study. Image thresholding analysis demonstrated a negative correlation to GPR features based on the mean of the signal amplitude and positive correlations based on the standard deviation of the signal amplitude [47].

The R values of up to -0.51 compare well with the R = 0.65 and 0.79 observed in cassava [40,48], considering the much smaller belowground biomass characteristic of peanut. Despite the moderate correlations, the variables gave reasonable ranking by pod weight for the population when compared with the manual data, with the top CSSLs and check varieties consistently featuring at the top. A sticking anomaly was the observation of CSSL 55, which has a lower pod weight among the top genotypes based on GPR ranks, which contributed to lowering the correlations. This may be attributed to the fact that GPR detects reflections of electromagnetic frequencies that are converted to an image from which the volumetric mass is estimated as pod weight. On the other hand, the manual pod weight measures only mass. Generally, genotypes with bigger pod sizes tend to have more pod weight mass; however, this is not always the case since other factors, such as the number of pods per plant, pod filling, and maturity at harvest, also affect the final pod weight. An example is CSSL 100, which generally has smaller pods but is among the best-performing genotypes by weight. With more research towards the improvement of GPR methods and analysis for fine biomass detection, it is conceivable that belowground HTP will be adopted with higher frequency in peanut research.

5. Conclusions

The findings of this study highlight novel phenomic approaches by which the peanut breeding pipeline can maintain and improve its current state. Examination of the CSSL population shows that alleles from peanut wild relatives can confer agronomically beneficial traits to the cultivated. The use of both aerial and belowground phenomic techniques has the potential to radically transform the peanut breeding pipeline by increasing the speed and precision of phenotype data acquisition. These techniques will facilitate the identification and speedy release of novel and better-adapted peanut varieties, hence improving the process of peanut breeding.

Supplementary Materials: The following supporting information can be downloaded at: https://www.mdpi.com/article/10.3390/agronomy13051223/s1. Table S1: GPR DFT raw data based on pipeline 1 for all plots; Table S2: GPR DFT raw data based on pipeline 2 for all plots; Table S3: GPR DFT raw data based on pipeline 3 for all plots; Table S4: Summary analysis of interactions between samples and field for manually collected data and spectral indices; Table S5: Ranking of the chromosome segment substitution lines (CSSLs) based on Green Normalized Difference Vegetation

Index (GNDVI), Visible Atmospherically Resistant Index (VARI), and Green Vegetation Index-R (GRV). The indices generally rank the lines similarly to the manual data, showing their capacity for selection in this population; Table S6: The effects of introgressions on the traits that were manually collected. Lines with significant introgression effects are presented. For TSWV late season, CSSL 10, CSSL 84, and CSSL 111, which are not statistically significant, are highlighted because of their high numeric effects. The introgression effects were calculated as percentages relative to the cultivated Fleur 11; Table S7: Summary statistics of all GPR variables; Table S8: Comparison of CSSL ranking ability between manual pod weight data and GPR variables that showed significant correlations with pod weight at Bowen and Gibbs. The CSSLs are ranked based on pod weight performance. The ranking by the GPR variables is indicated in the respective columns for each variable. Green font indicates the top 10 CSSLs as ranked by pod weight, while purple font indicates the bottom 10 CSSLs as ranked by pod weight.

Author Contributions: Conceptualization, D.G. and P.O.-A.; methodology, D.G., Y.C., W.P. and B.T.; software, H.R.-G., I.D. and D.H.; formal analysis, D.G., I.D. and H.R.-G.; resources, C.C.H., D.F., W.P., D.H. and P.O.-A.; data curation, D.G. and I.D.; original draft preparation, D.G.; supervision, P.O.-A.; funding acquisition, P.O.-A. and D.H. All authors have read and agreed to the published version of the manuscript.

Funding: The project was funded by the USAID Feed the Future Innovation Lab for Peanut and the NSF BREAD PHENO: High Throughput Phenotyping Early Stage Root Bulking in Cassava using Ground Penetrating Radar.

Data Availability Statement: Not applicable.

Acknowledgments: The authors would like to thank Jason Golden, Shannon Atkinson, Betty Tyler, and Kathy Marchant for their technical assistance.

Conflicts of Interest: The authors declare no conflict of interest.

References

1. Fonceka, D.; Tossim, H.-A.; Rivallan, R.; Vignes, H.; Lacut, E.; de Bellis, F.; Faye, I.; Ndoye, O.; Leal-Bertioli, S.C.M.; Valls, J.F.M.; et al. Construction of chromosome segment substitution lines in peanut (*Arachis hypogaea* L.) using a wild synthetic and QTL mapping for plant morphology. *PLoS ONE* **2012**, *7*, e48642. [CrossRef] [PubMed]
2. Seijo, J.G.; Lavia, G.I.; Fernández, A.; Krapovickas, A.; Ducasse, D.; Moscone, E.A. Physical mapping of the 5S and 18S-25S rRNA genes by FISH as evidence that *Arachis duranensis* and *A. ipaensis* are the wild diploid progenitors of *A. hypogaea* (Leguminosae). *Am. J. Bot.* **2004**, *91*, 1294–1303. [CrossRef] [PubMed]
3. Moretzsohn, M.C.; Gouvea, E.G.; Inglis, P.W.; Leal-Bertioli, S.C.M.; Valls, J.F.M.; Bertioli, D.J. A study of the relationships of cultivated peanut (*Arachis hypogaea*) and its most closely related wild species using intron sequences and microsatellite markers. *Ann. Bot.* **2013**, *111*, 113–126. [CrossRef]
4. Kochert, G.; Stalker, T.; Gimenes, M.; Galgaro, L.; Lopes, C.R.; Moore, K. RFLP and cytogenetic evidence on the origin and evolution of allotetraploid domesticated peanut, *Arachis hypogaea* (Leguminosae). *Am. J. Bot.* **1996**, *83*, 1282–1291. [CrossRef]
5. Foncéka, D.; Hodo-Abalo, T.; Rivallan, R.; Faye, I.; Sall, M.; Ndoye, O.; Fávero, A.P.; Bertioli, D.J.; Glaszmann, J.-C.; Courtois, B.; et al. Genetic mapping of wild introgressions into cultivated peanut: A way toward enlarging the genetic basis of a recent allotetraploid. *BMC Plant Biol.* **2009**, *9*, 103. [CrossRef]
6. Hopkins, M.S.; Casa, A.M.; Wang, T.; Mitchell, S.E.; Dean, R.E.; Kochert, G.D.; Kresovich, S. Discovery and Characterization of Polymorphic Simple Sequence Repeats (SSRs) in Peanut. *Crop Sci.* **1999**, *39*, 1243. [CrossRef]
7. Varshney, R.K.; Mahendar, T.; Aruna, R.; Nigam, S.N.; Neelima, K.; Vadez, V.; Hoisington, D.A. High level of natural variation in a groundnut (*Arachis hypogaea* L.) germplasm collection assayed by selected informative SSR markers. *Plant Breed.* **2009**, *128*, 486–494. [CrossRef]
8. Stalker, H.T.; Tallury, S.P.; Ozias-Akins, P.; Bertioli, D.; Bertioli, S.C.L. The value of diploid peanut relatives for breeding and genomics. *Peanut Sci.* **2013**, *40*, 70–88. [CrossRef]
9. Stalker, H.T. Utilizing wild species for peanut improvement. *Crop Sci.* **2017**, *57*, 1102–1120. [CrossRef]
10. Fávero, A.P.; Simpson, C.E.; Valls, J.F.M.; Vello, N.A. Study of the evolution of cultivated peanut through crossability studies among *Arachis ipaensis*, *A. duranensis*, and *A. hypogaea*. *Crop Sci.* **2006**, *46*, 1546–1552. [CrossRef]
11. Bertioli, D.J.; Jenkins, J.; Clevenger, J.; Dudchenko, O.; Gao, D.; Seijo, G.; Leal-Bertioli, S.C.M.; Ren, L.; Farmer, A.D.; Pandey, M.K.; et al. The genome sequence of segmental allotetraploid peanut *Arachis hypogaea*. *Nat. Genet.* **2019**, *51*, 877–884. [CrossRef] [PubMed]
12. Bertioli, D.J.; Cannon, S.B.; Froenicke, L.; Huang, G.; Farmer, A.D.; Cannon, E.K.S.; Liu, X.; Gao, D.; Clevenger, J.; Dash, S.; et al. The genome sequences of *Arachis duranensis* and *Arachis ipaensis*, the diploid ancestors of cultivated peanut. *Nat. Genet.* **2016**, *48*, 438–446. [CrossRef]

13. Clevenger, J.; Chu, Y.; Chavarro, C.; Agarwal, G.; Bertioli, D.J.; Leal-Bertioli, S.C.M.; Pandey, M.K.; Vaughn, J.; Abernathy, B.; Barkley, N.A.; et al. Genome-wide SNP genotyping resolves signatures of selection and tetrasomic recombination in Peanut. *Mol. Plant* **2017**, *10*, 309–322. [CrossRef] [PubMed]
14. Pandey, M.K.; Agarwal, G.; Kale, S.M.; Clevenger, J.; Nayak, S.N.; Sriswathi, M.; Chitikineni, A.; Chavarro, C.; Chen, X.; Upadhyaya, H.D.; et al. Development and evaluation of a high density genotyping "Axiom-Arachis" Array with 58 K SNPs for accelerating genetics and breeding in groundnut. *Sci. Rep.* **2017**, *7*, 40577. [CrossRef] [PubMed]
15. Clevenger, J.P.; Korani, W.; Ozias-Akins, P.; Jackson, S. Haplotype-based genotyping in polyploids. *Front. Plant Sci.* **2018**, *9*, 564. [CrossRef]
16. Korani, W.; Clevenger, J.P.; Chu, Y.; Ozias-Akins, P. Machine learning as an effective method for identifying true single nucleotide polymorphisms in polyploid plants. *Plant Genome* **2019**, *12*, 180023. [CrossRef]
17. Clevenger, J.; Chu, Y.; Scheffler, B.; Ozias-Akins, P. A developmental transcriptome map for allotetraploid *Arachis hypogaea*. *Front. Plant Sci.* **2016**, *7*, 1–18. [CrossRef]
18. Ozias-akins, P.; Cannon, E.K.S.; Cannon, S.B. Genomic Resources for Peanut Improvement. In *The Peanut Genome; Compendium of Plant Genomes Book Series*; Varshney, R., Pandey, M., Puppala, N., Eds.; Springer: Berlin/Heidelberg, Germany, 2017; pp. 69–91. [CrossRef]
19. Ozias-Akins, P. The orphan legume genome whose time has come: Symposium highlights from the american peanut research education society annual meeting. *Peanut Sci.* **2013**, *40*, 66–69. [CrossRef]
20. Svensgaard, J.; Roitsch, T.; Christensen, S. Development of a mobile multispectral imaging platform for precise field phenotyping. *Agronomy* **2014**, *4*, 322–336. [CrossRef]
21. De Witt, C. *On Competition*; 66.8.; Versl.; Landbouwk Underz: Wagenigen, The Netherlands, 1960.
22. Furbank, R.T.; Tester, M. Phenomics—Technologies to relieve the phenotyping bottleneck. *Trends Plant Sci.* **2011**, *16*, 635–644. [CrossRef]
23. Chen, D.; Neumann, K.; Friedel, S.; Kilian, B.; Chen, M.; Altmann, T.; Klukas, C. Dissecting the phenotypic components of crop plant growthand drought responses based on high-throughput image analysis w open. *Plant Cell* **2014**, *26*, 4636–4655. [CrossRef] [PubMed]
24. Fiorani, F.; Schurr, U. Future scenarios for plant phenotyping. *Annu. Rev. Plant Biol.* **2013**, *64*, 267–291. [CrossRef] [PubMed]
25. Ge, Y.; Bai, G.; Stoerger, V.; Schnable, J.C. Temporal dynamics of maize plant growth, water use, and leaf water content using automated high throughput RGB and hyperspectral imaging. *Comput. Electron. Agric.* **2016**, *127*, 625–632. [CrossRef]
26. Olson, D.; Anderson, J. Review on unmanned aerial vehicles, remote sensors, imagery processing, and their applications in agriculture. *Agron. J.* **2021**, *113*, 971–992. [CrossRef]
27. Cheng, Q.; Xu, H.; Fei, S.; Li, Z.; Chen, Z. Estimation of maize LAI using ensemble learning and UAV multispectral imagery under different water and fertilizer treatments. *Agriculture* **2022**, *12*, 1267. [CrossRef]
28. Rodene, E.; Xu, G.; Delen, S.P.; Zhao, X.; Smith, C.; Ge, Y.; Schnable, J.; Yang, J. A UAV-based high-throughput phenotyping approach to assess time-series nitrogen responses and identify trait-associated genetic components in maize. *Plant Phenome J.* **2022**, *5*, e20030. [CrossRef]
29. Jiang, R.; Sanchez-Azofeifa, A.; Laakso, K.; Wang, P.; Xu, Y.; Zhou, Z.; Luo, X.; Lan, Y.; Zhao, G.; Chen, X. UAV-based partially sampling system for rapid NDVI mapping in the evaluation of rice nitrogen use efficiency. *J. Clean. Prod.* **2021**, *289*, 125705. [CrossRef]
30. Jin, H.; Köppl, C.J.; Fischer, B.M.C.; Rojas-Conejo, J.; Johnson, M.S.; Morillas, L.; Lyon, S.W.; Durán-Quesada, A.M.; Suárez-Serrano, A.; Manzoni, S.; et al. Drone-based hyperspectral and thermal imagery for quantifying upland rice productivity and water use efficiency after biochar application. *Remote Sens.* **2021**, *13*, 1866. [CrossRef]
31. Alabi, T.R.; Abebe, A.T.; Chigeza, G.; Fowobaje, K.R. Estimation of soybean grain yield from multispectral high-resolution UAV data with machine learning models in West Africa. *Remote Sens. Appl. Soc. Environ.* **2022**, *27*, 100782. [CrossRef]
32. Marston, Z.P.D.; Cira, T.M.; Hodgson, E.W.; Knight, J.F.; MacRae, I.V.; Koch, R.L.; Rondon, S. Detection of stress induced by soybean aphid (Hemiptera: Aphididae) using multispectral imagery from unmanned aerial vehicles. *J. Econ. Entomol.* **2020**, *113*, 779–786. [CrossRef]
33. Balota, M.; Oakes, J. Exploratory use of a UAV platform for variety selection in peanut. In *Autonomous Air and Ground Sensing Systems for Agricultural Optimization and Phenotyping*; Society of Photo-Optical Instrumentation Engineers (SPIE) Conference Series; International Society for Optics and Photonics: Washington, DC, USA, 2016; Volume 9866, p. 98660F. [CrossRef]
34. Balota, M.; Oakes, J. UAV remote sensing for phenotyping drought tolerance in peanuts. In *Autonomous Air and Ground Sensing Systems for Agricultural Optimization and Phenotyping II*; Society of Photo-Optical Instrumentation Engineers (SPIE) Conference Series; International Society for Optics and Photonics: Washington, DC, USA, 2017; Volume 10218, p. 102180C. [CrossRef]
35. Abd-El Monsef, H.; Smith, S.E.; Rowland, D.L.; Abd El Rasol, N. Using multispectral imagery to extract a pure spectral canopy signature for predicting peanut maturity. *Comput. Electron. Agric.* **2019**, *162*, 561–572. [CrossRef]
36. Patrick, A.; Pelham, S.; Culbreath, A.; Corely Holbrook, C.; De Godoy, I.J.; Li, C. High throughput phenotyping of tomato spot wilt disease in peanuts using unmanned aerial systems and multispectral imaging. *IEEE Instrum. Meas. Mag.* **2017**, *20*, 4–12. [CrossRef]
37. Sarkar, S.; Ramsey, A.F.; Cazenave, A.B.; Balota, M. Peanut leaf wilting estimation from RGB color indices and logistic models. *Front. Plant Sci.* **2021**, *12*, 713. [CrossRef]

38. Sarkar, S.; Oakes, J.; Cazenave, A.B.; Burow, M.D.; Bennett, R.S.; Chamberlin, K.D.; Wang, N.; White, M.; Payton, P.; Mahan, J.; et al. Evaluation of the U.S. peanut germplasm mini-core collection in the Virginia-Carolina region using traditional and new high-throughput methods. *Agronomy* **2022**, *12*, 1945. [CrossRef]
39. Bagherian, K.; Puhl, R.B.; Bao, Y.; Zhang, Q.; Sanz-Saez, A.; Chen, C.; Dang, P. Phenotyping agronomic traits of peanuts using UAV-based hyperspectral imaging and deep learning. In Proceedings of the ASABE 2022 Annual International Meeting, Houston, TX, USA, 17–20 July 2022. [CrossRef]
40. Delgado, A.; Hays, D.B.; Bruton, R.K.; Ceballos, H.; Novo, A.; Boi, E.; Selvaraj, M.G. Ground penetrating radar: A case study for estimating root bulking rate in cassava (*Manihot esculenta* Crantz). *Plant Methods* **2017**, *13*, 65. [CrossRef]
41. Butnor, J.R.; Doolittle, J.A.; Kress, L.; Cohen, S.; Johnsen, K.H. Use of ground-penetrating radar to study tree roots in the southeastern United States. *Tree Physiol.* **2001**, *21*, 1269–1278. [CrossRef]
42. Butnor, J.R.; Doolittle, J.A.; Johnsen, K.H.; Samuelson, L.; Stokes, T.; Kress, L. Utility of ground-penetrating radar as a root biomass survey tool in forest systems. *Soil Sci. Soc. Am. J.* **2003**, *67*, 1607. [CrossRef]
43. Borden, K.A.; Isaac, M.E.; Thevathasan, N.V.; Gordon, A.M.; Thomas, S.C. Estimating coarse root biomass with ground penetrating radar in a tree-based intercropping system. *Agrofor. Syst.* **2014**, *88*, 657–669. [CrossRef]
44. Borden, K.A.; Thomas, S.C.; Isaac, M.E. Interspecific variation of tree root architecture in a temperate agroforestry system characterized using ground-penetrating radar. *Plant Soil* **2017**, *410*, 323–334. [CrossRef]
45. Borden, K.A.; Anglaaere, L.C.N.; Adu-Bredu, S.; Isaac, M.E. Root biomass variation of cocoa and implications for carbon stocks in agroforestry systems. *Agrofor. Syst.* **2019**, *93*, 369–381. [CrossRef]
46. Liu, X.; Dong, X.; Xue, Q.; Leskovar, D.I.; Jifon, J.; Butnor, J.R.; Marek, T. Ground penetrating radar (GPR) detects fine roots of agricultural crops in the field. *Plant Soil* **2018**, *423*, 517–531. [CrossRef]
47. Dobreva, I.D.; Ruiz-Guzman, H.A.; Barrios-Perez, I.; Adams, T.; Teare, B.L.; Payton, P.; Everett, M.E.; Burow, M.D.; Hays, D.B. Thresholding analysis and feature extraction from 3D ground penetrating radar data for noninvasive assessment of peanut yield. *Remote Sens.* **2021**, *13*, 1896. [CrossRef]
48. Agbona, A.; Teare, B.; Ruiz-Guzman, H.; Dobreva, I.D.; Everett, M.E.; Adams, T.; Montesinos-Lopez, O.A.; Kulakow, P.A.; Hays, D.B. Prediction of root biomass in cassava based on ground penetrating radar phenomics. *Remote Sens.* **2021**, *13*, 4908. [CrossRef]
49. Simpson, C.E. Pathways for introgression of pest resistance into *Arachis hypogaea* L. *Peanut Sci.* **1991**, *18*, 22–26. [CrossRef]
50. ESRI. *ArcGIS Desktop*; Environmental Systems Research Institute: Redlands, CA, USA, 2011.
51. R Core Team. *R: A Language and Environment for Statistical Computing*; R Foundation for Statistical Computing: Vienna, Austria, 2021.
52. Kuznetsova, A.; Brockhoff, P.B.; Christensen, R.H.B. LmerTest package: Tests in linear mixed effects models. *J. Stat. Softw.* **2017**, *82*, 1–26. [CrossRef]
53. Dunnett, C.W. A multiple comparison procedure for comparing several treatments with a control. *J. Am. Stat. Assoc.* **1955**, *50*, 1096–1121. [CrossRef]
54. Gitelson, A.A.; Kaufman, Y.J.; Stark, R.; Rundquist, D. Novel algorithms for remote estimation of vegetation fraction. *Remote Sens. Environ.* **2002**, *80*, 76–87. [CrossRef]
55. Tucker, C.J. Red and photographic infrared linear combinations for monitoring vegetation. *Remote Sens. Environ.* **1979**, *8*, 127–150. [CrossRef]
56. Louhaichi, M.; Borman, M.M.; Johnson, D.E. Spatially located platform and aerial photography for documentation of grazing impacts on wheat. *Geocarto Int.* **2001**, *16*, 65–70. [CrossRef]
57. Gitelson, A.A.; Viña, A.; Ciganda, V.; Rundquist, D.C.; Arkebauer, T.J. Remote estimation of canopy chlorophyll content in crops. *Geophys. Res. Lett.* **2005**, *32*. [CrossRef]
58. Richardson, A.J.; Wiegand, C.L. Distinguishing vegetation from soil background information. *Photogramm. Eng. Rem. S* **1977**, *43*, 1541–1552.
59. Gitelson, A.A.; Kaufman, Y.J.; Merzlyak, M.N. Use of a green channel in remote sensing of global vegetation from EOS-MODIS. *Remote Sens. Environ.* **1996**, *58*, 289–298. [CrossRef]
60. Sripada, R.P.; Heiniger, R.W.; White, J.G.; Weisz, R. Aerial color infrared photography for determining late-season nitrogen requirements in corn. *Agron. J.* **2005**, *97*, 1443–1451. [CrossRef]
61. Jordan, C.F. Derivation of leaf-area index from quality of light on the forest floor. *Ecology* **1969**, *50*, 663–666. [CrossRef]
62. Rouse, J.W., Jr.; Haas, R.H.; Schell, J.A.; Deering, D.W. Monitoring vegetation systems in the great plains with erts. *NASA Spec Publ.* **1974**, *351*, 309–317.
63. Baret, F.; Jacquemoud, S.; Hanocq, J.F. The soil line concept in remote sensing. *Remote Sens. Rev.* **1993**, *7*, 65–82. [CrossRef]
64. Huete, A.R. A soil-adjusted vegetation index (SAVI). *Remote Sens. Environ.* **1988**, *25*, 295–309. [CrossRef]
65. Broge, N.H.; Leblanc, E. Comparing prediction power and stability of broadband and hyperspectral vegetation indices for estimation of green leaf area index and canopy chlorophyll density. *Remote Sens. Environ.* **2001**, *76*, 156–172. [CrossRef]
66. Holbrook, C.C.; Timper, P.; Culbreath, A.K.; Kvien, C.K. Registration of 'Tifguard' Peanut. *J. Plant Regist.* **2008**, *2*, 92. [CrossRef]
67. Anderson, W.F.; Holbrook, C.C.; Culbreath, A.K. Screening the peanut core collection for resistance to tomato spotted wilt virus. *Peanut Sci.* **1996**, *23*, 57–61. [CrossRef]
68. Knauft, D.A.; Gorbet, D.W. Genetic diversity among peanut cultivars. *Crop Sci.* **1989**, *29*, 1417. [CrossRef]

69. Culbreath, A.K.; Todd, J.W.; Brown, S.L. Epidemiology and management of tomato spotted wilt in peanut. *Annu. Rev. Phytopathol.* **2003**, *41*, 53–75. [CrossRef] [PubMed]
70. Culbreath, A.K.; Todd, J.W.; Demski, J.W. Productivity of florunner peanut infected with tomato spotted wilt virus. *Peanut Sci.* **1992**, *19*, 11–14. [CrossRef]
71. Culbreath, A.K.; Todd, J.W.; Gorbet, D.W.; Brown, S.L.; Baldwin, J.; Pappu, H.R.; Shokes, F.M. Reaction of peanut cultivars to spotted wilt. *Peanut Sci.* **2000**, *27*, 35–39. [CrossRef]
72. Bertrand, P.F. 1997 Georgia plant disease loss estimates. *Univ. Ga. Coop. Ext. Pub. Pathol.* **1998**, *81*, 98–107.
73. Srinivasan, R.; Abney, M.R.; Culbreath, A.K.; Kemerait, R.C.; Tubbs, R.S.; Monfort, W.S.; Pappu, H.R. Three decades of managing Tomato spotted wilt virus in peanut in southeastern United States. *Virus Res.* **2017**, *106*, 203–212. [CrossRef] [PubMed]
74. Todd, J.W.; Culbreath, A.K.; Brown, S.L. Dynamics of vector populations and progress of spotted wilt disease relative to insecticide use in peanuts. *Acta Hortic.* **1996**, *431*, 483–490. [CrossRef]
75. Gorbet, D.W.; Norden, A.J.; Shokes, F.M.; Knauft, D.A. Registration of 'Southern Runner' peanut. *Crop Sci.* **1987**, *27*, 817. [CrossRef]
76. Branch, W.D. Registration of 'Georgia Green' peanut. *Crop Sci.* **1996**, *36*, 806. [CrossRef]
77. Branch, W.D. Registration of 'Georgia-06G' peanut. *J. Plant Regist.* **2007**, *1*, 120. [CrossRef]
78. Branch, W.D. Registration of "Georgia-09B" peanut. *J. Plant Regist.* **2010**, *4*, 175–178. [CrossRef]
79. Smith, H.D.; Littrell, H.R. Management of peanut foliar diseases with fungicides. *Am. Phytopathol. Soc.* **1980**, *64*, 356–361.
80. Chu, Y.; Holbrook, C.C.; Isleib, T.G.; Burow, M.; Culbreath, A.K.; Tillman, B.; Chen, J.; Clevenger, J.; Ozias-Akins, P. Phenotyping and genotyping parents of sixteen recombinant inbred peanut populations. *Peanut Sci.* **2018**, *45*, 1–11. [CrossRef]
81. Chu, Y.; Chee, P.; Culbreath, A.; Isleib, T.G.; Holbrook, C.C.; Ozias-Akins, P. Major QTLs for resistance to early and late leaf spot diseases are identified on chromosomes 3 and 5 in peanut (*Arachis hypogaea*). *Front. Plant Sci.* **2019**, *10*, 883. [CrossRef] [PubMed]
82. Woodward, J.E.; Brenneman, T.B.; Kemerait, R.C.; Smith, N.B.; Culbreath, A.K.; Stevenson, K.L. Use of resistant cultivars and reduced fungicide programs to manage peanut diseases in irrigated and nonirrigated fields. *Plant Dis.* **2008**, *92*, 896–902. [CrossRef]
83. Woodward, J.E.; Brenneman, T.B.; Kemerait, R.C.; Culbreath, A.K.; Smith, N.B. Management of peanut diseases with reduced input fungicide programs in fields with varying levels of disease risk. *Crop Prot.* **2010**, *29*, 222–229. [CrossRef]
84. Holbrook, C.C.; Culbreath, A.K. Registration of "Tifrunner" peanut. *J. Plant Regist.* **2007**, *1*, 124. [CrossRef]
85. Holbrook, C.C.; Culbreath, A.K. Registration of "Georganic" Peanut. *J. Plant Regist.* **2008**, *2*, 10–3198. [CrossRef]
86. Company, M.; Stalker, H.T.; Wynne, J.C. Cytology and leafspot resistance in *Arachis hypogaea* x wild species hybrids. *Euphytica* **1982**, *31*, 885–893. [CrossRef]
87. Tallury, S.P.; Isleib, T.G.; Copeland, S.C.; Rosas-Anderson, P.; Balota, M.; Singh, D.; Stalker, H.T. Registration of two multiple disease-resistant peanut germplasm lines derived from *Arachis cardenasii* Krapov. & W.C. Gregory, GKP 10017. *J. Plant Regist.* **2014**, *8*, 86–89. [CrossRef]
88. Holbrook, C.C.; Isleib, T.G.; Ozias-Akins, P.; Chu, Y.; Knapp, S.J.; Tillman, B.; Guo, B.; Gill, R.; Burow, M.D. Development and phenotyping of recombinant inbred line (RIL) populations for peanut (*Arachis hypogaea*). *Peanut Sci.* **2013**, *40*, 89–94. [CrossRef]
89. Moss, J.P.; Singh, A.K.; Reddy, L.J.; Nigam, S.N.; Subrahmanyam, P.; McDonald, D.; Reddy, A.G.S. Registration of ICGV 87165 peanut germplasm line with multiple resistance. *Crop Sci.* **1997**, *37*, 1028. [CrossRef]
90. Gowda, M.; Motagi, B.; Naidu, G.; Diddimani, S.; Sheshagiri, R. GPBD 4: A spanish bunch groundnut genotype resistant to rust and late leaf spot. *Int. Arachis Newsl.* **2002**, *22*, 29–32.
91. Godoy, I.J.; Santos, J.F.; De Carvalho Moretzsohn, M.; Rocha, A.; Moraes, A.; Michelotto, M.D.; Bolonhezi, D.; Nakayama, F.; Soares De Freitas, R.; Bertioli, D.J.; et al. 'IAC SEMPRE VERDE': A wild-derived peanut cultivar highly resistant to foliar diseases. *Crop Breed. Appl. Biotechnol.* **2022**, *22*, e41252232. [CrossRef]
92. Leal-Bertioli, S.C.M.; Godoy, I.J.; Santos, J.F.; Doyle, J.J.; Guimarães, P.M.; Abernathy, B.L.; Jackson, S.A.; Moretzsohn, M.C.; Bertioli, D.J. Segmental allopolyploidy in action: Increasing diversity through polyploid hybridization and homoeologous recombination. *Am. J. Bot.* **2018**, *105*, 1053–1066. [CrossRef] [PubMed]
93. Bertioli, D.J.; Seijo, G.; Freitas, F.O.; Valls, J.F.M.; Leal-Bertioli, S.C.M.; Moretzsohn, M.C. An overview of peanut and its wild relatives. *Plant Genet. Resour. Characterisation Util.* **2011**, *9*, 134–149. [CrossRef]
94. Galvão, L.S.; Roberts, D.A.; Formaggio, A.R.; Numata, I.; Breunig, F.M. View angle effects on the discrimination of soybean varieties and on the relationships between vegetation indices and yield using off-nadir Hyperion data. *Remote Sens. Environ.* **2009**, *113*, 846–856. [CrossRef]
95. Sakamoto, T.; Gitelson, A.A.; Wardlow, B.D.; Arkebauer, T.J.; Verma, S.B.; Suyker, A.E.; Shibayama, M. Application of day and night digital photographs for estimating maize biophysical characteristics. *Precis. Agric.* **2012**, *13*, 285–301. [CrossRef]
96. Perry, E.M.; Roberts, D.A. Sensitivity of narrow-band and broad-band indices for assessing nitrogen availability and water stress in an annual crop. *Agron. J.* **2008**, *100*, 1211–1219. [CrossRef]
97. Meng, J.; Xu, J.; You, X. Optimizing soybean harvest date using HJ-1 satellite imagery. *Precis. Agric.* **2015**, *16*, 164–179. [CrossRef]
98. Mullan, D. Spectral radiometry. In *Physiological Breeding: Interdisciplinary Approaches to Improve Crop Adaptation*; Reynolds, M., Pask, A., Mullan, D., Eds.; CIMMYT: Batan, Mexico, 2012; pp. 69–80.
99. Araus, J.; Casadesus, J.; Bort, J. Recent tools for the screening of physiological traits determining yield. In *Application of Physiology in Wheat Breeding*; Reynolds, M., Ortiz-Monasterio, J., McNab, A., Eds.; CIMMYT: Batan, Mexico, 2001; pp. 59–77.

100. Zarco-Tejada, P.J.; Morales, A.; Testi, L.; Villalobos, F.J. Spatio-temporal patterns of chlorophyll fluorescence and physiological and structural indices acquired from hyperspectral imagery as compared with carbon fluxes measured with eddy covariance. *Remote Sens. Environ.* **2013**, *133*, 102–115. [CrossRef]

101. Tossim, H.A.; Nguepjop, J.R.; Diatta, C.; Sambou, A.; Seye, M.; Sane, D.; Rami, J.F.; Fonceka, D. Assessment of 16 peanut (*Arachis hypogaea* L.) CSSLs derived from an interspecific cross for yield and yield component traits: QTL validation. *Agronomy* **2020**, *10*, 583. [CrossRef]

102. Guo, L.; Chen, J.; Cui, X.; Fan, B.; Lin, H. Application of ground penetrating radar for coarse root detection and quantification: A review. *Plant Soil* **2013**, *362*, 1–23. [CrossRef]

103. Butnor, J.R.; Samuelson, L.J.; Stokes, T.A.; Johnsen, K.H.; Anderson, P.H.; González-Benecke, C.A. Surface-based GPR underestimates below-stump root biomass. *Plant Soil* **2016**, *402*, 47–62. [CrossRef]

104. Yeung, S.W.; Yan, W.M.; Hau, C.H.B. Performance of ground penetrating radar in root detection and its application in root diameter estimation under controlled conditions. *Sci. China Earth Sci.* **2015**, *59*, 145–155. [CrossRef]

105. Hirano, Y.; Dannoura, M.; Aono, K.; Igarashi, T.; Ishii, M.; Yamase, K.; Makita, N.; Kanazawa, Y. Limiting factors in the detection of tree roots using ground-penetrating radar. *Plant Soil* **2008**, *319*, 15–24. [CrossRef]

106. Pauli, D.; Chapman, S.C.; Bart, R.; Topp, C.N.; Lawrence-Dill, C.J.; Poland, J.; Gore, M.A. The quest for understanding phenotypic variation via integrated approaches in the field environment. *Plant Physiol.* **2016**, *172*, 622–634. [CrossRef]

107. Liu, X.; Dong, X.; Leskovar, D.I. Ground penetrating radar for underground sensing in agriculture: A review. *Int. Agrophys.* **2016**, *30*, 533–543. [CrossRef]

MDPI AG
Grosspeteranlage 5
4052 Basel
Switzerland
Tel.: +41 61 683 77 34

Agronomy Editorial Office
E-mail: agronomy@mdpi.com
www.mdpi.com/journal/agronomy